T0178273

LONDON MATHEMATICAL SOCIETY LECTURE NOTE SERIES

Managing Editor: Professor N. J. Hitchin, Mathematical Institute, University of Oxford, 24-29 St Giles, Oxford OX1 3LB, United Kingdom

The titles below are available from booksellers, or from Cambridge University Press at www.cambridge.org

London Mathematical Society Lecture Note Series. 330

Non-Commutative Localization in Algebra and Topology

Edited by
ANDREW RANICKI
University of Edinburgh

CAMBRIDGE
UNIVERSITY PRESS

Shaftesbury Road, Cambridge CB2 8EA, United Kingdom

One Liberty Plaza, 20th Floor, New York, NY 10006, USA

477 Williamstown Road, Port Melbourne, VIC 3207, Australia

314–321, 3rd Floor, Plot 3, Splendor Forum, Jasola District Centre, New Delhi – 110025, India

103 Penang Road, #05–06/07, Visioncrest Commercial, Singapore 238467

Cambridge University Press is part of Cambridge University Press & Assessment,
a department of the University of Cambridge.

We share the University's mission to contribute to society through the pursuit of
education, learning and research at the highest international levels of excellence.

www.cambridge.org
Information on this title: www.cambridge.org/9780521681605

© Cambridge University Press & Assessment 2006

First published 2006

A catalogue record for this publication is available from the British Library

ISBN   978-0-521-68160-5   Paperback

Contents

Dedicated to the memory of

Desmond Sheiham

(13th November 1974 − 25th March 2005)

- Cambridge University (Trinity College), 1993–1997
  B.A. Hons. Mathematics 1st Class, 1996
  Part III Mathematics, Passed with Distinction, 1997

- University of Edinburgh, 1997–2001
  Ph.D. *Invariants of Boundary Link Cobordism*, 2001

- Visiting Assistant Professor, Mathematics Department,
  University of California at Riverside, 2001–2003

- Research Instructor, International University Bremen (IUB),
  2003–2005

Publications:

1. *Non-commutative Characteristic Polynomials and Cohn Localization*
   Journal of the London Mathematical Society (2) Vol. 64, 13–28 (2001)
   http://arXiv.org/abs/math.RA/0104158

2. *Invariants of Boundary Link Cobordism*
   Memoirs of the American Mathematical Society, Vol. 165 (2003)
   http://arXiv.org/abs/math.AT/0110249

3. *Whitehead Groups of Localizations and the Endomorphism Class Group*
   Journal of Algebra, Vol. 270, 261–280 (2003)
   http://arXiv.org/abs/math.KT/0209311

4. *Invariants of Boundary Link Cobordism II. The Blanchfield-Duval Form*
   in this volume, pages 142–218, http://arXiv.org/abs/math.AT/0404229

5. *Universal localization of triangular matrix rings*
   to appear in the Proceedings of the American Mathematical Society
   http://arXiv.org/abs/math.RA/0407497

6. (with A.R.) *Blanchfield and Seifert algebra in high-dimensional boundary links I. Algebraic K-theory*, http://arXiv.org/abs/math.AT/0508405

I was Des's Ph.D. supervisor at the University of Edinburgh. I prefer the German name "Doktorvater", and Des was in fact very much a mathematical son to me. We quickly found that we shared the same passion for mathematics. I suggested that he work on the computation of high-dimensional boundary link cobordism. Des's Ph.D. thesis was a complete solution of this 30-year old problem. The solution was beautiful, clever and original – as indeed was Des himself.

Des and I kept in close touch in the years after he left Edinburgh, and the relationship naturally evolved into one of mathematical brothers and collaborators. Des was not only a wonderful student but also a wonderful teacher. I certainly learned a lot from him myself! The students of his topology course at IUB were especially influenced by his teaching. Des made such an impression at IUB that he was invited to apply for a more permanent position there. In my letter of recommendation I wrote: "Dr. Sheiham is committed to both teaching and research in mathematics. He is a highly talented young mathematician who has a bright future ahead of him." It is a tragedy that he denied himself this future.

A.R.

Preface

Noncommutative localization is a powerful algebraic technique for constructing new rings by inverting elements, matrices and more generally morphisms of modules. The applications to topology are via the noncommutative localizations of the fundamental group rings.

The volume is the proceedings of a workshop on 'Noncommutative localization in algebra and topology' held at the International Centre for the Mathematical Sciences in Edinburgh on April 29 and 30, 2002, with 25 participants. The collection could be used as an introduction to noncommutative localization and its applications, but it is not an encyclopedia. Neither is it just a record of the talks at the meeting: the papers submitted by the participants are much more extensive than their talks, and in addition the volume contains papers commissioned from non-participants. The 'Historical Perspective' overleaf sets the papers in a historical context.

The meeting and the proceedings have the aim of bringing together the algebraists and topologists interested in noncommutative localization. I was particularly pleased that Professor Paul Cohn attended the meeting and contributed to the proceedings. He invented the universal 'Cohn noncommutative localization' technique which has turned out to be so useful both in its original algebraic setting and in its applications to topology.

The meeting was supported by a Scheme 1 Grant of the London Mathematical Society, and was an activity of the European Union TMR Network ERB FMRX CT-97-0107 'Algebraic K-Theory, Linear Algebraic Groups and Related Structures'. The papers in the proceedings were refereed individually. I am grateful to all the participants, to the speakers, to the authors of the papers, to the referees, to the LMS and the EU, and to the staff at ICMS, for their contributions to the success of both the meeting and the proceedings.

Additional material (such as errata) will be posted on

http://www.maths.ed.ac.uk/~aar/books

Andrew Ranicki

Edinburgh, July 2005

Historical Perspective

The localization of commutative rings is a classic technique of commutative algebra, notably in the construction of the quotient field of an integral domain. The localization of noncommutative rings was pioneered by Ore in 1931, who introduced the 'Ore condition' which allows noncommutative fractions. However, the first general reference to noncommutative localization in Mathematical Reviews only dates back to 1967; since then, the topic has grown in importance, and by now there are 525 references in total.

The history of noncommutative localization in algebra is outlined in the article of Cohn in this volume, including both the Ore localization and the more general universal localization technique he himself introduced in 1971. The article of Beachy characterizes Ore localizations as flat universal localizations. In the last 10 years the universal localization of group rings has been used to investigate the rationality of the L^2-Betti numbers and the Isomorphism Conjecture in algebraic K-theory, as described in the articles of Linnell and Reich.

Commutative localization entered algebraic topology with Serre's fundamental contributions to homotopy theory in the early 1950's. Some 20 years later the work of Sullivan, Quillen, Kan and Bousfield developed the localization of spaces, which is now a standard method. The article of Dwyer gives a homotopy-theoretic interpretation of the Cohn universal localization, making a direct link between localization in algebra and topology.

In the last 20 years noncommutative localization has been applied to the topology of manifolds via the Cappell-Shaneson homology version (1974) of the Browder-Novikov-Sullivan-Wall surgery theory (1962-1970), as well as to the circle-valued Morse-Novikov theory, and to codimension 1 splitting obstruction theory. The article of Ranicki surveys these applications. The article of Sheiham uses noncommutative localization to give a new interpretation of the high-dimensional boundary link cobordism invariants obtained in his 2001 Edinburgh Ph.D. thesis.

Noncommutative localization is closely related to the quotient construction of categories, particularly the Verdier quotient of triangulated categories developed in the 1960's. The article of Neeman makes this relation quite explicit, and also explains the connections with algebraic geometry via the work of Thomason.

The Ore noncommutative localization is a useful tool in the recent development of noncommutative geometry, as described in the article of Škoda.

Conference Participants

1. Pere Ara (Universitat Autonoma de Barcelona)

2. Andrew Baker (Glasgow University)

3. John Beachy (Northern Illinois University)

4. Tom Bridgeland (University of Edinburgh)

5. Jeremy Brookman (University of Edinburgh)

6. Ken Brown (Glasgow University)

7. Paul Cohn (University College London)

8. Diarmuid Crowley (Max Planck Institute, Bonn)

9. Matyas Domokos (University of Edinburgh)

10. Eivind Eriksen (University of Warwick)

11. Ben Franklin (Bristol University)

12. Michael Hudson (Bristol University)

13. Thomas Huettemann (Aberdeen University)

14. Colin Ingalls (University of Warwick)

15. Olav Arnfinn Laudal (University of Oslo)

16. Tom Lenagan (University of Edinburgh)

17. Amnon Neeman (Australian National University)

18. Francesc Perera (Queen's University Belfast)

19. Andrew Ranicki (University of Edinburgh)

20. Holger Reich (Universität Münster)

21. Jeremy Rickard (Bristol University)

22. Aidan Schofield (Bristol University)

23. Des Sheiham (University of California, Riverside)

24. James Shepherd (Bristol University)

25. Marco Varisco (Universität Münster)

Conference Photo

ICMS, 14 India Street, Edinburgh

Bridgeland, Huettemann, Varisco, Crowley
Ara, Domokos, Laudal, Eriksen, Brown, Beachy, Shepherd
Perera, Neeman, Ranicki, Cohn, Lenagan, Rickard
Brookman, Schofield, Ingalls, Sheiham, Hudson, Franklin, Reich

Conference Timetable

Monday, 29st April		
10.00 – 11.00	P. M. Cohn	Localization in general rings
11.30 – 12.30	H. Reich	Noncommutative localization, L^2–Betti numbers, and group von Neumann algebras
15.00 – 15.30	O. Laudal	Localizations and the structure of finitely generated algebras
16.30 – 17.30	A. Schofield	Universal localization and algebraic geometry
Tuesday, 30th April		
10.00 – 11.00	D. Sheiham	Whitehead groups of localizations
11.30 – 12.30	A. Neeman	Generalising Thomason's localization theorem to noncommutative rings
15.00 – 15.30	J. Beachy	Construction of quotient modules
16.30 – 17.30	A. Ranicki	Noncommutative localization and manifold transversality

On flatness and the Ore condition

John A. Beachy

In the standard theory of localization of a commutative Noetherian ring R at a prime ideal P, it is well-known that the localization R_P is a flat R-module. In the case of a prime ideal of a noncommutative Noetherian ring, it is not always possible to obtain a similar ring of fractions. An exposition of the standard theory in this more general situation can be found in [5]. The largest set in $R \backslash P$ that we can hope to invert is

$$\mathcal{C}(P) = \{c \in R \backslash P \mid cr \in P \text{ or } rc \in P \text{ implies } r \in P\} \,.$$

It is well-known that there exists a ring of left fractions R_P in which each element of $\mathcal{C}(P)$ is invertible if and only if $\mathcal{C}(P)$ satisfies the left Ore condition; that is, if and only if for each $a \in R$ and $c \in \mathcal{C}(P)$ there exist $b \in R$ and $d \in \mathcal{C}(P)$ with $da = bc$. In this case R_P is flat as a right R-module, as shown in Proposition II.3.5 of [5].

Even if $\mathcal{C}(P)$ does not satisfy the left Ore condition, it was shown by Cohn in [4] that it is still possible to obtain a localization at P, by inverting matrices rather than elements. Let $\Gamma(P)$ be the set of all square matrices C such that C is not a divisor of zero in the matrix ring $\mathrm{Mat}_n(R/P)$ (where C is an $n \times n$ matrix). The universal localization $R_{\Gamma(P)}$ of R at $\Gamma(P)$ is defined to be the universal $\Gamma(P)$-inverting ring. As shown in [4], the universal localization always exists, and has the desirable property that the canonical $\Gamma(P)$-inverting homomorphism $\theta : R \to R_{\Gamma(P)}$ is an epimorphism of rings. Furthermore, $R_{\Gamma(P)}/\mathrm{J}(R_{\Gamma(P)})$ is naturally isomorphic to $Q_{cl}(R/P)$, which is a simple Artinian ring by Goldie's theorem. (Here $\mathrm{J}(R_{\Gamma(P)})$ is the Jacobson radical of $R_{\Gamma(P)}$ and $Q_{cl}(R/P)$ is the classical ring of left quotients of R/P.)

It was shown by the author in [1] that $R_{\Gamma(P)}$ is flat as a right R-module only when the Ore condition is satisfied, in which case $R_{\Gamma(P)}$ coincides with R_P, the Ore ring of left fractions with denominators in $\mathcal{C}(P)$. There are similar results due to Braun [3] and Teichner [6] (see Corollaries 3 and 4, respectively). The goal of this paper is to find a general setting in which it is possible to give a common proof.

We will use the characterization of flat modules given in Proposition 10.7 of Chapter I of [5], which can be written in vector notation in the following way. The module M_R is flat $\iff$ if $\mathbf{m} \cdot \mathbf{r}^t = 0$ for $\mathbf{m} = (m_1, \ldots, m_n) \in M^n$ and $\mathbf{r} = (r_1, \ldots, r_n) \in R^n$, then there exist $A = (a_{ij}) \in \mathrm{Mat}_{k,n}(R)$ and $\mathbf{x} = (x_1, \ldots, x_k) \in M^k$ with $A\mathbf{r}^t = \mathbf{0}$ and $\mathbf{x}A = \mathbf{m}$. As a consequence, if $\theta : R \to T$ is a ring homomorphism, then θ induces on T the structure of a flat right R-module $\iff$ if $\mathbf{t} \cdot \theta(\mathbf{r})^t = 0$ for $\mathbf{t} = (t_1, \ldots, t_n) \in T^n$ and $\mathbf{r} = (r_1, \ldots, r_n) \in R^n$, then there exist $A = (a_{ij}) \in \mathrm{Mat}_{k,n}(R)$ and $\mathbf{u} = (u_1, \ldots, u_k) \in T^k$ with $A\mathbf{r}^t = \mathbf{0}$ and $\mathbf{u}\theta(A) = \mathbf{m}$.

This brief discussion brings us to the main theorem. Note that the statement of the theorem is independent of any chain conditions on the ring R.

Theorem 1. *Let $\phi : R \to Q$ be a ring homomorphism such that for all $q_1, \ldots, q_n \in Q$ there exists a unit $u \in Q$ with $uq_i \in \phi(R)$, for $1 \leq i \leq n$. Let $S \subseteq R$ be the set of elements inverted by ϕ. If there exists an S-inverting ring homomorphism $\theta : R \to T$ such that*

    (i) there exists a ring homomorphism $\eta : T \to Q$ with $\eta\theta = \phi$, and

    (ii) T is flat as a right R-module,

then S satisfies the left Ore condition.

Proof. Given $a \in R$ and $c \in S$, we must find $b \in R$ and $d \in S$ with $da = bc$. To clarify the situation, we give the following commutative diagram.

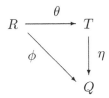

Since $c \in S$ and θ is S-inverting, it follows that $\theta(c)$ is invertible in T. If we let $\mathbf{t} = (\theta(a)\theta(c)^{-1}, 1)$ and $\mathbf{r} = (c, -a)$, then $\mathbf{t} \cdot \theta(\mathbf{r})^t = \theta(a)\theta(c)^{-1}\theta(c) - \theta(a) = 0$. As in the comments preceding the theorem, by Proposition I.10.7 of [5] there exist $\mathbf{u} \in T^k$ and $A \in \mathrm{Mat}_{2,k}(R)$ such that $A\mathbf{r} = \mathbf{0}$ and $\mathbf{u}\theta(A) = \mathbf{t}$.

From the second component of $\mathbf{t}$ we obtain $\sum_{i=1}^{k} u_i\theta(a_{i2}) = 1$. By assumption there exists a unit $u \in Q$ with $u\eta(u_i) \in \phi(R)$, for $1 \leq i \leq k$. Thus there exist $b_1, \ldots, b_k \in R$ with $u\eta(u_i) = \phi(b_i)$, for $1 \leq i \leq k$. If we let

$d = \sum_{i=1}^{k} b_i a_{i2}$, then

$$\begin{aligned}
\phi(d) &= \sum_{i=1}^{k} \phi(b_i)\phi(a_{i2}) = \sum_{i=1}^{k} u\eta(u_i)\eta\theta(a_{i2}) \\
&= u\eta(\sum_{i=1}^{k} u_i\theta(a_{i2})) = u\eta(1) = u \, ,
\end{aligned}$$

and so $d \in S$.

From the equation $\mathbf{Ar} = \mathbf{0}$ we obtain $a_{i2}a = a_{i1}c$, for $1 \leq i \leq k$. If we let $b = \sum_{i=1}^{k} b_i a_{i1}$, then

$$da = (\sum_{i=1}^{k} b_i a_{i2})a = \sum_{i=1}^{k} b_i(a_{i2}a) = \sum_{i=1}^{k} b_i(a_{i1}c) = (\sum_{i=1}^{k} b_i a_{i1})c = bc \, .$$

Thus the left Ore condition holds in S, completing the proof. End of proof.

Corollary 2 ([1], Corollary 3.2). *Let I be a semiprime left Goldie ideal of the ring R, and let $R_{\Gamma(I)}$ be the universal localization of R at I. If $R_{\Gamma(I)}$ is a flat right R-module, then $\mathcal{C}(I)$ is a left Ore set.*

Proof. Since I is assumed to be a semiprime left Goldie ideal, the ring R/I has a semisimple Artinian classical ring of left quotients $Q_{cl}(R/I)$. Let $Q = Q_{cl}(R/I)$, and let ϕ be the projection $R \to R/I$ followed by the canonical embedding $R/I \to Q$. An element $c \in R$ is in $\mathcal{C}(I)$ if and only if it is inverted by ϕ. It is well-known that the classical ring of left quotients has common denominators, so for $q_1, \ldots, q_k \in Q$ there exists an element $d \in R$ such that $d \in \mathcal{C}(I)$ and $\phi(d)q_i \in R/I$, for $1 \leq i \leq k$. Note that $\phi(d)$ is a unit of Q since $d \in \mathcal{C}(I)$. The canonical $\Gamma(I)$-inverting homomorphism $\theta : R \to R_{\Gamma(I)}$ inverts the subset $\mathcal{C}(I) \subseteq \Gamma(I)$, and the universality of θ guarantees the existence of $\eta : R_{\Gamma(I)} \to Q$ with $\eta\theta = \phi$, so the conditions of the theorem are satisfied. End of proof.

Corollary 3 ([3], Theorem 16). *Let I be a semiprime left Goldie ideal of the ring R, and let T be an extension ring of R such that $R \cap J(T) = I$ and $T/J(T)$ is naturally isomorphic to the left classical ring of quotients $Q_{cl}(R/I)$. If T is a flat right R-module, then $\mathcal{C}(I)$ satisfies the left Ore condition.*

Proof. Define Q and $\phi : R \to Q$ as in the proof of Corollary 2. If $\theta : R \to T$ is the inclusion mapping, and $\eta : T \to T/J(T)$ is the canonical projection, then $\eta\theta = \phi$ since we have assumed that the given isomorphism $T/J(T) \cong Q_{cl}(R/I)$ is natural. It follows that θ is a $\mathcal{C}(I)$-inverting homomorphism, since any element that is invertible modulo the Jacobson radical of a ring is invertible in the ring. Thus the conditions of the theorem are satisfied. End of proof.

Corollary 4 ([6], Main Theorem). *Let I be an ideal of the ring R, let $\phi : R \to R/I$ be the canonical projection, and let S be the set of elements inverted by ϕ. If the universal S-inverting ring is a flat right R-module, then S satisfies the left Ore condition.*

Proof. If R_S is the universal S-inverting ring, and $\theta : R \to R_S$ is the canonical S-inverting homomorphism, then by the universality of R_S there exists a homomorphism $\eta : R_S \to R/I$ with $\eta\theta = \phi$. Since ϕ maps R onto R/I, the remaining hypotheses of the theorem are certainly satisfied. End of proof.

References

[1] J. A. Beachy, Inversive localization at semiprime Goldie ideals, *Manuscripta Math.* **34** (1981) 211–239.

[2] J. A. Beachy, On universal localization at semiprime Goldie ideals, *Ring Theory, Proceedings of the Biennial Ohio State–Denison Conference, May, 1992*, S.K. Jain and S. Tariq Rizvi, ed., World Scientific: Singapore, New Jersey, London, Hong Kong, 1993, pp. 41–57.

[3] A. Braun, Completions of Noetherian PI rings, *J. Algebra* **133** (1990) 340–350.

[4] P. M. Cohn, Inversive localisation in Noetherian rings, *Commun. Pure Appl. Math.* **26** (1973) 679-691.

[5] B. Stenström, *Rings of Quotients*, Springer-Verlag: Berlin–Heidelberg–New York, 1975.

[6] P. Teichner, Flatness and the Ore condition, *Proc. Amer. Math. Soc.* **131**, 1977–1980 (2003).

The author would like to thank the Mathematics Department of the University of Glasgow for its hospitality during an extended visit in the spring of 2002.

Department of Mathematics
University of Glasgow, Glasgow, Scotland G12 8QW
and
Department of Mathematical Sciences
Northern Illinois University, DeKalb, IL 60115
U.S.A.

e-mail: beachy@math.niu.edu

Localization in general rings, a historical survey

P. M. Cohn

1 Introduction

The process of introducing fractions in a ring, or localization, has been applied in many different ways in algebra and geometry, and more recently it has also been used for noncommutative rings. Our object here is to survey the different methods of forming fractions, with particular emphasis on the noncommutative case. After a statement of the problem in §2 we look in §3 at different classes of rings that permit the introduction of fractions but are not embeddable in skew fields, and in §4 describe some topological methods. §5 deals with fractions in a general ring, including a statement of the necessary and sufficient conditions for embeddability in a skew field. Various classes of rings are considered in §6 and specific examples of such rings are given in §7. I should like to thank George Bergman, whose careful reading provided comments which resulted in a number of improvements. I am also indebted to a referee whose comments helped to clarify the text.

Throughout, all rings are associative, with a unit element, denoted by 1, which is inherited by subrings, preserved by homomorphisms and which acts unitally on modules. If $1 \neq 0$ and every non-zero element has an inverse, we speak of a *skew field*, but we shall frequently omit the prefix "skew", so that a "field" will mean a not necessarily commutative division ring.

2   The embedding problem

It is well known that a commutative ring is embeddable in a field if and only if it is an integral domain, i.e. $1 \neq 0$ and $a, b \neq 0 \implies ab \neq 0$.

The constructive proof of the field of fractions is well known: write $R^\times = R \backslash \{0\}$ and take the set of all formal expressions as^{-1}, where $a \in R$, $s \in R^\times$ and define $as^{-1} = a_1 s_1^{-1}$ whenever $as_1 = a_1 s$; now it is easy to verify that the result is a field of fractions for R.

For rings that are not necessarily commutative we have the following general result. Given a ring R and a subset S, a homomorphism $f : R \to R'$ will be called *S-inverting* if every element of Sf is invertible in R'. The homomorphism f is called *universal S-inverting* if it is S-inverting and every S-inverting homomorphism to a ring can be factored uniquely by f. Now we can state:

Theorem 2.1. *For every ring R and subset S of R there exists a ring R_S and a homomorphism $f : R \to R_S$ which is universal S-inverting.*

The proof is almost immediate: we take R and for each $s \in S$ adjoin an element s' with relations $ss' = s's = 1$. The resulting ring R_S with the obvious map $f : R \to R_S$ clearly satisfies the conditions.   □

The same method allows us to adjoin formal inverses for any set of matrices. The only drawback of this result is that it gives us no indication when f is injective; in fact R_S might even be the zero ring. When R is an integral domain and $S = R^\times$, f need not be injective and even if it is, the ring $R_{R^\times}$ need not be a field. In fact van der Waerden had written in 1930 (in [29]): "Die Möglichkeit der Einbettung nichtkommutativer Ringe ohne Nullteiler in einen sie enthaltenden Schiefkörper bildet ein ungelöstes Problem, außer in ganz speziellen Fällen."

In the following year O. Ore [26] posed a very natural question and gave a complete answer. He asked under what conditions a (not necessarily commutative) ring R had a field of fractions in which each element had the form as^{-1}, where $a \in R$, $s \in R^\times$. A necessary condition is that every expression of the form $s^{-1}a$ can also be written as $a's'^{-1}$, or multiplying up, $as' = sa'$, for some $a' \in R$, $s' \in R^\times$. It turns out that this condition (and the absence of zero-divisors) is also sufficient; thus we have the following result:

Theorem 2.2. (O. Ore [26]) *A ring R can be embedded in a field in which every element has the form as^{-1} ($a \in R, s \in R^{\times}$) if and only if*

O.1. R is an integral domain,

O.2. Given $a \in R$, $s \in R^{\times}$, there exist $a' \in R$, $s' \in R^{\times}$ such that $as' = sa'$.

The proof is a straightforward verification, along the lines of the proof in the commutative case. □

We note that O.2 may be expressed as: $aR \cap bR \neq 0$ for any $a, b \in R^{\times}$.

A ring satisfying O.1-2 is called a *right Ore domain*. Left Ore domains are defined similarly and for a right and left Ore domain the qualifiers are omitted. For example, any principal ideal domain is an Ore domain; this is easily verified and also follows from the next result, first noted by Goldie [16]:

Theorem 2.3. *Any right Noetherian domain satisfies O.2; hence every Noetherian domain is an Ore domain.*

Proof. Let R be right Noetherian; for $a, b \in R^{\times}$ we have to show that $aR \cap bR \neq 0$. Consider the elements $b, ab, a^2 b, \ldots$. The right ideal generated by them is finitely generated, so we have an equation

$$bc_0 + abc_1 + a^2 bc_2 + \cdots + a^{n-1} bc_{n-1} - a^n b = 0 .$$

Let c_r be the first non-zero c_i; cancelling a^r, we obtain the equation

$$bc_r + abc_{r+1} + \cdots + a^{n-r-1} bc_{n-1} - a^{n-r} b = 0 ,$$

which shows that $aR \cap bR \neq 0$. □

3 Rings that are almost embeddable

To clarify the situation we shall introduce the following classes of rings (following Bokut [6]). We denote by $\mathcal{D}_0$ the class of all integral domains (not necessarily commutative), by $\mathcal{D}_1$ the class of all rings R such that $R^\times$ is embeddable in a group, by $\mathcal{D}_2$ the class of all rings such that the universal $R^\times$-inverting map is injective, and by $\mathcal{E}$ the class of all rings embeddable in (skew) fields. Then it is clear that

$$\mathcal{D}_0 \supseteq \mathcal{D}_1 \supseteq \mathcal{D}_2 \supseteq \mathcal{E} \ .$$

The question raised by van der Waerden was whether $\mathcal{D}_0 = \mathcal{E}$. It was answered in 1937 by A. I. Malcev [20], who showed that $\mathcal{D}_0 \neq \mathcal{D}_1$ (essentially the example below) and asked whether $\mathcal{D}_1 = \mathcal{E}$. This question was answered in 1967 by three people independently: A.J. Bowtell [7] and A. A. Klein [18] gave examples showing that $\mathcal{D}_2 \neq \mathcal{E}$, and L. A. Bokut [4, 5] gave examples showing that $\mathcal{D}_1 \neq \mathcal{D}_2$. The proofs of Bowtell and Klein are fairly short and an example to show that $\mathcal{D}_1 \neq \mathcal{E}$ follows easily by general theory (see §5 below), but Bokut's proof has not been simplified.

A. I. Malcev's example was of a semigroup with cancellation which was not embeddable in a group. Its group ring over $\mathbb{Z}$ is an integral domain which is not embeddable in a field. His example may be described as follows. Let R be a ring on 8 generators, arranged as two 2×2 matrices A, B, with defining relations (expressed in matrix form)

$$AB = \begin{pmatrix} 0 & * \\ 0 & 0 \end{pmatrix} \tag{1}$$

This ring is an integral domain (by [9], Th.2.11.2), but if we adjoin inverses of all non-zero elements of R, then we can reduce the (1,1)-entry of A to 0 by row operations and the (2,2)-entry of B to 0 by column operations, all of which leaves the right-hand side of (1) unchanged, so the (1,2)-entry on the right is $\neq 0$ whereas on the left-hand side it is now equal to 0. This shows that the adjunction of inverses has mapped some elements to zero. Two years later Malcev in [21] gave necessary and sufficient conditions for a semigroup to be embeddable in a group, but this had no consequences for rings.

4 Topological methods

Some embedding methods are adapted to particular types of rings. Here we shall give two examples, group rings and filtered rings, exemplified by the universal envelopes of Lie algebras; for both there is an embedding method which is essentially topological.

Let G be a group and k a commutative field. Then the group algebra kG of G may be described as the k-algebra with the elements of G as basis and the multiplication given by the group law in G. Of course this algebra is not always an integral domain; if G has elements of finite order, say $a^n = 1$, then $a - 1$ is a zero-divisor, as is easily seen. Even for torsion-free groups it is not known whether the group algebra is an integral domain. One case of ensuring that we have an integral domain is to take an ordered group. We recall that a group G is said to be (totally) *ordered* if there is a total ordering on G such that for all $a, b, c, d \in G$, $a \leqslant b$, $c \leqslant d$ implies $ac \leqslant bd$. For an ordered group the group algebra can be embedded in a field, by the procedure of forming series. For any element $f = \sum a_g g$ ($g \in G, a_g \in k$) of the group algebra its *support* is defined as

$$\mathrm{sup}(f) = \{g \in G \mid a_g \neq 0\} .$$

Thus the group algebra may be described as the set of all expressions $f = \sum a_g g$ which have finite support. We now consider the set $k((G))$ of all series $f = \sum a_g g$ whose support is well-ordered. This set admits multiplication, for if $f_1 = \sum a_g g$, $f_2 = \sum b_g g$, then in the product $\sum a_h b_{h^{-1}g} g$ the coefficient of each g is a finite (possibly empty) sum. This algebra $k((G))$ could be called the *ordered series ring* of G over k; clearly it contains kG as a subalgebra.

Theorem 4.1. *Let G be a totally ordered group. Then the ordered series ring $k((G))$ over any field k is again a field; hence the group algebra of any totally ordered group can be embedded in a field.*

For the proof one takes a general element $a_0 g_0 + a_1 g_1 + \ldots$ and writes it as $a_0 g_0 (1 - f)$, where all elements in the support of f are > 1. Its inverse is $(1 + f + f^2 + \ldots) a_0^{-1} g_0^{-1}$ and now the well-ordering of the support is used to show that only finitely many terms $f, f^2, \ldots$ contribute to the coefficient of a given group element. For the details see [9], Th.8.7.5, or [10], Th.2.4.5; the method is due to Malcev [22] and Neumann [25] and is usually called the *Malcev-Neumann construction.* It shows that for example the group algebra of any free group can be embedded in a field, since the free group can be totally ordered by writing its elements as infinite power products of basic commutators and taking the lexicographic ordering of the exponents (see [17]).

A second method applies to certain classes of filtered rings. We recall that by a *filtered ring* one understands a ring R with a chain of submodules

$$R = R_0 \supseteq R_1 \supseteq R_2 \supseteq \dots$$

such that $R_i R_j \subseteq R_{i+j}$, $\cap R_n = 0$. With every filtered ring one can associate a graded ring: $G(R) = \bigoplus \mathrm{gr}_n(R)$, where $\mathrm{gr}_n(R) = R_n/R_{n+1}$. Now we have:

Theorem 4.2. *Let R be a filtered ring whose associated graded ring is an Ore domain. Then R can be embedded in a field.*

There have been a number of different proofs; without giving complete details, we shall mention the methods used. It is easy to see that R is an integral domain, and $R^\times$, the multiplicative semigroup of non-zero elements has a system of congruences, whose intersection is the trivial (diagonal) congruence. The quotients are embeddable in groups, and by taking their inverse limit one embeds $R^\times$ in a group. Now it is just a matter of phrasing the definition of a skew field as a group with an additional operation $x \mapsto 1 - x$ and showing that the group obtained satisfies this definition. This was the original proof in [8]. A variant of this method, using a different method to define the additive structure on a 'group with 0': $G \cup \{0\}$, was given by Dauns in [13].

Another proof, by Wehrfritz [30], uses inverse limits of quotient groups of p-jets, and a proof using the matrix ideals described in §5 below was given by Valitskas [28]; he extends the filtration to matrices and constructs a prime matrix ideal disjoint from the set of non-zero scalar matrices. The shortest proof so far is due to Lichtman [19] (see [10], 2.6); he describes the filtration by means of a variable t, leading to a t-adic valuation and then forming the localization at the multiplicative set generated by t.

5   Matrix localization

The examples of the last section show that in general the conditions for embeddability are likely to be quite complicated, and it seems natural to phrase the question more generally by asking for homomorphisms into a skew field, not necessarily injective. Even this may have a negative answer: a 2×2 matrix ring over a field, F_2, has no homomorphisms to a skew field, for F_2 is a simple ring with zero-divisors.

Let us for a moment look at the commutative case. Given a homomorphism $f : R \to K$, where K is a field and R, K are commutative, the kernel is a prime ideal $\mathfrak{p}$, and we have the following analysis of f:

$$
\begin{array}{ccc}
R & \longrightarrow & R/\mathfrak{p} \\
\downarrow & & \downarrow \\
R_{\mathfrak{p}} & \longrightarrow & K
\end{array}
$$

Suppose that K is the field generated by the image of f; then we can form K either by putting all the elements in $\mathfrak{p}$ equal to zero, thus obtaining an integral domain $R/\mathfrak{p}$, and forming the field of fractions gives us K (up to isomorphism). Or we can make all the elements outside $\mathfrak{p}$ invertible, which gives us a local ring $R_{\mathfrak{p}}$, whose residue class field is again K.

When R is noncommutative, the rings $R/\mathfrak{p}$, $R_{\mathfrak{p}}$ are not of much help, but we can try to replace $\mathfrak{p}$ by the set $\mathcal{P}$ of all square matrices mapping to singular matrices; this set $\mathcal{P}$ will be called the *singular kernel* of f. The elements of K may then be described as components of the solutions of matrix equations

$$ Au = b \, , \text{ where } A \notin \mathcal{P} \, . \tag{1} $$

Here A is a matrix over R but we shall not distinguish between the entries of A as elements of R and their images in K. It will be more convenient to combine A and b to a single matrix $(-b \quad A)$, replacing b by $-A_0$ and A by $(A_* \quad A_\infty)$ and renaming $(A_0 \quad A_* \quad A_\infty)$ as A, we can write the equation for $x \in K$

$$ Au = (A_0 \quad A_* \quad A_\infty) \begin{pmatrix} 1 \\ u_* \\ x \end{pmatrix} = 0 \, , \tag{2} $$

where A_0, A_∞ are columns and A_* is an $n \times (n-1)$ matrix. The square matrix $(A_0 \quad A_*)$ is called the *numerator* and $(A_* \quad A_\infty)$ the *denominator* of the equation (2), while A_* is the *core*. We have an analogue of Cramer's rule:

$$ (A_* \quad -A_0) = (A_* \quad A_\infty) \begin{pmatrix} I & u_* \\ 0 & x \end{pmatrix} . \tag{3} $$

In the commutative case consider (1) and let U_i be the matrix obtained from the unit matrix by replacing the i-th column by u; then $AU_i = A^{(i)}$ is the matrix obtained from A by replacing the i-th column by b. Thus $\det A^{(i)} = \det A . \det U_i = (\det A).u_i$. In the general case we no longer have determinants, but (3) can be used for many of the purposes served by Cramer's rule in the classical case.

When inverting elements we had the problem that in general a sum $as^{-1} + bt^{-1}$ or product $as^{-1}.bt^{-1}$ cannot be written in the form cw^{-1}, but when we invert matrices, this problem is overcome, for if the equations for p and q are

$$
\begin{pmatrix} A_0 & A_* & A_\infty \end{pmatrix} \begin{pmatrix} 1 \\ u \\ p \end{pmatrix} = 0 \,, \quad \begin{pmatrix} B_0 & B_* & B_\infty \end{pmatrix} \begin{pmatrix} 1 \\ v \\ q \end{pmatrix} = 0
$$

then the equations for $p - q$, pq are easily verified to be

$$
\begin{pmatrix} B_0 & B_* & B_\infty & 0 & 0 \\ A_0 & 0 & A_\infty & A_* & A_\infty \end{pmatrix} \begin{pmatrix} 1 \\ v \\ q \\ u \\ p - q \end{pmatrix} = 0 \,,
$$

$$
\begin{pmatrix} B_0 & B_* & B_\infty & 0 & 0 \\ 0 & 0 & A_0 & A_* & A_\infty \end{pmatrix} \begin{pmatrix} 1 \\ v \\ q \\ uq \\ pq \end{pmatrix} = 0 \,.
$$

We note that the denominators of $p - q$ and pq are each a triangular matrix sum of the denominators of p and q.

Our aim will be to study the properties of the set $\mathcal{P}$; we shall find that it bears close resemblance to an ideal. Its properties are the following, where by a *full* matrix we understand a matrix A which is square, say $n \times n$, and in any factorization

$$
A = PQ \,, \text{ where } P \text{ is } n \times r \text{ and } Q \text{ is } r \times n \,,
$$

we have $r \geqslant n$. We shall also abbreviate the diagonal sum $\begin{pmatrix} A & 0 \\ 0 & B \end{pmatrix}$ as $A \oplus B$.

P.1. Every non-full (square) matrix is in $\mathcal{P}$.

P.2. $A \in \mathcal{P} \Longrightarrow A \oplus B \in \mathcal{P}$ for all B.

P.3. $1 \notin \mathcal{P}$ and if $A, B \notin \mathcal{P}$, then $A \oplus B \notin \mathcal{P}$.

P.4. Suppose that $A = (A_1, A_2, \ldots, A_n)$, $A' = (A_1', A_2, \ldots, A_n) \in \mathcal{P}$; then

$$A \nabla A' = (A_1 + A_1', A_2, \ldots, A_n) \in \mathcal{P} .$$

This operation $A \nabla A'$ is called the *determinantal sum*; of course a similar rule holds for other columns and for rows; which column or row is intended will usually be clear from the context.

Given a homomorphism $f : R \to K$ to a field, with singular kernel $\mathcal{P}$, we cannot form $R/\mathcal{P}$ as in the commutative case, but the localization $R_{\mathcal{P}}$ still exists; it is formed by adjoining formal inverses of all the square matrices over R that are not in $\mathcal{P}$. As in the commutative case, here it is important to distinguish R_{Σ}, the ring obtained by formally inverting all the matrices in Σ and $R_{\mathcal{P}}$ which is R_{Σ}, where Σ is the set of all square matrices over R that are not in $\mathcal{P}$. Moreover, $R_{\mathcal{P}}$ is a local ring.

Theorem 5.1. *Let R be a ring with a homomorphism $f : R \to K$ to a field, with singular kernel $\mathcal{P}$. Then $R_{\mathcal{P}}$ is a local ring.*

Proof. The equation for $x \in R_{\mathcal{P}}$ has the form

$$Au = \begin{pmatrix} A_0 & A_* & A_\infty \end{pmatrix} \begin{pmatrix} 1 \\ u_* \\ x \end{pmatrix} = 0 ,$$

and Cramer's rule states

$$\begin{pmatrix} A_* & -A_0 \end{pmatrix} = \begin{pmatrix} A_* & A_\infty \end{pmatrix} \begin{pmatrix} I & u_* \\ 0 & x \end{pmatrix} . \tag{4}$$

Since the denominator is non-singular, we have $\begin{pmatrix} A_* & A_\infty \end{pmatrix} \notin \mathcal{P}$. To prove that $R_{\mathcal{P}}$ is a local ring we show that if x is a non-unit in $R_{\mathcal{P}}$, then $1 - x$ is a unit. If x is a non-unit, then so is the left-hand side of (4), hence $\begin{pmatrix} A_* & -A_0 \end{pmatrix} \in \mathcal{P}$. We have:

$$\begin{pmatrix} A_* & A_\infty \end{pmatrix} = \begin{pmatrix} A_* & A_\infty + A_0 \end{pmatrix} \nabla \begin{pmatrix} A_* & -A_0 \end{pmatrix} \notin \mathcal{P}$$

and it follows that the first term on the right is a unit over R. Now the equation

$$\left(A_0 + A_\infty \quad A_* \quad A_\infty \right) \begin{pmatrix} 1 \\ u_* \\ x - 1 \end{pmatrix} = 0$$

shows that $1 - x$ is a unit in $R_{\mathcal{P}}$.   □

This result suggests that we use the above properties of $\mathcal{P}$ to define it abstractly, as a certain class of matrices and try to prove that we still get a local ring. We use $\mathfrak{M}(R)$ to denote the set of all square matrices over R and make the following definitions.

Definition 1. In any ring R a *matrix ideal* is a subset $\mathcal{A}$ of $\mathfrak{M}(R)$ such that

I.1. $\mathcal{A}$ contains all non-full matrices,

I.2. If $C \in \mathcal{A}$ and $B \in \mathfrak{M}(R)$ then $C \oplus B \in \mathcal{A}$,

I.3. Any determinantal sum of matrices in $\mathcal{A}$, when defined, lies in $\mathcal{A}$.

Definition 2. A *prime* matrix ideal for R is a matrix ideal $\mathcal{P}$ such that $1 \notin \mathcal{P}$ and $A, B \notin \mathcal{P} \Longrightarrow A \oplus B \notin \mathcal{P}$.

Just as in the commutative case one can now prove:

Theorem 5.2. *Every proper matrix ideal is contained in a maximal matrix ideal, and every maximal matrix ideal is prime.*

We can also prove a result which justifies the introduction of matrix ideals:

Theorem 5.3. *Let R be a ring with a prime matrix ideal $\mathcal{P}$. Then the localization $R_{\mathcal{P}}$ is a local ring and if K is the residue-class field, then the natural map $R \to R_{\mathcal{P}} \to K$ has the singular kernel $\mathcal{P}$.*

The proof is not difficult, but quite long (see [9], Th.7.4.8 or [10], Th 4.3.1). It leads to a criterion for embeddability:

Theorem 5.4. *Let R be any ring and denote by $\mathcal{N}$ the least matrix ideal (generated by all non-full matrices). Then*

(i) *R has a homomorphism to a field if and only if $\mathcal{N}$ is proper,*

(ii) *R can be embedded in a field if and only if $\mathcal{N}$ contains no diagonal matrix with non-zero elements along the main diagonal.*

Proof (Sketch). It is clear from Th.5.3 that to find a homomorphism from R to a field we must find a prime matrix ideal, and by Th.5.2 this is possible if there is a proper matrix ideal, i.e. if the least matrix ideal is proper, so (i) is established.

The condition in (ii) is clearly necessary, for any diagonal matrix with non-zero elements along the diagonal cannot lie in the singular kernel for an embedding in a field. When (ii) is satisfied, then the set $\mathcal{S}$ of all these diagonal matrices is disjoint from $\mathcal{N}$, and an extension of Th.5.2 will show that there is a matrix ideal containing $\mathcal{N}$ and disjoint from $\mathcal{S}$ and maximal subject to these conditions, and further any such matrix ideal is prime. Clearly it will lead to a field K with a homomorphism from R to K, which is injective, by construction. □

Above we have been concerned with inverting elements of R and more generally, certain full matrices over R. Even more generally one can formally invert any matrices, even rectangular ones. Such matrices may be regarded as describing homomorphisms between free modules. A further generalization allows us to invert homomorphisms between finitely generated projective R-modules. We recall that a pair of such modules is specified by a pair of idempotent matrices E, F and a mapping from one to the other is given by a matrix A such that $AE = A = FA$. An inverse is given by a matrix X such that

$$EX \ = \ X \ = \ XF \ , \ XA \ = \ E \ , \ AX = F \ . \tag{5}$$

Thus to invert A formally we introduce indeterminates forming the matrix X with (5) as defining relations. Bergman and Dicks in [3] proved that the property of being hereditary is preserved by such localizations (recall that a ring is called hereditary if all its ideals are projective).

Another way of looking at the embedding of a ring R in a field K is to define for each R-module its rank as the dimension, considered as a K-module. This allows one to define a dependence relation which may be regarded as a matroid with the exchange property. It has been used in an alternative approach to construct the field K by Bergman in [2].

Sometimes it is convenient to express the information specifying an element of R_Σ in a single matrix. An element of R_Σ is determined by a 'matrix block' $\begin{pmatrix} A & q \\ p & c \end{pmatrix}$ where $A \in \Sigma$, $c \in R$, p is a row and q a column over R (see [15], [23]). The kernel of the natural mapping $\lambda : R \to R_\Sigma$ can be shown to consist of all $c \in R$ such that

$$\begin{pmatrix} A & 0 & 0 \\ 0 & B & q \\ p & 0 & c \end{pmatrix} = \begin{pmatrix} U \\ u \end{pmatrix} \begin{pmatrix} V & v \end{pmatrix}$$

where $A, B, U, V \in \Sigma$, p, u are rows and q, v are columns over R. This is known as Malcolmson's criterion and in certain cases it leads to a precise description of the kernel of λ (see Th.6.4 below).

6   Specializations, Semifirs and Sylvester Domains

For any ring R we understand by an R-field a field K with a homomorphism $f : R \to K$. If K is the field generated by im f, it is called an *epic R-field*. We shall denote the singular kernel of f (defined in §5) by Ker f or also Ker K. Given epic R-fields K, L such that Ker $K \subseteq$ Ker L, let us take R and localize at Ker$L$; this gives a local ring R_0 say, with L as residue-class field (by Theorem 5.3). Since Ker $K \subseteq$ Ker L, there is a homomorphism $R_0 \to K$; the image R_1, say, is a local ring; hence L is a homomorphic image of R_1. Thus we have a subring of K, a local ring generated as local ring by the image of R, with a homomorphism to L. Any homomorphism from a local subring containing R to L will be called a *specialization* from K to L over R. The set of all epic R-fields is partially ordered with respect to specialization by regarding K as greater than or equal to L if there is a specialization from K to L over R; a greatest element, if one exists, will be called a *universal R-field*. Clearly it is unique if it does exist; it is the epic R-field which has every other epic R-field as specialization. If R is embedded in its universal R-field U, then U will be called a *universal field of fractions* for R. The concept of specialization has been generalized to a situation involving more than two R-fields by Bergman [1] (see also [10], Ch.7).

Our first task is to define a class of rings which have a universal field of fractions. A ring in which every right ideal is free, of unique rank, is called a *free right ideal ring*, or *right fir* for short. Left firs are defined similarly and by a *fir* we understand a left and right fir. A ring in which every right ideal on at most n generators is free, of unique rank, is called an *n-fir*. Strictly speaking this should be called a "right n-fir", but the definition can be shown to be left-right symmetric. Let $\mathcal{F}_n$ denote the class of n-firs; then $\bigcap \mathcal{F}_n = \mathcal{F}$ is the class of rings in which every finitely generated (left or) right ideal is free of unique rank; such a ring will be called a *semifir*. In an integral domain every non-zero principal (left or) right ideal is free on one generator, and conversely, if every non-zero principal right ideal is free of unique rank 1, then we have an integral domain. Thus a 1-fir is just an integral domain, and as n increases, the class of n-firs shrinks, so

$$\mathcal{F}_1 \supset \mathcal{F}_2 \supset \mathcal{F}_3 \supset \dots , \quad \cap \mathcal{F}_n = \mathcal{F} .$$

Here all the inclusions are proper, again by Th.2.11.2 of [9], and we observe that a ring in $\mathcal{F}_2 \backslash \mathcal{F}_3$ shows that $\mathcal{D}_1 \neq \mathcal{E}$ (see §3). Now we have:

Theorem 6.1. *In every semifir the set of non-full matrices is a prime matrix ideal and the corresponding field is a universal field of fractions.* □

We shall not give the proof here (see [9], Cor. 7.5.11 or [10], Cor.4.5.9), but remark that an essential part of the proof is the fact that Sylvester's law of nullity holds in semifirs:

Given any two $n \times n$ matrices A, B

$$\rho(AB) \geqslant \rho A + \rho B - n . \tag{1}$$

This is well known to hold for matrices over a field (even skew); for more general cases we first have to extend the notion of rank. Given an $n \times n$ matrix A, consider a factorization

$$A = PQ \tag{2}$$

where P is $n \times r$ and Q is $r \times n$. If for a given A, r is minimal among all such factorizations, then (2) will be called a *rank factorization* of A and r the *inner rank* of A, denoted by ρA. This defines the rank of a matrix over any ring, and it is easily verified that over a field it reduces to the usual notion. We also note that a full matrix is just an $n \times n$ matrix of inner rank n.

Proposition 6.2. *Every semifir satisfies Sylvester's law of nullity.*

Proof. We first prove the following special case of (1):

$$\text{For any } n \times n \text{ matrices } A, B, \text{ if } AB = 0, \text{ then } \rho A + \rho B \leqslant n. \qquad (3)$$

An easy induction shows that in a free module over a semifir, any finitely generated submodule is again free, of unique rank. Let A, B be $n \times n$ matrices over a semifir R, such that $AB = 0$. If $B \neq 0$, this tells us that the submodule of R^n (as right R-module) generated by the columns of A is not free on these columns. Since this submodule is free, of rank r say, the induced map $R^n \to R^r$ splits, so the kernel is free of rank $n - r$ (by the uniqueness of the rank in semifirs). The resulting isomorphism $R^n \to R^r \oplus R^{n-r}$ yields an invertible matrix P such that AP has its first r columns linearly independent and the remainder zero. The relation $AB = 0$ now becomes $AP.P^{-1}B = 0$, so PB has its first r rows zero. It follows that $\rho A \leqslant r$, $\rho B \leqslant n - r$ and (3) follows. To deduce (1), suppose that $\rho(AB) = r$ and write $AB = PQ$, where P has r columns. Then

$$(A \quad P) \begin{pmatrix} B \\ -Q \end{pmatrix} = 0 \,,$$

hence

$$\rho A + \rho B \leqslant \rho (A \quad P) + \rho \begin{pmatrix} B \\ -Q \end{pmatrix} \leqslant n + r \,,$$

and therefore $\rho A + \rho B - n \leqslant r = \rho(AB)$, as we had to show. $\qquad \square$

The law of nullity is so useful that it is worthwhile studying the class of rings defined by it. Following Dicks and Sontag [14] we call a ring ($\neq 0$) satisfying the law of nullity a *Sylvester domain*. Clearly the inner rank of the zero element, regarded as a 1×1 matrix, is 0, for we can write 0 as product of two "empty" matrices, i.e. a 1×0 by a 0×1 matrix. Any non-zero element clearly has rank 1, hence if a, b are any elements such that $ab = 0$, then $\rho a + \rho b \leqslant 1$, so either $a = 0$ or $b = 0$. It follows that a Sylvester domain is indeed an integral domain. Further, by Prop.6.2, every semifir is a Sylvester domain.

In Th.5.4 we had a criterion for the existence of R-fields, which was not very explicit. That can now be remedied with the help of Sylvester domains. We recall that for any $S \subseteq R$, R_S denotes the universal S-inverting ring; this should not be confused with $R_{\mathcal{P}}$, an ambiguity familiar from the commutative case.

Theorem 6.3. *Let R be any ring and Φ the set of all full matrices over R. Then the localization R_Φ is a field, the universal field of fractions of R, if and only if R is a Sylvester domain.*

A complete proof can be found in [9], Th.7.5.10. As a sample we shall show that for a Sylvester domain R_Φ is a field. Thus assume that R is a Sylvester domain and let $x \in R_\Phi$ be a non-unit; if we can show that $x = 0$, then R_Φ is a field, which is what is claimed. Let

$$Au \;=\; 0 \tag{4}$$

be an equation for x. Since x is a non-unit, the numerator $N = \begin{pmatrix} A_0 & A_* \end{pmatrix}$ is not invertible over R_Φ and hence is not full over R. Thus $N = PQ$, where P is $n \times (n-1)$ and Q is $(n-1) \times n$. Therefore we have

$$A \;=\; \begin{pmatrix} A_0 & A_* & A_\infty \end{pmatrix} \;=\; \begin{pmatrix} PQ & A_\infty \end{pmatrix} \;=\; \begin{pmatrix} P & A_\infty \end{pmatrix} \begin{pmatrix} Q & 0 \\ 0 & 1 \end{pmatrix} .$$

Here the denominator is full, hence $\begin{pmatrix} P & A_\infty \end{pmatrix}$, as a left factor of the denominator, is also full, and so is invertible over R_Φ. Substituting this expression for A into (4) we can cancel the left factor $\begin{pmatrix} P & A_\infty \end{pmatrix}$ in (4) and conclude that $x = 0$. □

For a Sylvester domain Malcolmson's criterion leads to the following explicit description of the kernel:

Theorem 6.4. *Let R be a Sylvester domain. Then a matrix block over R represents 0 if and only if it is not full.*

The proof depends on the fact that for a Sylvester domain the product (or diagonal sum) of full matrices is again full (see [10], Th.4.5.11).

7 Examples

It only remains to give some actual examples of firs, semifirs and Sylvester domains. As already noted, every principal ideal domain is a fir, hence the polynomial ring in one variable over a field, $k[x]$, is a fir. For more than one variable this is no longer true, but if we do not allow the variables to commute, we obtain the free associative algebra $k\langle x_1, \ldots, x_n \rangle$, and this is a fir, even when the number of variables is infinite. The proof is by a weakened form of the Euclidean algorithm, called the *weak algorithm*, and can be found in [9], Cor.2.4.3, [10], Th.5.4.1, [11], Th.5.15, 17). More generally, any family of firs having a common subfield has a coproduct, which is again a fir ([10], Th.5.3.9). This shows in particular that any family of fields with a common subfield has a coproduct which is a fir. An example of a Sylvester domain (apart from the firs and semifirs already mentioned) is the polynomial ring $k[x, y]$ in two variables over a field. However, when the number of variables exceeds two, the ring obtained is no longer a Sylvester domain. For 3 variables this is most easily seen by considering the matrix

$$\begin{pmatrix} 0 & z & -y \\ -z & 0 & x \\ y & -x & 0 \end{pmatrix}$$

which is full (as some trials will show) and yet a zero-divisor:

$$\begin{pmatrix} 0 & z & -y \\ -z & 0 & x \\ y & -x & 0 \end{pmatrix} \begin{pmatrix} x \\ y \\ z \end{pmatrix} = 0 .$$

We also note that Sylvester's law of nullity fails even for full matrices:

$$\begin{pmatrix} 0 & z & -y \\ -z & 0 & x \\ y & -x & 0 \end{pmatrix} \begin{pmatrix} 1 & 0 & 0 \\ 0 & 1 & 0 \\ 0 & 0 & z \end{pmatrix} = \begin{pmatrix} 0 & z \\ -z & 0 \\ y & -x \end{pmatrix} \begin{pmatrix} 1 & 0 & -x \\ 0 & 1 & -y \end{pmatrix} .$$

Of course the polynomial ring in any number of variables over a field has a universal field of fractions; all the above argument shows is that over a polynomial ring in more than two variables some full matrices cannot be inverted.

References

[1] Bergman, G. M., *Rational relations and rational identities in division rings. II*, J. Algebra 43 (1976), 267–297.

[2] ———, *Constructing division rings as module-theoretic direct limits*, Trans. Amer. Math. Soc. 354 (2002), 2079–2114.

[3] _____ and Dicks, W., *Universal derivations and universal ring constructions*, Pacif. J. Math. 79 (1978), 293–337.

[4] Bokut, L. A., *The embedding of rings in skew fields* (Russian), Dokl. Akad. Nauk SSSR 175 (1967), 755–758.

[5] _____, *On Malcev's problem* (Russian), Sibirsk. Mat. Zh. 10 (1969), 965–1005.

[6] _____, *Associative Rings 1, 2* (Russian). NGU Novosibirsk 1981.

[7] Bowtell, A. J., *On a question of Malcev*, J. Algebra 9 (1967), 126–139.

[8] Cohn, P. M., *On the embedding of rings in skew fields*, Proc. London Math. Soc. (3) 11 (1961), 511–530.

[9] _____, *Free Rings and their Relations*, 2nd Ed. LMS Monographs No.19, Academic Press 1985.

[10] _____, *Skew Fields, Theory of General Division Rings*, Encyclopedia of Mathematics and its Applications, Vol.57, Cambridge University Press 1995.

[11] _____, *Introduction to Ring Theory*, SUMS, Springer Verlag London 2000.

[12] _____, *Free ideal rings and localization in general rings*, Mathematical Monographs, Cambridge University Press (to appear).

[13] Dauns, J., *Embeddings in division rings*, Trans. Amer. Math. Soc. 150 (1970), 287–299.

[14] Dicks, W. and Sontag, E. D., *Sylvester domains*, J. Pure Appl. Algebra 13 (1978), 243–275.

[15] Gerasimov, V. N., *Localizations in associative rings* (Russian), Sibirsk. Mat. Zh. 23 (1982), 36–54.

[16] Goldie, A. W., *The structure of prime rings under ascending chain conditions*, Proc. London Math. Soc. (3) 8 (1958), 589–608.

[17] Hall, M. Jr., *Theory of Groups*, Macmillan, New York 1959.

[18] Klein, A. A., *Rings nonembeddable in fields with multiplicative semigroups embeddable in groups*, J. Algebra 7 (1967), 100–125.

[19] Lichtman, A. I., *Valuation methods in division rings*, J. Algebra 177 (1995), 870–898.

[20] Malcev, A. I., *On the immersion of an algebraic ring into a field*, Math. Ann. 113 (1937), 686–691.

[21] _____, *Über die Einbettung von assoziativen Systemen in Gruppen*, (Russian, German summary) I. Mat. Sb. 6(48) (1939), 331–336, II, 8(50) (1940), 251–264.

[22] _____, *On the embedding of group algebras in division algebras*, (Russian), Doklady Akad. Nauk SSSR 60 (1948), 1499–1501.

[23] Malcolmson, P., *Construction of universal matrix localizations*, Advances in non-comm. Ring theory. Proc. 12th G. H. Hudson Symp. Plattsburgh, 1981. Lecture Notes in Math. No. 951, Springer-Verlag, Berlin 1982, pp. 117–131.

[24] _____, *Matrix localizations of n-firs*, Trans. Amer. Math. Soc. 282 (1984), I 503–518, II 519–527.

[25] Neumann, B. H., *On ordered division rings*, Trans. Amer. Math. Soc. 66 (1949), 202–252.

[26] Ore, O., *Linear equations in non-commutative fields*, Ann. Math. 32 (1931), 463–477.

[27] Schofield, A. H., *Representations of Rings in Skew Fields*, LMS Lecture Notes No.92, Cambridge University Press 1985.

[28] Valitskas, A. I., *P.Cohn's method of embedding certain filtered rings in skew fields* (Russian), to appear.

[29] van der Waerden, B. L., *Moderne Algebra I*, Springer-Verlag, Leipzig 1930.

[30] Wehrfritz, B. A. F., *On Cohn's method of embedding an enveloping algebra into a division ring*, Ukrain. Mat. Zh. 44 (1992), 729–735.

Department of Mathematics
University College London
Gower Street
London WC1E 6BT
United Kingdom

e-mail: pmc@math.ucl.ac.uk

Noncommutative localization in homotopy theory

William G. Dwyer

1 Introduction

In a sense, noncommutative localization is at the center of homotopy theory, or even more accurately, one form of it *is* homotopy theory. After all, Gabriel and Zisman [9] and later Quillen [18] observed that the homotopy category of CW-complexes can be obtained from the category of topological spaces by formally inverting the maps which are weak homotopy equivalences. More generally, the homotopy category of any Quillen model category [6] [11] can be built by formally inverting maps. In a slightly different direction, the process of *localization with respect to a map* (§2) has recently developed into a powerful tool for making homotopy-theoretic constructions [2, §4]; roughly speaking, localizing with respect to f involves converting an object X into a new one, $L_f(X)$, with the property that, as far as mapping into $L_f(X)$ goes, f looks like an equivalence.

In this paper we will show how the Cohn noncommutative localization described in [19] can be interpreted as an instance of localization with respect to a map (3.2). Actually, we produce a derived form of the Cohn localization, and show that the circumstances in which the Cohn localization is most useful are exactly those in which the higher derived information vanishes (3.3). Finally, we sketch how the derived Cohn localization can sometimes be computed by using a derived form of the categorical localization construction from Gabriel and Zisman (§4).

1.1 The context. It is necessary to choose what to work with: algebraic objects, such as rings, chain complexes, and differential graded algebras (DGAs), or geometric ones, such as ring spectra and module spectra [7] [12]. Since this paper focuses mostly on Cohn localization, we've picked the algebraic option. If R is a ring, the term *R-module* will refer to an

This research was partially supported by NSF grant DMS02–04169 (unbounded) chain complex over R. See [21], [1], or [22, §10] for algebraic accounts of how to work with these complexes, and [13] for a topological approach. The differentials in our complexes always lower degree by one, and all unspecified modules are *left* modules. To maintain at least a little topological standing, we denote the i'th homology group of an R-module X by $\pi_i X$; this is in fact isomorphic to the i'th homotopy group of the Eilenberg-MacLane spectrum corresponding to X [7]. A map between R-modules which induces an isomorphism on π_* is called a *quasi-isomorphism* or *equivalence*; the *homotopy category of R-modules* (also knows as the *derived category of R*) is obtained from the category of R-modules by formally inverting the equivalences. A *cofibration sequence* of R-modules is one which becomes a distinguished triangle in the derived category. If $f : X \to Y$ is a map of R-modules, the *cofibre C* of f is the chain complex mapping cone of f [22, 1.2.8], and there is a cofibration sequence $X \to Y \to C$. We use Σ for the shift or suspension operator.

An ordinary module M over R gives rise to an R-module in our sense by treating M as a chain complex concentrated in degree 0; we refer to such an M as a *discrete* module over R, and we do not distinguish in notation between M and its associated complex.

We will sometimes work in a context which includes differential graded algebras (DGAs) [22, 4.5.2]. In this setting a ring is identified with the associated DGA concentrated in degree 0.

1.2 Tensor and Hom. The symbol $\otimes_R$ refers to the tensor product of two R-modules, and Hom_R to the complex of homomorphisms between two R-modules (for this last, see [22, 2.7.4], but reindex so that all of the differentials reduce degree by one). For our purposes, both $\otimes_R$ and Hom_R are always taken in the derived sense, so that modules are to be replaced by suitable resolutions before the tensor product or function object is formed. Along the same lines, $\mathrm{End}_R(X)$ denotes the DGA given by the derived endomorphism complex of the R-module X.

These conventions are such that if R is a ring, M a discrete right module over R, and N a discrete left module, then $M \otimes_R N$ is a complex with

$$\pi_i(M \otimes_R N) \cong \begin{cases} \mathrm{Tor}_i^R(M, N) & i \geq 0 \\ 0 & i < 0. \end{cases}$$

Similarly, if M and N are discrete left R-modules, $\mathrm{Hom}_R(M, N)$ is a complex with

$$\pi_i \mathrm{Hom}_R(M, N) \cong \begin{cases} \mathrm{Ext}_R^{-i}(M, N) & i \leq 0 \\ 0 & i > 0. \end{cases}$$

In particular, $\pi_i \mathrm{End}_R(M) \cong \mathrm{Ext}_R^{-i}(M, M)$.

2 Localization with respect to a map

Suppose that R is a ring and that $f : A \to B$ is a map of R-modules.

2.1 Definition. An R-module Y is said to be f-*local* if f induces an equivalence $\mathrm{Hom}_R(B, Y) \to \mathrm{Hom}_R(A, Y)$.

In other words, Y is f-local if, as far as mapping into Y is concerned, f looks like an equivalence.

2.2 Definition. A map $X \to X'$ of R-modules is said to be an f-*local equivalence* if it induces an equivalence $\mathrm{Hom}_R(X', Y) \to \mathrm{Hom}_R(X, Y)$ for every f-local R-module Y. An f-*localization* of X is a map $\epsilon : X \to L_f(X)$, such that $L_f(X)$ is f-local and ϵ is an f-local equivalence.

2.3 Remark. It is not hard to see that any two f-localizations of X are equivalent, so that we can speak loosely of *the* f-localization of X. For any map f and R-module X, the f-localization $L_f(X)$ of X exists, and the construction of $L_f(X)$ can be made functorial in X (see [10], or 2.12 below). The functor L_f preserves equivalences, is idempotent up to equivalence, and preserves cofibration sequences up to equivalence. An R-module X is f-local if and only if $X \to L_f(X)$ is an equivalence. A map g of R-modules is an f-local equivalence if and only if $L_f(g)$ is an equivalence.

2.4 Remark. Let C be the cofibre of f. For any R-module Y there is a cofibration sequence

$$\mathrm{Hom}_R(C, Y) \to \mathrm{Hom}_R(B, Y) \to \mathrm{Hom}_R(A, Y).$$

This shows that Y is f-local if and only $\mathrm{Hom}_R(C, Y)$ is contractible, i.e, if and only if Y is local with respect to $0 \to C$. This last condition is sometimes expressed by saying that Y is C-*null* [8, 1.A.4]. The f-localization functor L_f can also be interpreted as a C-nullification functor.

2.5 Proposition. *Up to equivalence, the R-module $L_f(R)$ is a DGA, in such a way that the localization map $R \to L_f(R)$ is a morphism of DGAs.*

Proof. Let $Y = L_f(R)$ and $\mathcal{E}$ be the endomorphism DGA $\mathrm{End}_R(Y)$. Since Y is f-local, the map $R \to Y$ induces an equivalence

$$\mathcal{E} = \mathrm{End}_R(Y) \xrightarrow{\sim} \mathrm{Hom}_R(R, Y) = Y.$$

The action of R on Y then gives a double commutator map

$$R \to \operatorname{End}_{\mathcal{E}}(Y) \sim \operatorname{End}_{\mathcal{E}}(\mathcal{E}) \sim \mathcal{E} \sim Y \, .$$

It is easy to see that this is essentially the localization map $R \to Y$. Identifying Y with $\operatorname{End}_{\mathcal{E}}(Y)$ gives the required DGA structure. $\qquad\square$

From now on we will treat $L_f(R)$ as a DGA and $R \to L_f(R)$ as a homomorphism of DGAs.

2.6 Definition. The localization functor L_f is *smashing* if for every R-module X the map $X \sim R \otimes_R X \to L_f(R) \otimes_R X$ is an f-localization map.

2.7 Remark. For any R-module X, the natural map $X \to L_f(R) \otimes_R X$ is an f-local equivalence; one way to see this is to pick an f-local Y and consider the chain of equivalences

$$\operatorname{Hom}_R(L_f(R) \otimes_R X, Y) \sim \operatorname{Hom}_R(X, \operatorname{Hom}_R(L_f(R), Y)) \sim \operatorname{Hom}_R(X, Y) \, .$$

The question of whether L_f is smashing, then, is the question of whether for every X the R-module $L_f(R) \otimes_R X$ is f-local.

2.8 Remark. If L_f is smashing then the category of f-local R-modules is equivalent, from a homotopy point of view, to the category of $L_f(R)$-modules. In particular, the homotopy category of f-local R-modules is equivalent to the homotopy category of $L_f(R)$-modules.

2.9 Examples. Let $R = \mathbb{Z}$, pick a prime p, and let f be the map $\mathbb{Z} \xrightarrow{p} \mathbb{Z}$. Then L_f is smashing, and $L_f(X) \sim \mathbb{Z}[1/p] \otimes_{\mathbb{Z}} X$.

On the other hand, if f is the map $\mathbb{Z}[1/p] \to 0$, then L_f is the Ext-p-completion functor [2, 2.5], which is the total left derived functor of the p-completion functor. In particular, $L_f(\mathbb{Z}) \sim \mathbb{Z}_p$ and $L_f(\mathbb{Z}/p^{\infty}) \sim \Sigma \mathbb{Z}_p$. Since $\Sigma \mathbb{Z}_p$ is not equivalent to $\mathbb{Z}_p \otimes_{\mathbb{Z}} \mathbb{Z}/p^{\infty}$, L_f is *not* smashing in this case.

The main positive result about smashing localizations is due to Miller. Recall that an R-module A is said to be *small* if $\operatorname{Hom}_R(A, -)$ commutes up to equivalence with arbitrary coproducts. This is the same as saying that A is finitely built from R, or that A is equivalent to a chain complex of finite length made up of finitely generated projective R-modules.

2.10 Proposition. *[15] Let $f : A \to B$ be a map of R-modules. If A and B are small, or more generally if the cofibre C of f is equivalent to a coproduct of small R-modules, then L_f is smashing.*

2.11 Lemma. *If the cofibre C of $f : A \to B$ is equivalent to a coproduct of small R-modules, then the class of f-local R-modules is closed under arbitrary coproducts.*

Proof. Write $C \sim \coprod_\alpha C_\alpha$, where each C_α is small. Then Y is f-local if and only if Y is C-null (2.4), which is the case if and only if Y is C_α-null for each α. The lemma now follows from the fact that $\mathrm{Hom}_R(C_\alpha, -)$ commutes up to equivalence with coproducts. $\square$

Proof of 2.10. Consider the class of R-modules X for which $L_f(R) \otimes_R X$ is f-local. We have to show that this is the class of all R-modules (2.7). However, the class contains R itself, is closed under cofibration sequences (2.3), is closed under arbitrary coproducts (2.11), and is closed under equivalences. The usual method for constructing resolutions shows that this is enough to give the desired result. $\square$

2.12 Construction of $L_f(X)$. We will sketch an explicit description of $L_f(X)$, at least up to equivalence, in the case in which the cofibre C of f is equivalent to a coproduct of small R-modules. Actually, we will assume that C itself is small, since the adjustments to handle the general case are mostly notational.

Recall that the *homotopy colimit* of a sequence $X_0 \xrightarrow{\sigma_0} X_1 \xrightarrow{\sigma_1} \cdots$ of R-modules is the cofibre of the map $\sigma : \coprod X_i \to \coprod X_i$ given by $\sigma_i(x_i) = \sigma_i(x_i) - x_i$. The description of $L_f(X)$ depends on two observations.

1. $\mathrm{Hom}_R(C, -)$ commutes up to equivalence with sequential homotopy colimits.

2. If U is a coproduct of copies of suspensions of C, $g : U \to X$ is a map of R-modules, and X' is the cofibre of g, then $X \to X'$ is an f-local equivalence.

Item (1) is clear from the description above of sequential homotopy colimits. For (2), pick an f-local Y, consider the cofibre sequence

$$\mathrm{Hom}_R(X', Y) \to \mathrm{Hom}_R(X, Y) \to \mathrm{Hom}_R(U, Y),$$

and observe that the term on the right is trivial (2.4).

Consider a set of representatives $g_\alpha : \Sigma^{n_\alpha} C \to X$ for all nontrivial homotopy classes of maps from suspensions of C to X. Let $U = \coprod_\alpha \Sigma^{n_\alpha} C$, let $g : U \to X$ be the sum of the maps $\{g_\alpha\}$, and let $\Phi(X)$ denote the cofibre of g. There is a natural map $X \to \Phi(X)$. Iterate the process to construct a sequential diagram

$$X \to \Phi(X) \to \Phi^2(X) \to \cdots \to \Phi^n(X) \to \cdots , \qquad (2.13)$$

and let $\Phi^\infty(X) = \mathrm{hocolim}_n \, \Phi^n(X)$. We claim that $X \to \Phi^\infty(X)$ is an f-localization map, so that $\Phi^\infty(X) \sim L_f(X)$. The fact that $\Phi^\infty(X)$ is f-local follows from 2.4 and (1) above, since every map from a suspension of C into $\Phi^\infty(X)$ factors up to homotopy through $\Phi^n(X)$ for some n, and so is null homotopic, since it becomes null homotopic by construction in $\Phi^{n+1}(X)$. To see that $X \to \Phi^\infty(X)$ is an f-local equivalence, observe that by (2) above and induction the map $X \to \Phi^n(X)$ is an f-local equivalence for each $n \geq 1$. For an f-local Y it is now possible to compute

$$\mathrm{Hom}_R(\Phi^\infty(X), Y) \sim \mathrm{holim}_n \, \mathrm{Hom}_R(\Phi^n(X), Y)$$
$$\sim \mathrm{holim}_n \, \mathrm{Hom}_R(X, Y)$$
$$\sim \mathrm{Hom}_R(X, Y).$$

2.14 Remark. The standard construction of $L_f(X)$ is similar to the above, but slightly more complicated [10, 4.3]. To make the construction functorial, and not just functorial up to equivalence, it is necessary to build $\Phi(X)$ by using *all* maps from suspensions of C to X, not just a set of representatives of nontrivial homotopy classes. We neglected to mention above that C should have been replaced up to equivalence by a projective complex (cofibrant model); in the general setting there's also a slight adjustment [10, 4.2.2] to deal with the fact that X might not be fibrant, in other words, to deal with the fact that not every map $\Sigma^n C \to X$ in the homotopy category is necessarily represented by an actual map $\Sigma^n C \to X$. Finally, if C is not small the countable homotopy colimit in 2.13 has to be replaced by a parallel transfinite construction [10, 10.5].

2.15 Other structure. There is more that can be said if the cofibre C of f is small. Let $C^\# = \mathrm{Hom}_R(C, R)$. There is a "homology theory" on the category of R-modules determined by the functor $X \mapsto \pi_*(C^\# \otimes_R X)$; let $X \to \hat{X}$ denote Bousfield localization with respect to this theory [8, 1.E.4] [10, xi]. Then for any X there is a homotopy fibre square

$$
\begin{array}{ccc}
X & \longrightarrow & \hat{X} \\
\downarrow & & \downarrow \\
L_f(X) & \longrightarrow & L_f(\hat{X}).
\end{array}
$$

In the case in which $R = \mathbb{Z}$ and f is the map $\mathbb{Z} \xrightarrow{p} \mathbb{Z}$, this is the arithmetic square

$$
\begin{array}{ccc}
X & \longrightarrow & X_{\hat{p}} \\
\downarrow & & \downarrow \\
\mathbb{Z}[1/p] \otimes_{\mathbb{Z}} X & \longrightarrow & \mathbb{Z}[1/p] \otimes_{\mathbb{Z}} (X_{\hat{p}}) .
\end{array}
$$

See [3] for other results along these lines and for some (mostly commutative) examples.

3   The Cohn localization

In this section we construct the Cohn localization from the point of view of §2. Let R be a ring and let $\{f_\alpha : P_\alpha \to Q_\alpha\}$ be some set of maps between discrete (1.1) finitely generated projective R-modules. Let f denote $\coprod_\alpha f_\alpha$. The main results are as follows.

3.1 Proposition. *The DGA $L_f(R)$ is (-1)-connected, i.e., $\pi_i L_f(R)$ vanishes for $i < 0$.*

If P is a discrete R-module, let $P^{\#} = \mathrm{Ext}^0_R(P, R)$ denote its usual dual, and note that $P^{\#}$ is a discrete right R-module. For each $f_\alpha : P_\alpha \to Q_\alpha$, let $f_\alpha^{\#} : Q_\alpha^{\#} \to P_\alpha^{\#}$ be the dual of f_α, and let S denote the set $\{f_\alpha^{\#}\}$. Recall that a ring homomorphism $R \to R'$ is said to be *S-inverting* if for each α the map $\mathrm{Tor}_0^R(f_\alpha^{\#}, R')$ is an isomorphism. A *Cohn localization of R with respect to S* is an initial object $R \to S^{-1}R$ in the category of S-inverting ring homomorphisms $R \to R'$ [19, Part 1].

The map $R \to L_f(R)$ of DGAs (2.5) induces a ring homomorphism $R = \pi_0 R \to \pi_0 L_f(R)$.

3.2 Proposition. *The map $R \to \pi_0 L_f(R)$ is a Cohn localization of R with respect to S.*

From now on we will denote $\pi_0 L_f(R)$ by $\mathcal{L}$. In light of 3.2, we think of the DGA $L_f(R)$ as a *derived Cohn localization of R* with respect to S.

Recall [17] that the ring homomorphism $R \to \mathcal{L}$ is said to be *stably flat* if $\mathrm{Tor}_i^R(\mathcal{L}, \mathcal{L}) = 0$ for $i > 0$. It is in the stably flat case that Cohn localization leads to K-theory localization sequences.

3.3 Proposition. *The map $R \to \mathcal{L}$ is stably flat if and only if the groups $\pi_i L_f(R)$ vanish for $i > 0$.*

In other words, $R \to \mathcal{L}$ is stably flat if and only if $L_f(R)$ is equivalent as a DGA to $\mathcal{L}$, or if and only if the "higher derived functors" of Cohn localization, given by $\pi_i L_f(R)$, $i > 0$, vanish.

The rest of this section is taken up with proofs. Observe to begin with that the objects P_α and Q_α are small (§2) as R-modules, and so the cofibre of f is a coproduct of small objects. It follows that L_f is smashing (2.10) and that up to equivalence there is a relatively simple construction for $L_f(X)$ (2.12).

3.4 Proposition. *An R-module X is f-local if and only if each group $\pi_i X$ is f-local.*

3.5 Remark. It might be useful to spell out the meaning of this. The object R is an ordinary ring and the object X is a chain complex over R. Each group $\pi_i X$ is a discrete R-module, which can be treated as a chain complex over R concentrated in degree 0. The proposition states that X is f-local if and only if, for each $i \in \mathbb{Z}$, the R-module obtained from $\pi_i X$ is f-local.

Proof of 3.4. If P is a discrete projective module over R, it is not hard to see that there are natural isomorphisms

$$\pi_i \operatorname{Hom}_R(P, X) \cong \operatorname{Ext}^0_R(P, \pi_i X).$$

This is clearly true if P is free, and follows in general from a retract argument. The proposition then follows from definition 2.1.   □

3.6 Lemma. *Suppose that $h : P \to Q$ is a map of discrete finitely generated projective R-modules, W is the cofibre of h, X is an R-module which is (-1)-connected, $g : \Sigma^n W \to X$ is a map which is not null homotopic, and X' is the cofibre of g. Then X' is (-1)-connected.*

Proof. By a retract argument, we can assume that P and Q are free, so that $P \cong R^i$ and $Q \cong R^j$. In view of the definition of W, there is a cofibration sequence

$$\operatorname{Hom}_R(W, X) \to X^j \to X^i.$$

The corresponding long exact homotopy sequence shows that $\pi_k \operatorname{Hom}_R(W, X)$ vanishes for $k < -1$. Since $g : \Sigma^n W \to X$ is essential, it follows that $n \geq -1$. This gives a cofibration sequence

$$X \to X' \to \Sigma^{n+1} W \tag{3.7}$$

with $n + 1 \geq 0$. It is clear that there are isomorphisms

$$\pi_k W = \begin{cases} \operatorname{coker}(h) & k = 0 \\ \ker(h) & k = 1 \\ 0 & \text{otherwise} \end{cases}$$

and so in particular that $\pi_k W = 0$ for $k < 0$. The proof is completed by looking at the long exact homotopy sequence of 3.7. $\qquad\square$

Proof of 3.1. This follows from 3.6 and the construction of $L_f(R)$ sketched in 2.12. $\qquad\square$

Proof of 3.2. If P is a discrete finitely generated projective R-module, then for any discrete R-module M, there is a natural isomorphism

$$\operatorname{Tor}_0^R(P^\#, M) \cong \operatorname{Ext}_R^0(P, M) \,.$$

In particular, as in the proof of 3.4, the map $\operatorname{Tor}_0^R(f_\alpha^\#, M)$ is an isomorphism for all α if and only if M is f-local. By 3.4, $\mathcal{L}$ is f-local, and it follows that $R \to \mathcal{L}$ is S-inverting.

Now, suppose that $R \to R'$ is an arbitrary S-inverting ring homomorphism. As above, the ring R' is f-local as an R-module, and this implies that the map $R \to L_f(R)$ induces an equivalence

$$\operatorname{Hom}_R(L_f(R), R') \to \operatorname{Hom}_R(R, R') \sim R' \,. \tag{3.8}$$

In conjunction with 3.1, the universal coefficient spectral sequence

$$\operatorname{Ext}_R^i(\pi_j L_f(R), R') \Rightarrow \pi_{-i-j} \operatorname{Hom}_R(L_f(R), R')$$

shows that $\pi_0 \operatorname{Hom}_R(L_f(R), R')$ is isomorphic to $\operatorname{Ext}_R^0(\mathcal{L}, R')$. Applying π_0 to the equivalence 3.8 thus shows that every homomorphism $R \to R'$ of discrete modules over R extends uniquely to a homomorphism $\mathcal{L} \to R'$. In particular, the given ring homomorphism $u : R \to R'$ extends uniquely to $v : \mathcal{L} \to R'$. To show that v is a ring homomorphism, it is enough to show that for each element λ of $\mathcal{L}$, the two maps $a, b : \mathcal{L} \to R'$ given by $a(x) = v(x\lambda)$ and $b(x) = v(x)v(\lambda)$ are the same. Both a and b are maps of discrete R-modules, and so it is in fact enough to show that a and b agree when composed with the map $R \to \mathcal{L}$. But this is just the statement that v is a map of discrete R-modules. $\qquad\square$

3.9 Lemma. *Suppose that X and Y are respectively right and left R-modules such that $\pi_i X$ and $\pi_i Y$ vanish for $i < 0$. Then there are natural isomorphisms*

$$\pi_i(X \otimes_R Y) \cong \begin{cases} \operatorname{Tor}_0^R(\pi_0 X, \pi_0 Y) & i = 0 \\ 0 & i < 0. \end{cases}$$

Proof. This follows from the Künneth spectral sequences

$$\pi_i((\pi_j X) \otimes_R Y) \Rightarrow \pi_{i+j}(X \otimes_R Y)$$
$$\operatorname{Tor}_i^R(M, \pi_j Y) \Rightarrow \pi_{i+j}(M \otimes_R Y).$$

In the second spectral sequence, M is a discrete right R-module (e.g., $\pi_k X$ for some $k \geq 0$). $\square$

3.10 Lemma. *The natural map*

$$L_f(R) \sim R \otimes_R L_f(R) \to L_f(R) \otimes_R L_f(R)$$

is an equivalence. The natural map $\mathcal{L} \cong \operatorname{Tor}_0^R(R, \mathcal{L}) \to \operatorname{Tor}_0^R(\mathcal{L}, \mathcal{L})$ is an isomorphism.

Proof. Since $L_f(L_f(R)) \sim L_f(R)$, the first statement follows from the fact that L_f is smashing (2.10). The second then follows from 3.1 and 3.9. $\square$

Proof of 3.3. Suppose that $\pi_i L_f(R) = 0$ for $i > 0$, or in other words (3.1), that $L_f(R) \sim \mathcal{L}$. It follows from 3.10 that $\mathcal{L} \otimes_R \mathcal{L} \sim \mathcal{L}$; applying π_* then gives isomorphisms

$$\operatorname{Tor}_i^R(\mathcal{L}, \mathcal{L}) \cong \pi_i(\mathcal{L} \otimes_R \mathcal{L}) \cong \begin{cases} \mathcal{L} & i = 0 \\ 0 & \text{otherwise.} \end{cases}$$

Suppose on the other hand that $\pi_i(\mathcal{L} \otimes_R \mathcal{L}) \cong \operatorname{Tor}_i^R(\mathcal{L}, \mathcal{L})$ vanishes for $i > 0$. Consider the class of all $\mathcal{L}$-modules X with the property that the natural map

$$X \sim R \otimes_R X \to \mathcal{L} \otimes_R X$$

is an equivalence. This class includes $\mathcal{L}$ (3.9, 3.10), is closed under equivalences, is closed under cofibration sequences, and is closed under arbitrary coproducts. As in the proof of 2.10, this is enough to show that the class contains all $\mathcal{L}$-modules. In particular, for any discrete $\mathcal{L}$-module M there are isomorphisms

$$\operatorname{Tor}_i^R(\mathcal{L}, M) \cong \begin{cases} M & i = 0 \\ 0 & i > 0. \end{cases}$$

Each group $\pi_j L_f(R)$ is a module over $\mathcal{L} = \pi_0 L_f(R)$, and so it follows from the Künneth spectral sequence

$$\mathrm{Tor}_i^R(\mathcal{L}, \pi_j L_f(R)) \Rightarrow \pi_{i+j}(\mathcal{L} \otimes_R L_f(R))$$

that the natural map $L_f(R) \to \mathcal{L} \otimes_R L_f(R)$ is an equivalence, and in particular that the R-module structure on $L_f(R)$ extends to an $\mathcal{L}$-module structure. This structure can be used to factor the natural map $R \to L_f(R)$ as a composite $R \to \mathcal{L} \to L_f(R)$. Applying L_f to this composite gives a diagram

$$L_f(R) \to \mathcal{L} \to L_f^2(R)$$

in which we have used 3.4 to identify $L_f(\mathcal{L}) \sim \mathcal{L}$. The composite map $L_f(R) \to L_f^2(R)$ is an equivalence, since $R \to L_f(R)$ is an f-local equivalence. Applying π_* shows that $\pi_i L_f(R) \cong 0$ for $i > 0$. $\square$

4   Localization of categories

In this section we sketch without proof a connection between the Cohn localization of a ring and the process of forming the derived localization of a category. The connecting link between the two is the notion of *ring with several objects*.

4.1 Derived localization of categories. Suppose that $\mathbf{C}$ is a small category and $\mathbf{W}$ a subcategory which contains all the objects of $\mathbf{C}$. The *localization of $\mathbf{C}$ with respect to $\mathbf{W}$* is a functor $\mathbf{C} \to \mathbf{W}^{-1}\mathbf{C}$ which is initial in the category of all functors with domain $\mathbf{C}$ which take the arrows in $\mathbf{W}$ into isomorphisms. A derived form of this localization can be constructed by forming a free simplicial resolution $(F\mathbf{C}, F\mathbf{W})$ of the pair $(\mathbf{C}, \mathbf{W})$ and taking the dimensionwise localization $(F\mathbf{W})^{-1}F\mathbf{C}$ [5]. This results in a category $L(\mathbf{C}, \mathbf{W})$ with the same objects as $\mathbf{C}$, but enriched over simplicial sets. Up to an enriched analog of categorical equivalence, $L(\mathbf{C}, \mathbf{W})$ is the same as the hammock localization of [4], and from this point of view there is a natural functor $\mathbf{C} \to L(\mathbf{C}, \mathbf{W})$. This functor is universal, in an appropriate sense, among functors from $\mathbf{C}$ to categories enriched over simplicial sets which send the arrows of $\mathbf{W}$ into maps which are invertible up to homotopy.

4.2 Examples. The following examples do not involve small categories, but it is still possible to make sense of them. Let $\mathbf{C}$ be the category of topological spaces and $\mathbf{W}$ the subcategory of weak homotopy equivalences. Let X and Y be spaces with CW-approximations X' and Y'. Then the set of maps $X \to Y$ in $\mathbf{W}^{-1}\mathbf{C}$ is isomorphic to the set of homotopy classes of maps

$X' \to Y'$; the simplicial set of maps $X \to Y$ in $L(\mathbf{C}, \mathbf{W})$ is equivalent to the singular complex of the mapping space $\mathrm{Map}(X', Y')$.

Let R be a ring, $\mathbf{C}$ the category of unbounded chain complexes over R, i.e., the category of R-modules, and $\mathbf{W} \subset \mathbf{C}$ the subcategory of quasi-isomorphisms. Then $\mathbf{W}^{-1}\mathbf{C}$ is the derived category of R. If X and Y are R-modules, then the homotopy groups of the simplicial set of maps $X \to Y$ in $L(\mathbf{C}, \mathbf{W})$ are $\pi_i \mathrm{Hom}_R(X, Y)$, $i \geq 0$.

4.3 Rings with several objects. A *ring T with several objects* is a small additive category [16]; a discrete T-module is an additive functor from T to abelian groups. There is a category of discrete T-modules in which the morphisms are natural transformations between functors. Define a *T-module* to be a chain complex of discrete T-modules, i.e., an additive functor from T to the category of chain complexes over $\mathbb{Z}$. One can build a homotopy theory of T-modules in which the weak equivalences are natural transformations which are objectwise quasi-isomorphisms (see [20] for geometric versions of this). We use the notation $\mathrm{Hom}_T(X, Y)$ for the derived chain complex of maps between two T-modules X, Y.

Every object $x \in T$ gives rise to a discrete small projective T-module P_x, where P_x assigns to y the group of maps $x \to y$ in T. Suppose that $\{f_\alpha : P_{x_\alpha} \to P_{y_\alpha}\}$ is a set of maps between such projectives, and let $f = \coprod_\alpha f_\alpha$. The ideas in §2 give for any T-module X an f-local module $L_f(X)$ and an f-local equivalence $X \to L_f(X)$. There is an associated category $L_f(T)$ enriched over chain complexes (in other words, $L_f(T)$ is "a DGA with several objects"); this has the same objects as T, and the function complex of maps x to y in $L_f(T)$ is given by $\mathrm{Hom}_T(L_f(y), L_f(x))$. There is a functor $i : T \to L_f(T)$ and by the same smallness argument used in the proof of 2.10, the functor L_f can be identified as (derived) left Kan extension along i.

If $\mathbf{C}$ is a simplicial category, let $\mathbb{Z}\mathbf{C}$ denote the simplicial additive category obtained by applying the free abelian group functor dimensionwise to the morphism sets of $\mathbf{C}$. There is an associated category $N\mathbb{Z}\mathbf{C}$ enriched over chain complexes, formed by normalizing the simplicial abelian groups which appear as morphism objects in $\mathbb{Z}\mathbf{C}$ and using the Eilenberg-Zilber formula [22, 6.5.11] to define composition. If $\mathbf{C}$ is an ordinary category treated as a simplicial category with discrete morphism sets, then $N\mathbb{Z}\mathbf{C}$ is the additive category obtained by taking free abelian groups on the morphism sets of $\mathbf{C}$, so at what we hope is minimal risk of confusion we will just denote it $\mathbb{Z}\mathbf{C}$.

If $(\mathbf{C}, \mathbf{W})$ is a pair of categories as above (4.1), then for each morphism $w : x \to y$ in $\mathbf{W}$ let $f_w : P_y \to P_x$ be the corresponding map between projective $\mathbb{Z}\mathbf{C}$-modules and let $f_{\mathbf{W}} = \coprod f_w$.

4.4 Proposition. *Let* $(\mathbf{C}, \mathbf{W})$ *and* $f = f_{\mathbf{W}}$ *be as above. Then in an appropriate enriched sense the two categories* $N\mathbb{Z}L(\mathbf{C}, \mathbf{W})$ *and* $L_f(\mathbb{Z}\mathbf{C})$ *are equivalent.*

"Equivalence" here means that the two categories are related by a zigzag of morphisms between enriched categories with the property that these morphisms give the identity map on object sets and induce quasi-isomorphisms on function complexes.

4.5 Rings. Suppose that T is a ring with a *finite* number of objects, in other words, a small additive category with a finite number of objects. Let $P = \coprod_x P_x$, where the coproduct runs through all of the objects in T and P_x is the projective from 4.3. Let $\mathcal{E}$ be the endomorphism ring of P in the category of discrete T-modules, and $\mathcal{P}(T)$ the ring $\mathcal{E}^{\mathrm{op}}$. The notation $\mathcal{P}(T)$ is meant to suggest that this is a kind of path algebra of T. As an abelian group, $\mathcal{P}(T)$ is isomorphic to the sum $\coprod_{x,y} T(x,y)$ of all of the morphism groups of T; products are defined by using the composition in T to the extent possible and otherwise setting the products equal to 0. Since P is a small projective generator for the category of discrete T-modules, ordinary Morita theory shows that $\mathrm{Ext}_T^0(P, -)$ gives an equivalence between the category of discrete T-modules and the category of discrete $\mathcal{P}(T)$-modules. Not surprisingly, this extends to a homotopy-theoretic equivalence between the category of T-modules and the category of $\mathcal{P}(T)$-modules.

The construction $\mathcal{P}(-)$ can be extended to categories enriched over chain complexes; if T' is such a category, then $\mathcal{P}(T')$ is a DGA.

4.6 Proposition. *Suppose that* T *is a ring with a finite number of objects,* $f : P \to Q$ *is a map between discrete small projective* T-modules, and $g : P' \to Q'$ *the corresponding map between discrete finitely-generated* $\mathcal{P}(T)$-*modules. Then the DGA* $\mathcal{P}(L_f(T))$ *is, in an appropriate sense, equivalent to the DGA* $L_g(\mathcal{P}(T))$.

4.7 Remark. The word "equivalent" in the proposition signifies that the DGA $L_g(\mathcal{P}(T))$ is related to $\mathcal{P}(L_f(T))$ by a zigzag of quasi-isomorphisms between DGAs.

4.8 An example. This is along the lines of [19, 2.4]. Let $H \leftarrow G \to K$ be a two-source of groups. Form a category $\mathbf{C}$ with three objects, x, y, and z and the following pattern of morphisms

$$
\begin{array}{ccccc}
y & \overset{H}{\leftarrow} & x & \overset{K}{\longrightarrow} & z \\
| & & | & & | \\
H & & G & & K
\end{array}
$$

This signifies, for instance, that H is the set of maps $x \to y$, and G is the monoid of endomorphisms of x. The action of G on H by composition is then the translation action determined by the given homomorphism $G \to H$.

Let ξ denote the pushout of the diagram $H \leftarrow G \to K$ of groups and X the homotopy pushout of the diagram $BH \leftarrow BG \to BK$ of spaces. By the van Kampen theorem, $\pi_1 X \cong \xi$.

4.9 Lemma. *The nerve of* **C** *is equivalent to* X.

Let $\mathbf{W} \subset \mathbf{C}$ be the subcategory whose nonidentity morphisms are the maps $x \to y$ and $x \to z$ corresponding to the identity elements of H and K, respectively. Note that all of the morphisms in $\mathbf{W}^{-1}\mathbf{C}$ are invertible; in fact, $\mathbf{W}^{-1}\mathbf{C}$ is isomorphic to a connected groupoid with three objects x, y, and z and vertex groups isomorphic to ξ. Let ΩX denote the simplicial loop group of X.

4.10 Proposition. *[5] The simplicial category* $L(\mathbf{C}, \mathbf{W})$ *is weakly equivalent to a connected simplicial groupoid with three objects* x, y, *and* z *and vertex group* ΩX.

Let $f = f_{\mathbf{W}}$ be map between projective $\mathbb{Z}\mathbf{C}$ modules determined as above by $\mathbf{W}$, and g the corresponding map between projective $\mathcal{P}(\mathbb{Z}\mathbf{C})$-modules. Note that $\mathcal{P}(\mathbb{Z}\mathbf{C})$ is the matrix ring associated in [19, 2.4] to the amalgamated product $\mathbb{Z}H *_{\mathbb{Z}G} \mathbb{Z}K$ and that g is the sum of the two maps σ_1 and σ_2 described there. Concatenating 4.4 with 4.6 gives the following result.

4.11 Proposition. *The DGA* $L_g\mathcal{P}(\mathbb{Z}\mathbf{C})$ *is equivalent in an appropriate sense to the* 3×3-*matrix algebra on the chain algebra* $C_*(\Omega X; \mathbb{Z})$. *In particular, for* $i \geq 0$ *there are natural isomorphisms*

$$\pi_i L_g \mathcal{P}(\mathbb{Z}\mathbf{C}) \cong H_i(\Omega X; \mathbb{Z}) \oplus \cdots \oplus H_i(\Omega X; \mathbb{Z}) \quad (9 \text{ times}).$$

This Cohn localization is stably flat (3.3) if and only if the universal cover of X is acyclic and X itself is equivalent to $B\xi$; this occurs, for instance, if the maps $G \to H$ and $G \to K$ are injective. It does not occur if $G = \mathbb{Z}$ and H and K are the trivial group.

Here's another example, which can be treated along the same lines. Let X be a connected space and M the monoid constructed by McDuff [14] with BM weakly equivalent to X. Let R be the monoid ring $\mathbb{Z}M$, and for each $m \in M$ let $f_m : R \to R$ be given by right multiplication by m. Denote the sum $\coprod_m f_m$ by f. Then for $i \geq 0$ there are natural isomorphisms

$$\pi_i L_f(R) \cong H_i(\Omega X; \mathbb{Z}).$$

References

[1] Luchezar L. Avramov and Hans-Bjørn Foxby, *Homological dimensions of unbounded complexes*, J. Pure Appl. Algebra **71** (1991), no. 2-3, 129–155. MR 93g:18017

[2] W. G. Dwyer, *Localizations*, Axiomatic, Enriched and Motivic Homotopy Theory (J. P. C. Greenlees, ed.), Kluwer, 2004, Proceedings of the NATO ASI, pp. 3–28.

[3] W. G. Dwyer and J. P. C. Greenlees, *Complete modules and torsion modules*, Amer. J. Math. **124** (2002), no. 1, 199–220. MR 2003g:16010

[4] W. G. Dwyer and D. M. Kan, *Calculating simplicial localizations*, J. Pure Appl. Algebra **18** (1980), no. 1, 17–35. MR 81h:55019

[5] ———, *Simplicial localizations of categories*, J. Pure Appl. Algebra **17** (1980), no. 3, 267–284. MR 81h:55018

[6] W. G. Dwyer and J. Spaliński, *Homotopy theories and model categories*, Handbook of algebraic topology, North-Holland, Amsterdam, 1995, pp. 73–126. MR 96h:55014

[7] A. D. Elmendorf, I. Kříž, M. A. Mandell, and J. P. May, *Rings, modules, and algebras in stable homotopy theory*, American Mathematical Society, Providence, RI, 1997, With an appendix by M. Cole. MR 97h:55006

[8] Emmanuel Dror Farjoun, *Cellular spaces, null spaces and homotopy localization*, Lecture Notes in Mathematics, vol. 1622, Springer-Verlag, Berlin, 1996. MR 98f:55010

[9] P. Gabriel and M. Zisman, *Calculus of fractions and homotopy theory*, Ergebnisse der Mathematik und ihrer Grenzgebiete, Band 35, Springer-Verlag New York, Inc., New York, 1967. MR 35 #1019

[10] Philip S. Hirschhorn, *Model categories and their localizations*, Mathematical Surveys and Monographs, vol. 99, American Mathematical Society, Providence, RI, 2003. MR 2003j:18018

[11] Mark Hovey, *Model categories*, Mathematical Surveys and Monographs, vol. 63, American Mathematical Society, Providence, RI, 1999. MR 99h:55031

[12] Mark Hovey, Brooke Shipley, and Jeff Smith, *Symmetric spectra*, J. Amer. Math. Soc. **13** (2000), no. 1, 149–208. MR 2000h:55016

[13] Igor Kříž and J. P. May, *Operads, algebras, modules and motives*, Astérisque (1995), no. 233, iv+145pp. MR 96j:18006

[14] Dusa McDuff, *On the classifying spaces of discrete monoids*, Topology **18** (1979), no. 4, 313–320. MR 81d:55020

[15] Haynes Miller, *Finite localizations*, Bol. Soc. Mat. Mexicana (2) **37** (1992), no. 1-2, 383–389, Papers in honor of José Adem (Spanish). MR 96h:55009

[16] Barry Mitchell, *Rings with several objects*, Advances in Math. **8** (1972), 1–161. MR 45 #3524

[17] A. Neeman and A. Ranicki, *Noncommutative localization and chain complexes I. Algebraic K- and L-theory*, http://arXiv.math.RA/0109118 Geometry and Topology **8** (2004), 1385–1425. MR 2119300

[18] Daniel G. Quillen, *Homotopical algebra*, Lecture Notes in Mathematics, No. 43, Springer-Verlag, Berlin, 1967. MR 36 #6480

[19] Andrew Ranicki, *Noncommutative localization in topology*, http://arXiv.math.AT/0303046 pp. 81–102 in this volume.

[20] Stefan Schwede and Brooke Shipley, *Stable model categories are categories of modules*, Topology **42** (2003), 103–153. MR 2003g:55034

[21] N. Spaltenstein, *Resolutions of unbounded complexes*, Compositio Math. **65** (1988), no. 2, 121–154. MR 89m:18013

[22] Charles A. Weibel, *An introduction to homological algebra*, Cambridge University Press, Cambridge, 1994. MR 95f:18001

Department of Mathematics
University of Notre Dame
Notre Dame
IN 46556 USA

e-mail: dwyer.1@nd.edu

Noncommutative localization in group rings

Peter A. Linnell

Abstract

This paper will briefly survey some recent methods of localization in group rings, which work in more general contexts than the classical Ore localization. In particular the Cohn localization using matrices will be described, but other methods will also be considered.

1 Introduction

Let R be a commutative ring and let $S = \{s \in R \mid sr \neq 0 \text{ for all } r \in R \setminus 0\}$, the set of non-zerodivisors of R. Then, in the same manner as one constructs $\mathbb{Q}$ from $\mathbb{Z}$, we can form the quotient ring RS^{-1} which consists of elements of the form r/s with $r \in R$ and $s \in S$, and in which $r_1/s_1 = r_2/s_2$ if and only if $r_1 s_2 = s_1 r_2$. We can consider R as a subring RS^{-1} by identifying $r \in R$ with $r/1 \in RS^{-1}$. Then RS^{-1} is a ring containing R with the property that every element is either a zerodivisor or invertible. Furthermore, every element of RS^{-1} can be written in the form rs^{-1} with $r \in R$ and $s \in S$ (though not uniquely so). In the case R is an integral domain, then RS^{-1} will be a field and will be generated as a field by R (i.e. if K is a subfield of RS^{-1} containing R, then $K = RS^{-1}$). Moreover if K is another field containing R which is generated by R, then K is isomorphic to RS^{-1} and in fact there is a ring isomorphism $RS^{-1} \to K$ which is the identity on R.

The question we will be concerned with here is what one can do with a noncommutative ring R; certainly many of the above results do not hold in general. In particular, Malcev [24] constructed domains which are not embeddable in division rings. We will concentrate on the case when our ring

Keywords Cohn localization, Ore condition, ring of quotients
AMS Classification Primary: 16S10; Secondary: 16U20, 20C07

is a crossed product $k*G$, where k is a division ring and G is a group [25], and in particular when the crossed product is the group ring kG with k a field. A field will always mean a commutative field, and we shall use the terminology "division ring" for the noncommutative case. Though our main interest is in group rings, often it is a trivial matter to extend results to crossed products. This has the advantage of facilitating induction arguments, because if $H \lhd G$ and $k * G$ is a crossed product, then $k * G$ can also be viewed as a crossed product $(k * H) * (G/H)$ [25, p. 2].

2   Ore Localization

We shall briefly recall the definition of a crossed product, and also establish some notational conventions for this paper. Let R be a ring with a 1 and let G be a group. Then a crossed product of G over R is an associative ring $R*G$ which is also a free left R-module with basis $\{\bar{g} \mid g \in G\}$. Multiplication is given by $\bar{x}\bar{y} = \tau(x,y)\overline{xy}$ where $\tau(x,y)$ is a unit of R for all $x,y \in G$. Furthermore we assume that $\bar{1}$ is the identity of $R * G$, and we identify R with $R\bar{1}$ via $r \mapsto r\bar{1}$. Finally $\bar{x}r = r^{\sigma(x)}\bar{x}$ where $\sigma(x)$ is an automorphism of R for all $x \in G$; see [25, p. 2] for further details.

We shall assume that all rings have a 1, subrings have the same 1, and ring homomorphisms preserve the 1. We say that the element s of R is a non-zerodivisor (sometimes called a regular element) if $sr \neq 0 \neq rs$ whenever $0 \neq r \in R$; otherwise s is called a zerodivisor. Let S denote the set of non-zerodivisors of the ring R. The simplest extension to noncommutative rings is when the ring R satisfies the *right Ore condition*, that is given $r \in R$ and $s \in S$, then there exists $r_1 \in R$ and $s_1 \in S$ such that $rs_1 = sr_1$. In this situation one can form the Ore localization RS^{-1}, which in the same way as above consists of elements of the form $\{rs^{-1} \mid r \in R, s \in S\}$. If $s_1 s = s_2 r$, then $r_1 s_1^{-1} = r_2 s_2^{-1}$ if and only if $r_1 s = r_2 r$; this does not depend on the choice of r and s. To define addition in RS^{-1}, note that any two elements can be written in the form $r_1 s^{-1}, r_2 s^{-1}$ (i.e. have the same common denominator), and then we set $r_1 s^{-1} + r_2 s^{-1} = (r_1 + r_2)s^{-1}$. To define multiplication, if $s_1 r = r_2 s$, we set $(r_1 s_1^{-1})(r_2 s_2^{-1}) = r_1 r(s_2 s)^{-1}$. Then RS^{-1} is a ring with $1 = 11^{-1}$ and $0 = 01^{-1}$, and $\{r1^{-1} \mid r \in R\}$ is a subring isomorphic to R via the map $r \mapsto r1^{-1}$. Furthermore RS^{-1} has the following properties:

- Every element of S is invertible in RS^{-1}.

- Every element of RS^{-1} is either invertible or a zerodivisor.

- If $\theta\colon R \to K$ is a ring homomorphism such that θs is invertible for all $s \in S$, then there is a unique ring homomorphism $\theta'\colon RS^{-1} \to K$ such that $\theta'(r1^{-1}) = \theta r$ for all $r \in R$; in other words, θ can be extended in a unique way to RS^{-1}.

- RS^{-1} is a flat left R-module [32, Proposition II.3.5].

Of course one also has the left Ore condition, which means that given $r \in R$ and $s \in S$, one can find $r_1 \in R$ and $s_1 \in S$ such that $s_1 r = r_1 s$, and then one can form the ring $S^{-1}R$, which consists of elements of the form $s^{-1}r$ with $s \in S$ and $r \in R$. However in the case of the group ring kG for a field k and group G, they are equivalent by using the involution on kG induced by $g \mapsto g^{-1}$ for $g \in G$. When a ring satisfies both the left and right Ore condition, then the rings $S^{-1}R$ and RS^{-1} are isomorphic, and can be identified. In this situation, we say that RS^{-1} is a classical ring of quotients for R. When R is a domain, a classical ring of quotients will be a division ring. On the other hand if already every element of R is either invertible or a zerodivisor, then R is its own classical quotient ring. For more information on Ore localization, see [13, §9].

Problem 2.1. *Let k be a field. For which groups G does kG have a classical quotient ring?*

One could ask more generally given a division ring D, for which groups G does a crossed product $D * G$ always have a classical quotient ring? We have put in the "always" because D and G do not determine a crossed product $D * G$. One could equally consider the same question with "always" replaced by "never".

For a nonnegative integer n, let F_n denote the free group on n generators, which is nonabelian for $n \geq 2$. If G is abelian in Problem 2.1, then kG certainly has a classical quotient ring because kG is commutative in this case. On the other hand if G has a subgroup isomorphic to F_2, then $D * G$ cannot have a classical quotient ring. We give an elementary proof of this well-known statement, which is based on [18, Theorem 1].

Proposition 2.2. *Let G be a group which has a subgroup isomorphic to the free group F_2 on two generators, let D be a division ring, and let $D * G$ be a crossed product. Then $D * G$ does not satisfy the right Ore condition, and in particular does not have a classical quotient ring.*

Proof. First suppose G is free on a, b. We prove that $(\bar{a}-1)D*G\cap(\bar{b}-1)D* G = 0$. Write $A = \langle a \rangle$ and $B = \langle b \rangle$. Suppose $\alpha \in (\bar{a}-1)D*G\cap(\bar{b}-1)D*G$.

Then we may write

$$\alpha = \sum_i (u_i - 1)\overline{x_i}d_i = \sum_i (v_i - 1)\overline{y_i}e_i \tag{2.1}$$

where $u_i = \bar{a}^{q(i)}$ for some $q(i) \in \mathbb{Z}$, $v_i = \bar{b}^{r(i)}$ for some $r(i) \in \mathbb{Z}$, $d_i, e_i \in D$ and $x_i, y_i \in G$. The general element g of G can be written in a unique way $g_1 \ldots g_l$, where the g_i are alternately in A and B, and $g_i \neq 1$ for all i; we shall define the length $\lambda(g)$ of g to be l. Of course $\lambda(1) = 0$. Let L be the maximum of all $\lambda(x_i), \lambda(y_i)$, let s denote the number of x_i with $\lambda(x_i) = L$, and let t denote the number of y_i with $\lambda(y_i) = L$. We shall use induction on L and then on $s + t$, to show that $\alpha = 0$. If $L = 0$, then $x_i, y_i = 1$ for all i and the result is obvious. If $L > 0$, then without loss of generality, we may assume that $s > 0$. Suppose $\lambda(x_i) = L$ and x_i starts with an element from A, so $x_i = a^p h$ where $0 \neq p \in \mathbb{Z}$ and $\lambda(h) = L - 1$. Then

$$(u_i - 1)\overline{x_i}d_i = (\bar{a}^{q(i)} - 1)\bar{a}^p \bar{h}dd_i = (\bar{a}^{q(i)+p} - 1)\bar{h}dd_i - (\bar{a}^p - 1)\bar{h}dd_i$$

for some $d \in D$. This means that we have found an expression for α with smaller $s + t$, so all the x_i with $\lambda(x_i) = L$ start with an element from B. Therefore if $\beta = \sum_i u_i\overline{x_i}d_i$ where the sum is over all i such that $\lambda(x_i) = L$, then each x_i starts with an element of B and hence $\lambda(a^{q(i)}x_i) = L + 1$. We now see from (2.1) that $\beta = 0$. Since $s > 0$ by assumption, the expression for β above is nontrivial and therefore there exists $i \neq j$ such that $a^{q(i)}x_i = a^{q(j)}x_j$. This forces $q(i) = q(j)$ and $x_i = x_j$. Thus $u_i = u_j$ and we may replace $(u_i - 1)\overline{x_i}d_i + (u_j - 1)\overline{x_j}d_j$ with $(u_i - 1)\overline{x_i}(d_i + d_j)$, thereby reducing s by 1 and the proof that $(\bar{a} - 1)D * G \cap (\bar{b} - 1)D * G = 0$ is complete.

In general, suppose G has a subgroup H which is free on the elements x, y. Then the above shows that $(x - 1)D * H \cap (y - 1)D * H = 0$, and it follows that $(x - 1)D * G \cap (y - 1)D * G = 0$. Since $x - 1$ and $y - 1$ are non-zerodivisors in $D * G$, it follows that $D * G$ does not have the right Ore property. □

Recall that the class of elementary amenable groups is the smallest class of groups which contains all finite groups and the infinite cyclic group $\mathbb{Z}$, and is closed under taking group extensions and directed unions. It is not difficult to show that the class of elementary amenable groups is closed under taking subgroups and quotient groups, and contains all solvable-by-finite groups. Moreover every elementary amenable group is amenable, but F_2 is not amenable. Thus any group which has a subgroup isomorphic to F_2 is not elementary amenable. Also Thompson's group F [4, Theorem 4.10] and the Gupta-Sidki group [14] are not elementary amenable even though they do not contain F_2. The Gupta-Sidki has sub-exponential growth [10] and is therefore amenable [26, Proposition 6.8]. The following result follows from [17, Theorem 1.2]

Theorem 2.3. *Let G be an elementary amenable group, let D be a division ring, and let $D * G$ be a crossed product. If the finite subgroups of G have bounded order, then $D * G$ has a classical ring of quotients.*

It would seem plausible that Theorem 2.3 would remain true without the hypothesis that the finite subgroups have bounded order. After all, if G is a locally finite group and k is a field, then kG is a classical quotient ring for itself. However the lamplighter group, which we now describe, yields a counterexample. If A, C are groups, then $A \wr C$ will indicate the Wreath product with base group $B := A^{|C|}$, the direct sum of $|C|$ copies of A. Thus B is a normal subgroup of $A \wr C$ with corresponding quotient group isomorphic to C, and C permutes the $|C|$ copies of A regularly. The case $A = \mathbb{Z}/2\mathbb{Z}$ and $C = \mathbb{Z}$ is often called the lamplighter group. Then [21, Theorem 2] is

Theorem 2.4. *Let $H \neq 1$ be a finite group, let k be a field, and let G be a group containing $H \wr \mathbb{Z}$. Then kG does not have a classical ring of quotients.*

Thus we have the following problem.

Problem 2.5. *Let k be a field. Classify the elementary amenable groups G for which kG has a classical ring of quotients. If $H \leqslant G$ and kG has a classical ring of quotients, does kH also have a classical ring of quotients?*

The obstacle to constructing a classical quotient ring in the case of elementary amenable groups is the finite subgroups having unbounded order, so let us consider the case of torsion-free groups. In this situation it is unknown whether kG is a domain, so let us assume that this is the case. Then we have the following result of Tamari [33]; see [8, Theorem 6.3], also [23, Example 8.16], for a proof.

Theorem 2.6. *Let G be an amenable group, let D be a division ring, and let $D * G$ be a crossed product which is a domain. Then $D * G$ has a classical ring of quotients which is a division ring.*

What about torsion-free groups which do not contain F_2, yet are not amenable? Given such a group G and a division ring D, it is unknown whether a crossed product $D * G$ has a classical quotient ring. Thompson's group F is orderable [4, Theorem 4.11]; this means that it has a total order $\leq$ which is left and right invariant, so if $a \leq b$ and $g \in F$, then $ga \leq gb$ and $ag \leq bg$. Therefore if D is a division ring and $D * F$ is a crossed product, then by the Malcev-Neumann construction [5, Corollary 8.7.6] the power series ring $D((F))$ consisting of elements with well-ordered support is a division ring. It is still unknown whether Thompson's group is amenable. We state the following problem.

Problem 2.7. *Let F denote Thompson's orderable group and let D be a division ring. Does $D * F$ have a classical ring of quotients?*

If the answer is negative, then Theorem 2.6 would tell us that Thompson's group is not amenable. Since Thompson's group seems to be right on the borderline between amenability and nonamenability, one would expect the answer to be in the affirmative.

3 Cohn's Theory

What happens when the ring R does not have the Ore condition, in other words R does not have a classical ring of quotients? Trying to form a ring from R by inverting the non-zerodivisors of R does not seem very useful. The key idea here is due to Paul Cohn; instead of trying to invert just elements, one inverts matrices instead. Suppose Σ is any set of matrices over R (not necessarily square, though in practice Σ will consist only of square matrices) and $\theta \colon R \to S$ is a ring homomorphism. If M is a matrix with entries $m_{ij} \in R$, then θM will indicate the matrix over S which has entries $\theta(m_{ij})$. We say that θ is Σ-inverting if θM is invertible over S for all $M \in \Sigma$. We can now define the universal localization of R with respect to Σ, which consists of a ring R and a *universal* Σ-inverting ring homomorphism $\lambda \colon R \to R_\Sigma$. This means that given any other Σ-inverting homomorphism $\theta \colon R \to S$, then there is a unique ring homomorphism $\phi \colon R_\Sigma \to S$ such that $\theta = \phi \lambda$. The ring R_Σ always exists by [5, Theorem 7.2.1], and by the universal property is unique up to isomorphism. Furthermore λ is injective if and only if R can be embedded in a ring over which all the matrices in Σ become invertible.

A related concept is the Σ-rational closure. Given a set of matrices Σ over R and a Σ-inverting ring homomorphism $\theta \colon R \to S$, the Σ-rational closure $R_\Sigma(S)$ of R in S consists of all entries of inverses of matrices in $\theta(\Sigma)$. In general $R_\Sigma(S)$ will not be a subring of S. We say that Σ is upper multiplicative if given $A, B \in \Sigma$, then $\begin{pmatrix} A & C \\ 0 & B \end{pmatrix} \in \Sigma$ for any matrix C of the appropriate size. If in addition permuting the rows and columns of a matrix in Σ leaves it in Σ, then we say that Σ is multiplicative.

Suppose now that Σ is a set of matrices over R and $\theta\colon R \to S$ is a Σ-inverting ring homomorphism. If Σ is upper multiplicative, then $R_\Sigma(S)$ is a subring of S [5, Theorem 7.1.2]. Also if Φ is the set of matrices over R whose image under θ becomes invertible over S, then Φ is multiplicative [5, Proposition 7.1.1]. In this situation we call $R_\Phi(S)$ the rational closure $\mathcal{R}_S(R)$ of R in S. By the universal property of R_Φ, there is a ring homomorphism $R_\Phi \to R_\Phi(S) = \mathcal{R}_S(R)$ which is surjective. A very useful tool is the following consequence of [5, Proposition 7.1.3], which we shall call Cramer's rule; we shall let $\mathrm{M}_n(R)$ denote the $n \times n$ matrices over R.

Proposition 3.1. *Let Σ be an upper multiplicative set of matrices of R and let $\theta\colon R \to S$ be a Σ-inverting ring homomorphism. If $p \in R_\Sigma(S)$, then p is stably associated to a matrix with entries in $\theta(R)$. This means that there exists a positive integer n and invertible matrices $A, B \in \mathrm{M}_n(S)$ such that $A\operatorname{diag}(p, 1, \ldots, 1)B \in \mathrm{M}_n(\theta R)$.*

Given a ring homomorphism $\theta\colon R \to S$ and an upper multiplicative set of matrices Σ of R, the natural epimorphism $R_\Sigma \twoheadrightarrow R_\Sigma(S)$ will in general not be isomorphism, even if θ is injective, but there are interesting situations where it is; we describe one of them. Let k be a PID (principal ideal domain), let X be a set, let $k\langle X\rangle$ denote the free algebra on X, let $k\langle\langle X\rangle\rangle$ denote the noncommutative power series ring on X, and let Λ denote the subring of $k\langle\langle X\rangle\rangle$ generated by $k\langle X\rangle$ and $\{(1+x)^{-1} \mid x \in X\}$. Then $\Lambda \cong kF$ where F denotes the free group on X [5, p. 529]. Let Σ consist of those square matrices over Λ with constant term invertible over k, and let $\Sigma' = \Sigma \cap k\langle X\rangle$. If we identify Λ with kF by the above isomorphism, then Σ consist of those matrices over kF which become invertible under the augmentation map $kF \to k$. Since Σ and Σ' are precisely the matrices over Λ and $k\langle X\rangle$ which become invertible over $k\langle\langle X\rangle\rangle$ respectively, we see that $k\langle X\rangle_{\Sigma'}(k\langle\langle X\rangle\rangle) = \Lambda_\Sigma(k\langle\langle X\rangle\rangle) = \mathcal{R}_{k\langle\langle X\rangle\rangle}(k\langle X\rangle) = \mathcal{R}_{k\langle\langle X\rangle\rangle}(\Lambda)$. By universal properties, we have a sequence of natural maps

$$\Lambda \xrightarrow{\alpha} k\langle X\rangle_{\Sigma'} \xrightarrow{\beta} \Lambda_\Sigma \xrightarrow{\gamma} k\langle X\rangle_{\Sigma'}(k\langle\langle X\rangle\rangle).$$

The map $\gamma\beta$ is an isomorphism by [7, Theorem 24]. Therefore the image under α of every matrix in Σ becomes invertible in $k\langle X\rangle_{\Sigma'}$, hence there is a natural map $\phi\colon \Lambda_\Sigma \to k\langle X\rangle_{\Sigma'}$ such that $\beta\phi$ and $\phi\beta$ are the identity maps. We deduce that γ is also an isomorphism. It would be interesting to know if γ remains an isomorphism if k is assumed to be only an integral domain. We state the following problem.

Problem 3.2. *Let X be a set, let F denote the free group on X, and let k be an integral domain. Define a k-algebra monomorphism $\theta\colon kF \to k\langle\langle X\rangle\rangle$ by $\theta(a) = a$ for $a \in k$ and $\theta(x) = 1 + x$ for $x \in X$, let Σ be the set of matrices over kF which become invertible over $k\langle\langle X\rangle\rangle$ via θ, and let $\phi\colon kF_\Sigma \to k\langle\langle X\rangle\rangle$ be the uniquely defined associated ring homomorphism. Determine when ϕ is injective.*

If R is a subring of the ring T, then we define the division closure $\mathcal{D}_T(R)$ of R in T to be the smallest subring $\mathcal{D}_T(R)$ of T containing R which is closed under taking inverses, i.e. $x \in \mathcal{D}_T(R)$ and $x^{-1} \in T$ implies $x^{-1} \in \mathcal{D}_T(R)$. In general $\mathcal{D}_T(R) \subseteq \mathcal{R}_T(R)$, i.e. the division closure is contained in the rational closure [5, Exercise 7.1.1]. However if T is a division ring, then the rational closure is a division ring and is equal to the division closure.

It is clear that taking the division closure is an idempotent operation; in other words $\mathcal{D}_T(\mathcal{D}_T(R)) = \mathcal{D}_T(R)$. It is also true that taking the rational closure is an idempotent operation; we sketch the proof below.

Proposition 3.3. *Let R be a subring of the ring T and assume that R and T have the same 1. Then $\mathcal{R}_T(\mathcal{R}_T(R)) = \mathcal{R}_T(R)$.*

Proof. Write $R' = \mathcal{R}_T(R)$ and let M be a matrix over R' which is invertible over T; we need to prove that all the entries of M^{-1} are in R'. We may assume that $M \in \mathrm{M}_d(R')$ for some positive integer d. Cramer's rule, Proposition 3.1, applied to the inclusion $\mathrm{M}_d(R) \to \mathrm{M}_d(R')$ tells us that M is stably associated to a matrix with entries in $\mathrm{M}_d(R)$. This means that for some positive integer e, there exists a matrix $L \in \mathrm{M}_e(\mathrm{M}_d(R)) = \mathrm{M}_{de}(R)$ of the form $\mathrm{diag}(M, 1, \ldots, 1)$ and invertible matrices $A, B \in \mathrm{M}_{de}(R')$ such that ALB is a matrix $X \in \mathrm{M}_{de}(R)$.

Since A, L, B are all invertible in $\mathrm{M}_{de}(T)$, we see that X^{-1} has (by definition of rational closure) all its entries in $\mathrm{M}_{de}(R')$. But $L^{-1} = BX^{-1}A$, which shows that $L^{-1} \in \mathrm{M}_{de}(R')$. Therefore $M^{-1} \in \mathrm{M}_d(R')$ as required. $\square$

We also have the following useful result.

Proposition 3.4. *Let n be a positive integer, let R be a subring of the ring T, and assume that R and T have the same 1. Then $\mathcal{R}_{\mathrm{M}_n(T)}(\mathrm{M}_n(R)) = \mathrm{M}_n(\mathcal{R}_T(R))$.*

Proof. Write $R' = \mathcal{R}_T(R)$ and $S = \mathrm{M}_n(T)$. Suppose $M \in \mathcal{R}_S(\mathrm{M}_n(R))$. Then M appears as an entry of A^{-1}, where $A \in \mathrm{M}_d(\mathrm{M}_n(R))$ for some positive integer d is invertible in $\mathrm{M}_d(S)$. By definition all the entries of

A^{-1} (when viewed as a matrix in $\mathrm{M}_{dn}(T)$) are in R', which shows that $M \in \mathrm{M}_n(R')$.

Now let $M \in \mathrm{M}_n(R')$. We want to show that $M \in \mathcal{R}_S(\mathrm{M}_n(R))$. Since $\mathcal{R}_S(\mathrm{M}_n(R))$ is a ring, it is closed under addition, so we may assume that M has exactly one nonzero entry. Let a be this entry. Then a appears as an entry of A^{-1} where A is an invertible matrix in $\mathrm{M}_m(R)$ for some positive integer m which is a multiple of n. By permuting the rows and columns, we may assume that a is the $(1,1)$-entry. Now form the $p \times p$ matrix $B = \mathrm{diag}(1, \ldots, 1, A, 1, \ldots, 1)$, so that the $(1,1)$-entry of A is in the (n,n)-entry of B (thus there are $n-1$ ones on the main diagonal and then A) and m divides p. By considering B^{-1}, we see that $\mathrm{diag}(1, \ldots, 1, a) \in \mathcal{R}_S(\mathrm{M}_n(R))$. Since $\mathrm{diag}(1, \ldots, 1, 0) \in \mathcal{R}_S(\mathrm{M}_n(R))$, it follows that $\mathrm{diag}(0, \ldots, 0, a) \in \mathcal{R}_S(\mathrm{M}_n(R))$. By permuting the rows and columns, we conclude that $M \in \mathcal{R}_S(\mathrm{M}_n(R))$. $\qquad\square$

When one performs a localization, it would be good to end up with a local ring. We now describe a result of Sheiham [30, §2] which shows that this is often the case. For any ring R, we let $\mathrm{Jac}(R)$ indicate the Jacobson radical of R. Let $\theta \colon R \to S$ be a ring homomorphism, let Σ denote the set of all matrices A over R with the property that $\theta(A)$ is an invertible matrix over S, and let $\lambda \colon R \to R_\Sigma$ denote the associated map. Then we have a ring homomorphism $\phi \colon R_\Sigma \to S$ such that $\theta = \phi\lambda$, and Sheiham's result is

Theorem 3.5. *If S is a local ring, then $\phi^{-1}\mathrm{Jac}(S) = \mathrm{Jac}(R_\Sigma)$*

Thus in particular if S is a division ring, then R_Σ is a local ring.

4   Uniqueness of Division Closure and Unbounded Operators

If R is a domain and D is a division ring containing R such that $\mathcal{D}_D(R) = D$ (i.e. R generates D as a field), then we say that D is a division ring of fractions for R. If R is an integral domain and D, E are division rings of fractions for R, then D and E are fields and are just the Ore localizations of R with respect to the nonzero elements of R. In this case there exists a unique isomorphism $D \to E$ which is the identity on R. Furthermore any automorphism of D can be extended to an automorphism of R.

When D and E are not commutative, i.e. it is only assumed that they are division rings, then this is not the case; in fact D and E may not be isomorphic even just as rings. Therefore we would like to have a criterion for when two such division rings are isomorphic, and also a criterion for the closely related property of when an automorphism of R can be extended to an automorphism of D.

Consider now the complex group algebra $R = \mathbb{C}G$. Here we may embed $\mathbb{C}G$ into the ring of unbounded operators $\mathcal{U}(G)$ on $L^2(G)$ affiliated to $\mathbb{C}G$; see e.g. [20, §8] or [23, §8]. We briefly recall the construction and state some of the properties. Let $L^2(G)$ denote the Hilbert space with Hilbert basis the elements of G; thus $L^2(G)$ consists of all square summable formal sums $\sum_{g \in G} a_g g$ with $a_g \in \mathbb{C}$ and inner product $\langle \sum_g a_g g, \sum_h b_h h \rangle = \sum_{g,h} a_g \overline{b_h}$. We have a left and right action of G on $L^2(G)$ defined by the formulae $\sum_h a_h h \mapsto \sum_h a_h gh$ and $\sum_h a_h h \mapsto \sum_h a_h hg$ for $g \in G$. It follows that $\mathbb{C}G$ acts faithfully as bounded linear operators on the left of $L^2(G)$, in other words we may consider $\mathbb{C}G$ as a subspace of $\mathcal{B}(L^2(G))$, the bounded linear operators on $L^2(G)$. The weak closure of $\mathbb{C}G$ in $\mathcal{B}(L^2(G))$ is the group von Neumann algebra $\mathcal{N}(G)$ of G, and the unbounded operators affiliated to G, denoted $\mathcal{U}(G)$, are those closed densely defined unbounded operators which commute with the right action of G. We have a natural injective $\mathbb{C}$-linear map $\mathcal{N}(G) \to L^2(G)$ defined by $\theta \mapsto \theta 1$ (where 1 denotes the element 1_1 of $L^2(G)$), so we may identify $\mathcal{N}(G)$ with a subspace of $L^2(G)$. When $H \leqslant G$, we may consider $L^2(H)$ as a subspace of $L^2(G)$ and using the above identification, we may consider $\mathcal{N}(H)$ as a subring of $\mathcal{N}(G)$. Also given $\alpha \in L^2(G)$, we can define a $\mathbb{C}$-linear map $\hat{\alpha} \colon \mathbb{C}G \to L^2(G)$ by $\hat{\alpha}(\beta) = \alpha\beta$ for $\beta \in L^2(G)$. Since $\mathbb{C}G$ is a dense linear subspace of $L^2(G)$, it yields a densely defined unbounded operator on $L^2(G)$ which commutes with the right action of G, and it is not difficult to see that this defines a unique element of $\mathcal{U}(G)$, which we shall also call $\hat{\alpha}$. We now have $\mathcal{N}(G) \subseteq L^2(G) \subseteq \mathcal{U}(G)$. Obviously if G is finite, then $\mathcal{N}(G) = L^2(G) = \mathcal{U}(G)$, because all terms are equal to $\mathbb{C}G$. In fact the converse is true, that is if G is infinite, then we do not have equality: more precisely

Proposition 4.1. *Let G be an infinite group. Then $\mathcal{N}(G) \neq L^2(G) \neq \mathcal{U}(G)$.*

Proof. Given $p \in (1, \infty)$, the L^p-conjecture in the case of discrete groups states that if G is an infinite group, then $L^p(G)$ is not closed under multiplication. This was solved in the affirmative for $p \geq 2$ in [28, Theorem 3], and in general for not necessarily discrete groups in [29, Theorem 1]. Thus $L^2(G)$ is not closed under multiplication. Since $\mathcal{N}(G)$ and $\mathcal{U}(G)$ are closed under multiplication, the result follows. $\qquad\square$

Some results related to Proposition 4.1 on the various homological dimensions of $\mathcal{N}(G)$ and $\mathcal{U}(G)$ can be found in [34, §6].

At this stage it is less important to understand the construction of $\mathcal{U}(G)$ than to know its properties. Recall that R is a von Neumann regular ring means that given $r \in R$, there exists $x \in R$ such that $rxr = r$. All matrix rings over a von Neumann regular ring are also von Neumann regular [12, Lemma 1.6], and every element of a von Neumann regular ring is either invertible or a zerodivisor. We now have that $\mathcal{U}(G)$ is a von Neumann regular ring containing $\mathcal{N}(G)$, and is a classical ring of quotients for $\mathcal{N}(G)$ [2, proof of Theorem 10] or [23, Theorem 8.22(1)]. Thus the embedding of $\mathcal{N}(H)$ in $\mathcal{N}(G)$ for $H \leqslant G$ as described above extends to a natural embedding of $\mathcal{U}(H)$ in $\mathcal{U}(G)$. Also $\mathcal{U}(G)$ is rationally closed in any overing. Furthermore $\mathcal{U}(G)$ is a self injective unit-regular ring which is the maximal ring of quotients of $\mathcal{N}(G)$ [2, Lemma 1, Theorems 2 and 3]. Thus we have embedded $\mathbb{C}G$ in a ring, namely $\mathcal{U}(G)$, in which every element is either invertible or a zerodivisor. In fact every element of any matrix ring over $\mathcal{U}(G)$ is either invertible or a zerodivisor. Of course the same is true for any subfield k of $\mathbb{C}$, that is kG can be embedded in a ring in which every element is either invertible or a zerodivisor. Let us write $\mathcal{D}(kG) = \mathcal{D}_{\mathcal{U}(G)}(kG)$ and $\mathcal{R}(kG) = \mathcal{R}_{\mathcal{U}(G)}(kG)$. Then if $H \leqslant G$, we may by the above identify $\mathcal{D}(kH)$ with $\mathcal{D}_{\mathcal{U}(G)}(kH)$ and $\mathcal{R}(kH)$ with $\mathcal{R}_{\mathcal{U}(G)}(kH)$. More generally, we shall write $\mathcal{D}_n(kG) = \mathcal{D}_{M_n(\mathcal{U}(G))}(M_n(kG))$ and $\mathcal{R}_n(kG) = \mathcal{R}_{M_n(\mathcal{U}(G))}(M_n(kG))$. Thus $\mathcal{D}_1(kG) = \mathcal{D}(kG)$ and $\mathcal{R}_1(kG) = \mathcal{R}(kG)$. Also, we may identify $\mathcal{D}_n(kH)$ with $\mathcal{D}_{M_n(\mathcal{U}(G))}(M_n(kH))$ and $\mathcal{R}(kG)$ with $\mathcal{R}_{M_n(\mathcal{U}(G))}(M_n(kH))$.

Often $\mathcal{R}(kG)$ is a very nice ring. For example when G has a normal free subgroup with elementary amenable quotient, and also the finite subgroups of G have bounded order, it follows from [19, Theorem 1.5(ii)] that $\mathcal{R}(\mathbb{C}G)$ is a semisimple Artinian ring, i.e. a finite direct sum of matrix rings over division rings. Thus in particular every element of $\mathcal{R}(\mathbb{C}G)$ is either invertible or a zerodivisor. We state the following problem.

Problem 4.2. *Let G be a group and let k be a subfield of $\mathbb{C}$. Is every element of $\mathcal{R}_n(kG)$ either invertible or a zerodivisor for all positive integers n? Furthermore is $\mathcal{D}_n(kG) = \mathcal{R}_n(kG)$?*

The answer is certainly in the affirmative if G is amenable.

Proposition 4.3. *Let G be an amenable group, let n be a positive integer, and let k be a subfield of $\mathbb{C}$. Then every element of $\mathcal{R}_n(kG)$ is either a zerodivisor or invertible. Furthermore $\mathcal{D}_n(kG) = \mathcal{R}_n(kG)$.*

Proof. Write $R = \mathcal{R}_n(kG)$ and let $A \in R$. By Cramer's rule Proposition 3.1, there is a positive integer d and invertible matrices $X, Y \in M_d(R)$ such that

$B := X \operatorname{diag}(A, 1, \ldots, 1)Y \in \mathrm{M}_{dn}(kG)$. Suppose $ZA \neq 0 \neq AZ$ whenever $0 \neq Z \in \mathrm{M}_n(kG)$. Then B is a non-zerodivisor in $\mathrm{M}_{dn}(kG)$. We claim that B is also a non-zerodivisor in $\mathrm{M}_{dn}(\mathbb{C}G)$. If our claim is false, then either $BC = 0$ or $CB = 0$ for some nonzero $C \in \mathrm{M}_{dn}(\mathbb{C}G)$. Without loss of generality, we may assume that $BC = 0$. Then for some positive integer m, we may choose $e_1, \ldots, e_m \in \mathbb{C}$ which are linearly independent over k such that we may write $C = C_1 e_1 + \cdots + C_m e_m$, where $0 \neq C_i \in \mathrm{M}_{dn}(kG)$ for all i. The equation $BC = 0$ now yields $BC_1 = 0$, contradicting the fact that B is a non-zerodivisor in $\mathrm{M}_{dn}(kG)$, and the claim is established.

Now B induces by left multiplication a right $\mathbb{C}G$-monomorphism $\mathbb{C}G^{dn} \to \mathbb{C}G^{dn}$. This in turn induces a right $\mathcal{N}(G)$-map $\mathcal{N}(G)^{dn} \to \mathcal{N}(G)^{dn}$, and the kernel of this map has dimension 0 by [22, Theorem 5.1]. It now follows from the theory of [22, §2] that this kernel is 0, consequently B is a non-zerodivisor in $\mathrm{M}_{dn}(\mathcal{N}(G))$. Since $\mathcal{U}(G)$ is a classical ring of quotients for $\mathcal{N}(G)$, we see that B is invertible in $\mathrm{M}_{dn}(\mathcal{U}(G))$ and hence B is invertible in $\mathrm{M}_d(R)$. Therefore A is invertible in R and the result follows. $\qquad\square$

One could ask the following stronger problem.

Problem 4.4. *Let G be a group and let k be a subfield of $\mathbb{C}$. Is $\mathcal{R}(kG)$ a von Neumann regular ring?*

Since being von Neumann regular is preserved under Morita equivalence [12, Lemma 1.6] and $\mathcal{R}_n(kG)$ can be identified with $\mathrm{M}_n(\mathcal{R}(kG))$ by Proposition 3.4, we see that this is equivalent to asking whether $\mathcal{R}_n(kG)$ is a von Neumann regular ring. Especially interesting is the case of the lamplighter group, specifically

Problem 4.5. *Let G denote the lamplighter group. Is $\mathcal{R}(\mathbb{C}G)$ a von Neumann regular ring?*

Suppose $H \leq G$ and T is a right transversal for H in G. Then $\bigoplus_{t \in T} L^2(H)t$ is a dense linear subspace of $L^2(G)$, and $\mathcal{U}(H)$ is naturally a subring of $\mathcal{U}(G)$ as follows. If $u \in \mathcal{U}(H)$ is defined on the dense linear subspace D of $L^2(H)$, then we can extend u to the dense linear subspace $\bigoplus_{t \in T} Dt$ of $L^2(G)$ by the rule $u(dt) = (ud)t$ for $t \in T$, and the resulting unbounded operator commutes with the right action of G. It is not difficult to show that $u \in \mathcal{U}(G)$ and thus we have an embedding of $\mathcal{U}(H)$ into $\mathcal{U}(G)$, and this embedding does not depend on the choice of T. In fact it will be the same embedding as described previously. It follows that $\mathcal{R}(kH)$ is naturally a subring of $\mathcal{R}(kG)$. Clearly if $\alpha_1, \ldots, \alpha_n \in \mathcal{U}(H)$ and $t_1, \ldots, t_n \in T$, then $\alpha_1 t_1 + \cdots + \alpha_n t_n = 0$

if and only if $\alpha_i = 0$ for all i, and it follows that if $\beta_1, \ldots, \beta_n \in \mathcal{R}(kH)$ and $\beta_1 t_1 + \cdots + \beta_n t_n = 0$, then $\beta_i = 0$ for all i.

The above should be compared with the theorem of Hughes [15] which we state below. Recall that a group is *locally indicable* if every nontrivial finitely generated subgroup has an infinite cyclic quotient. Though locally indicable groups are left orderable [3, Theorem 7.3.1] and thus $k * G$ is certainly a domain whenever k is a division ring, G is a locally indicable group and $k * G$ is a crossed product, it is still unknown whether such crossed products can be embedded in a division ring. Suppose however $k * G$ has a division ring of fractions D. Then we say that D is *Hughes-free* if whenever $N \lhd H \leqslant G$, H/N is infinite cyclic, and $h_1, \ldots, h_n \in N$ are in distinct cosets of N, then the sum $\mathcal{D}_D(k * N)h_1 + \cdots + \mathcal{D}_D(k * N)h_n$ is direct.

Theorem 4.6. *Let G be a locally indicable group, let k be a division ring, let $k * G$ be a crossed product, and let D, E be Hughes-free division rings of fractions for $k * G$. Then there is an isomorphism $D \to E$ which is the identity on $k * G$.*

This result of Hughes is highly nontrivial, even though the paper [15] is only 8 pages long. This is because the proof given by Hughes in [15] is extremely condensed, and though all the steps are there and correct, it is difficult to follow. A much more detailed and somewhat different proof is given in [6].

Motivated by Theorem 4.6, we will extend the definition of Hughes free to a more general situation.

Definition. Let D be a division ring, let G be a group, let $D * G$ be a crossed product, and let Q be a ring containing $D * G$ such that $\mathcal{R}_Q(D * G) = Q$, and every element of Q is either a zerodivisor or invertible. In this situation we say that Q is *strongly Hughes free* if whenever $N \lhd H \leqslant G$, $h_1, \ldots, h_n \in H$ are in distinct cosets of N and $\alpha_1, \ldots, \alpha_n \in \mathcal{R}_Q(D * N)$, then $\alpha_1 h_1 + \cdots + \alpha_n h_n = 0$ implies $\alpha_i = 0$ for all i (i.e. the h_i are linearly independent over $\mathcal{R}_Q(D * N)$).

Then we would like to extend Theorem 4.6 to more general groups, so we state

Problem 4.7. *Let D be a division ring, let G be a group, let $D * G$ be a crossed product, and let Q be a ring containing $D * G$ such that $\mathcal{R}_Q(D * G) = Q$, and every element of Q is either a zerodivisor or invertible. Suppose P, Q are strongly Hughes free rings for $D * G$. Does there exists an isomorphism $P \to Q$ which is the identity on $D * G$?*

It is clear that if G is locally indicable and Q is a division ring of fractions for $D*G$, then Q is strongly Hughes free implies Q is Hughes free. We present the following problem.

Problem 4.8. *Let G be a locally indicable group, let D be a division ring, let $D * G$ be a crossed product, and let Q be a division ring of fractions for $D * G$ which is Hughes free. Is Q strongly Hughes free?*

It would seem likely that the answer is always "yes". Certainly if G is orderable, then $\mathcal{R}_{D((G))}(D*G)$, the rational closure (which is the same as the division closure in this case) of $D * G$ in the Malcev-Neumann power series ring $D((G))$ [5, Corollary 8.7.6] is a Hughes free division ring of fractions for $D * G$. Therefore by Theorem 4.6 of Hughes, all Hughes free division ring of fractions for $D * G$ are isomorphic to $\mathcal{R}_{D((G))}(D * G)$. It is easy to see that this division ring of fractions is strongly Hughes free and therefore all Hughes free division ring of fractions for $D * G$ are strongly Hughes free.

5   Other Methods

Embedding $\mathbb{C}G$ into $\mathcal{U}(G)$ has proved to be a very useful tool, but what about other group rings? In general we would like a similar construction when k is a field of nonzero characteristic. If D is a division ring, then we can always embed $D * G$ into a ring in which every element is either a unit or a zerodivisor, as follows. Let $V = D * G$ viewed as a right vector space over D, so V has basis $\{\bar{g} \mid g \in G\}$. Then $D * G$ acts by left multiplication on V and therefore can be considered as a subring of the ring of all linear transformations $\mathrm{End}_D(V)$ of V. This ring is von Neumann regular. However it is too large; it is not even directly finite (that is $xy = 1$ implies $yx = 1$) when G is infinite. Another standard method is to embed $D * G$ in its maximal ring of right quotients [11, §2.C]. If R is a right nonsingular ring, then its maximal ring of right quotients $Q(R)$ is a ring containing R which is a right injective von Neumann regular ring, and furthermore as a right R-module, $Q(R)$ is the injective hull of R [11, Corollary 2.31]. By [31, Theorem 4], when k is a field of characteristic zero, kG is right (and left) nonsingular, consequently $Q(kG)$ is a right self-injective von Neumann regular ring. However again it is too large in general. If G is a nonabelian free group, then kG is a domain which by Proposition 2.2 does not satisfy the Ore condition, so we see from [11, Exercise 6.B.14] that $Q(R)$ is not directly finite.

A very useful technique is that of ultrafilters, see [16, p. 76, §2.6] for example. We briefly illustrate this in an example. Let k be a field and let G be a group. Suppose G has a descending chain of normal subgroups $G = G_0 \geqslant G_1 \geqslant \cdots$ such that $k[G/G_n]$ is embeddable in a division ring for all n. Then can we embed kG in a division ring? It is easy to prove that kG is a domain, but to prove the stronger statement that G can be embedded in division ring seems to require the theory of ultrafilters. For most applications (or at least for what we are interested in), it is sufficient to consider ultrafilters on the natural numbers $\mathbb{N} = \{1, 2, \dots\}$. A filter on $\mathbb{N}$ is a subset ω of the power set $\mathcal{P}(\mathbb{N})$ of $\mathbb{N}$ such that if $X, Y \in \omega$ and $X \subseteq Z \subseteq \mathbb{N}$, then $X \cap Y \in \omega$ and $Z \in \omega$. A filter is proper if it does not contain the empty set $\emptyset$, and an ultrafilter is a maximal proper filter. By considering the maximal ideals in the Boolean algebra on $\mathcal{P}(\mathbb{N})$, it can be shown that any proper filter can be embedded in an ultrafilter (this requires Zorn's lemma), and an ultrafilter has the following properties.

- If $X, Y \in \omega$, then $X \cap Y \in \omega$.

- If $X \in \omega$ and $X \subseteq Y$, then $Y \in \omega$.

- If $X \in \mathcal{P}(\mathbb{N})$, then either X or its complement are in ω.

- $\emptyset \notin \omega$.

An easy example of an ultrafilter is the set of all subsets containing n for some fixed $n \in \mathbb{N}$; such an ultrafilter is called a *principal ultrafilter*. An ultrafilter not of this form is called a non-principal ultrafilter.

Given division rings D_n for $n \in \mathbb{N}$ and an ultrafilter ω on $\mathbb{N}$, we can define an equivalence relation $\sim$ on $\prod_n D_n$ by $(d_1, d_2, \dots) \sim (e_1, e_2, \dots)$ if and only if there exists $S \in \omega$ such that $d_n = e_n$ for all $n \in S$. Then the set of equivalence classes $(\prod_n D_n)/\sim$ is called the ultraproduct of the division rings D_i with respect to the ultrafilter ω, and is a division ring [16, p. 76, Proposition 2.1]. This can be applied when R is a ring with a descending sequence of ideals $I_1 \supseteq I_2 \supseteq \dots$ such that $\bigcap_n I_n = 0$ and R/I_n is a division ring. The set of all cofinite subsets of $\mathbb{N}$ is a filter, so here we let ω be any ultrafilter containing this filter. The corresponding ultraproduct D of the division rings R/I_n is a division ring. Furthermore the natural embedding of R into $\prod_n R/I_n$ defined by $r \mapsto (r + R/I_1, r + R/I_2, \dots)$ induces an embedding of R into D. This proves that R can be embedded in a division ring.

In their recent paper [9], Gábor Elek and Endre Szabó use these ideas to embed the group algebra kG over an arbitrary division ring k in a nice von Neumann regular ring for the class of *sofic* groups. The class of sofic groups is a large class of groups which contains all residually amenable groups and is closed under taking free products.

Suppose $\{a_n \mid n \in \mathbb{N}\}$ is a bounded sequence of real numbers and ω is a non-principal ultrafilter. Then there is a unique real number l with the property that given $\epsilon > 0$, then l is in the closure of $\{a_n \mid n \in S\}$ for all $S \in \omega$. We call this the ω-limit of $\{a_n\}$ and write $l = \lim_\omega a_n$.

Now let G be a countable amenable group. Then G satisfies the Følner condition and therefore there exist finite subsets X_i of G ($i \in \mathbb{N}$) such that

- $\bigcup_i X_i = G$.

- $|X_i| < |X_{i+1}|$ for all $i \in \mathbb{N}$.

- If $g \in G$, then $\lim_{i \to \infty} |gX_i \cap X_i|/|X_i| = 1$.

Let k be a division ring and let V_i denote the right k-vector space with basis X_i ($i \in \mathbb{N}$). The general element of $\prod_i \operatorname{End}_k(V_i)$ (Cartesian product) is of the form $\bigoplus_i \alpha_i$ where $\alpha_i \in \operatorname{End}_k(V_i)$ for all i. For $\beta \in \operatorname{End}_k(V_i)$, we define

$$\operatorname{rk}_i(\beta) = \frac{\dim_k(\beta V_i)}{\dim_k V_i},$$

a real number in $[0, 1]$. Now choose a non-principal ultrafilter ω for $\mathbb{N}$. Then for $\alpha \in \prod_i \operatorname{End}_k(V_i)$, we define $\operatorname{rk}(\alpha) = \lim_\omega \operatorname{rk}_n(\alpha_n)$ and

$$I = \{\alpha \in \prod_i \operatorname{End}_k(V_i) \mid \operatorname{rk}(\alpha) = 0\} .$$

It is not difficult to check that I is a two-sided ideal of $\prod_i \operatorname{End}_k(V_i)$. Now set

$$\operatorname{R}_k(G) = \frac{\prod_i \operatorname{End}_k(V_i)}{I}$$

and let $[\alpha]$ denote the image of α in $\operatorname{R}_k(G)$. Since $\operatorname{End}_k(V_i)$ is von Neumann regular and direct products of von Neumann regular rings are von Neumann regular, we see that $\prod_i \operatorname{End}_k(V_i)$ is von Neumann regular and we deduce that $\operatorname{R}_k(G)$ is also von Neumann regular. Next we define $\operatorname{rk}([\alpha]) = \operatorname{rk}(\alpha)$. It can be shown that rk is a well-defined rank function [12, p. 226, Chapter 16] and therefore $\operatorname{R}_k(G)$ is directly finite [12, Proposition 16.11].

For $g \in G$ and $x \in X_i$, we can define $\phi(g)x = gx$ if $gx \in X_i$ and $\phi(g)x = x$ if $gx \notin X_i$. This determines an embedding (which is not a homomorphism) of G into $\prod_i \operatorname{End}_k(V_i)$, and it is shown in [9] that the composition with the natural epimorphism $\prod_i \operatorname{End}_k(V_i) \twoheadrightarrow \operatorname{R}_k(G)$ yields a homomorphism $G \to \operatorname{R}_k(G)$. This homomorphism extends to a ring homomorphism $\theta \colon kG \to$

$R_k(G)$ and [9] shows that $\ker \theta = 0$. Thus we have embedded kG into $R_k(G)$; in particular this shows that kG is directly finite because $R_i(G)$ is. In fact this construction for G amenable can be extended to the case G is a sofic group, consequently kG is directly finite if k is a division ring and G is sofic. The direct finiteness of $k * G$ for k a division ring and G free-by-amenable had earlier been established in [1].

Another type of localization is considered in [27]. Recall that a monoid M is a semigroup with identity, that is M satisfies the axioms for a group except for the existence of inverses. If A is a monoid with identity 1, then M is an A-monoid means that there is an action of A on M satisfying $a(bm) = (ab)m$ and $1m = m$ for all $a, b \in A$ and $m \in M$. In the case A is a ring with identity 1 (so A is a monoid under multiplication) and M is a left A-module, then M is an A-monoid. Let $\text{End}(M)$ denote the monoid of all endomorphisms of the A-monoid M. Given a submonoid S of $\text{End}(M)$, Picavet constructs an A-monoid $S^{-1}M$ with the property that every endomorphism in S becomes an automorphism of M, in other words the elements of S become invertible. To achieve this, he requires that S is a *localizable* submonoid of $\text{End}(M)$. This means that the following Ore type conditions hold:

- For all $u, v \in S$, there exist $u', v' \in S$ such that $u'u = v'v$.

- For all $u, v, w \in S$ such that $uw = vw$, there is $s \in S$ such that $su = sv$.

The construction is similar to Ore localization. We describe this in the case R is ring, M is an R-module and $S = \{\theta^n \mid n \in \mathbb{N}\}$ where θ is an endomorphism of M. Clearly S is localizable. For $m, n \in \mathbb{N}$ with $m \leq n$, we set $M_n = M$ and $\theta_{mn} = \theta^{n-m} \colon M_m \to M_n$. Then (M_n, θ_{mn}) forms a direct system of R-modules, and $S^{-1}M$ is the direct limit of this system. Clearly θ^n induces an R-automorphism on $S^{-1}M$ for all n, so we have inverted θ. In the case R is a division ring, M is finitely generated and θ is a noninvertible nonnilpotent endomorphism of M, the sequence of R-modules $M\theta^n$ eventually stabilizes to a proper nonzero R-submodule of M, which is $S^{-1}M$. It would be interesting to see if this construction has applications to group rings.

References

[1] Pere Ara, Kevin C. O'Meara, and Francesc Perera. Stable finiteness of group rings in arbitrary characteristic. *Adv. Math.*, 170(2):224–238, 2002.

[2] S. K. Berberian. The maximal ring of quotients of a finite von Neumann algebra. *Rocky Mountain J. Math.*, 12(1):149–164, 1982.

[3] Roberta Botto Mura and Akbar Rhemtulla. *Orderable groups*. Marcel Dekker Inc., New York, 1977. Lecture Notes in Pure and Applied Mathematics, Vol. 27.

[4] J. W. Cannon, W. J. Floyd, and W. R. Parry. Introductory notes on Richard Thompson's groups. *Enseign. Math. (2)*, 42(3-4):215–256, 1996.

[5] P. M. Cohn. *Free rings and their relations*, volume 19 of *London Mathematical Society Monographs*. Academic Press Inc. [Harcourt Brace Jovanovich Publishers], London, second edition, 1985.

[6] Warren Dicks, Dolors Herbera, and Javier Sánchez. On a theorem of Ian Hughes about division rings of fractions. *Comm. Algebra*, 32(3):1127–1149, 2004.

[7] Warren Dicks and Eduardo D. Sontag. Sylvester domains. *J. Pure Appl. Algebra*, 13(3):243–275, 1978.

[8] Józef Dodziuk, Peter Linnell, Varghese Mathai, Thomas Schick, and Stuart Yates. Approximating L^2-invariants, and the Atiyah conjecture. *Comm. Pure Appl. Math.*, 56(7):839–873, 2003.

[9] Gábor Elek and Endre Szabó. Sofic groups and direct finiteness. *J. Algebra*, 280(2):426–434, 2004.

[10] Jacek Fabrykowski and Narain Gupta. On groups with sub-exponential growth functions. *J. Indian Math. Soc. (N.S.)*, 49(3-4):249–256 (1987), 1985.

[11] K. R. Goodearl. *Ring theory*. Marcel Dekker Inc., New York, 1976. Nonsingular rings and modules, Pure and Applied Mathematics, No. 33.

[12] K. R. Goodearl. *von Neumann regular rings*. Robert E. Krieger Publishing Co. Inc., Malabar, FL, second edition, 1991.

[13] K. R. Goodearl and R. B. Warfield, Jr. *An introduction to noncommutative Noetherian rings*, volume 16 of *London Mathematical Society Student Texts*. Cambridge University Press, Cambridge, 1989.

[14] Narain Gupta. On groups in which every element has finite order. *Amer. Math. Monthly*, 96(4):297–308, 1989.

[15] Ian Hughes. Division rings of fractions for group rings. *Comm. Pure Appl. Math.*, 23:181–188, 1970.

[16] Nathan Jacobson. *Basic algebra. II.* W. H. Freeman and Co., San Francisco, Calif., 1980.

[17] P. H. Kropholler, P. A. Linnell, and J. A. Moody. Applications of a new *K*-theoretic theorem to soluble group rings. *Proc. Amer. Math. Soc.*, 104(3):675–684, 1988.

[18] Jacques Lewin. On the intersection of augmentation ideals. *J. Algebra*, 16:519–522, 1970.

[19] Peter A. Linnell. Division rings and group von Neumann algebras. *Forum Math.*, 5(6):561–576, 1993.

[20] Peter A. Linnell. Analytic versions of the zero divisor conjecture. In *Geometry and cohomology in group theory (Durham, 1994)*, volume 252 of *London Math. Soc. Lecture Note Ser.*, pages 209–248. Cambridge Univ. Press, Cambridge, 1998.

[21] Peter A. Linnell, Wolfgang Lück, and Thomas Schick. The Ore condition, affiliated operators, and the lamplighter group. In *High-dimensional manifold topology*, pages 315–321. World Sci. Publishing, River Edge, NJ, 2003.

[22] Wolfgang Lück. Dimension theory of arbitrary modules over finite von Neumann algebras and L^2-Betti numbers. I. Foundations. *J. Reine Angew. Math.*, 495:135–162, 1998.

[23] Wolfgang Lück. L^2-*invariants: theory and applications to geometry and K-theory*, volume 44 of *Ergebnisse der Mathematik und ihrer Grenzgebiete. 3. Folge. A Series of Modern Surveys in Mathematics [Results in Mathematics and Related Areas. 3rd Series. A Series of Modern Surveys in Mathematics]*. Springer-Verlag, Berlin, 2002.

[24] A Malcev. On the immersion of an algebraic ring into a field. *Math. Ann.*, 113:686–691, 1937.

[25] Donald S. Passman. *Infinite crossed products*, volume 135 of *Pure and Applied Mathematics*. Academic Press Inc., Boston, MA, 1989.

[26] Alan L. T. Paterson. *Amenability*, volume 29 of *Mathematical Surveys and Monographs*. American Mathematical Society, Providence, RI, 1988.

[27] Gabriel Picavet. Localization with respect to endomorphisms. *Semigroup Forum*, 67(1):76–96, 2003.

[28] M. Rajagopalan. On the L^p-space of a locally compact group. *Colloq. Math.*, 10:49–52, 1963.

[29] Sadahiro Saeki. The L^p-conjecture and Young's inequality. *Illinois J. Math.*, 34(3):614–627, 1990.

[30] Desmond Sheiham. Whitehead groups of localizations and the endomorphism class group. *J. Algebra*, 270(1):261–280, 2003.

[31] Robert L. Snider. On the singular ideal of a group algebra. *Comm. Algebra*, 4(11):1087–1089, 1976.

[32] Bo Stenström. *Rings of quotients*. Springer-Verlag, New York, 1975. Die Grundlehren der Mathematischen Wissenschaften, Band 217, An introduction to methods of ring theory.

[33] D. Tamari. A refined classification of semi-groups leading to generalized polynomial rings with a generalized degree concept. In Johan C. H. Gerretsen and Johannes de Groot, editors, *Proceedings of the International Congress of Mathematicians, Amsterdam, 1954*, volume 3, pages 439–440. Groningen, 1957.

[34] Lia Vaš. *Torsion theories for group von Neumann algebras*. PhD thesis, University of Maryland at College Park, 2002.

Department of Mathematics, Virginia Tech
Blacksburg, VA 24061-0123
USA

e-mail: linnell@math.vt.edu
http://www.math.vt.edu/people/linnell/

A non-commutative generalisation of Thomason's localisation theorem

Amnon Neeman

Abstract

In this survey we remind the reader of Thomason's localisation theorem of [19]. Then we review the generalisation given in my article [9]. After this background, we explain how the article [11], by myself and Ranicki, applies the generalised Thomason localisation theorem of [9] to the universal localisation of a (non-commutative) ring R.

Introduction

This article attempts to give a brief survey of recent joint work by Ranicki and the author [11]. Since this is a survey article rather than a research one, we make an attempt to present the results in historical perspective.

Let us begin with an old theorem of Serre's. Suppose X is a quasi-projective, noetherian scheme, and U is an open subset. In his 1955 paper *Faisceaux algébriques cohérents* [17], Serre tells us that every coherent sheaf on U can be extended to a coherent sheaf on X. Given a vector bundle $\mathcal{V}$ on U, Serre's result permits us to extend $\mathcal{V}$ to a coherent sheaf $\mathcal{S}$ on all of X. If it so happens that X is smooth, then the coherent sheaf $\mathcal{S}$ admits a finite resolution by vector bundles. There is a finite complex $\mathcal{C}$ of vector bundles on X, whose restriction to $U \subset X$ is quasi–isomorphic to $\mathcal{V}$. Using this, Quillen obtained a long exact sequence in K–theory, of the form

$$K_{i+1}(U) \longrightarrow G_i(X - U) \longrightarrow K_i(X) \longrightarrow K_i(U) \longrightarrow G_{i-1}(X - U)$$

Here, $G_i(X - U)$ means the G–theory of $X - U$, that is the Quillen K–theory of the abelian category of coherent sheaves on $X - U$. See Quillen's article [12].

Keywords Scheme, derived category, localisation, K–theory

It is natural to ask if this procedure still works when X is singular, and the answer is No. It is easy to construct a pair $U \subset X$, with U an open subset of a (singular) scheme X, so that the map

$$K_0(X) \longrightarrow K_0(U)$$

is not surjective[1]. This means that there exists on U a vector bundle $\mathcal{V}$ which is not the formal difference of the restriction to U of any vector bundles from X. It certainly implies that $\mathcal{V}$ cannot be quasi–isomorphic to the restriction to U of a complex $\mathcal{C}$ of vector bundles on X.

One would still like to have a localisation long exact sequence in the K–theory of singular schemes. To keep the discussion short and clear, let us sacrifice some generality. In this survey, X will be a quasi–projective, noetherian scheme. We want to have a long exact sequence

$$K_{(i+1)}(U) \longrightarrow K_i^Z(X) \longrightarrow K_i(X) \longrightarrow K_i(U) \longrightarrow K_{(i-1)}^Z(X).$$

In other words, one has a map of spectra $K(X) \longrightarrow K(U)$, and one wants a simple description of the homotopy fiber, preferably not much more complicated than Quillen's $G(X - U)$. To be useful, the description must be "local" in a neighbourhood of $X - U \subset X$. It took 17 years between Quillen's paper [12], and Thomason's [19], which satisfactorily solves the problem.

Remark 0.1. Thomason proves his theorems in the generality of quasi–compact, quasi–separated schemes. But the special case where X is quasi–projective and noetherian is already quite interesting enough. In Remark 2.6, we will very briefly outline what modifications are needed for more general X.

Remark 0.2. In the 17 years between Quillen's paper [12] and Thomason's [19] there was some progress. There are results by Gersten [4], Levine [8] and Weibel [24] and [25].

The foundational tool which Thomason uses is Waldhausen's localisation theorem. But the point is that the straight-forward, obvious way to apply Waldhausen's theorem does not work. The obvious thing to try is the following.

---

[1]Take a normal variety X, and inside it a Weil divisor D which is not Cartier. Let U be the complement of the singular locus. Then on U the divisor D gives a line bundle (and a section). This line bundle gives a class in $K_0(U)$ which does not extend.

Let X be our quasi–projective, noetherian scheme, let U be a Zariski open subset, and let $Z = X - U$ be the (closed) complement. There is a map of derived categories

$$D^b(\text{Vect}/X) \longrightarrow D^b(\text{Vect}/U)$$

which takes a bounded chain complex of vector bundles on X and restricts it to U. Let the kernel of this functor be denoted $D^b_Z(\text{Vect}/X)$. The objects of $D^b_Z(\text{Vect}/X)$ are the bounded chain complexes of vector bundles on X, whose cohomology is supported on Z. In other words, if we restrict a chain complex in $D^b_Z(\text{Vect}/X)$ to $U = X - Z$, we get an acyclic complex.

The composite

$$D^b_Z(\text{Vect}/X) \longrightarrow D^b(\text{Vect}/X) \xrightarrow{\ \pi\ } D^b(\text{Vect}/U)$$

vanishes, and hence the map π factors through the quotient $\dfrac{D^b(\text{Vect}/X)}{D^b_Z(\text{Vect}/X)}$.
We have a diagram

Applying Waldhausen's K–theory to a suitable model for this diagram, we have a diagram in the category of spectra

The part of the diagram below

is a homotopy fibration. We would certainly be very happy if we knew that the map

$$K(i): \quad K\left(\frac{D^b(\mathrm{Vect}/X)}{D_Z^b(\mathrm{Vect}/X)}\right) \longrightarrow K(U)$$

is a homotopy equivalence. But the counterexample in the footnote on page 61 shows that already the map

$$K_0(i): \quad K_0\left(\frac{D^b(\mathrm{Vect}/X)}{D_Z^b(\mathrm{Vect}/X)}\right) \longrightarrow K_0(U)$$

is not surjective. So for many years, the prevailing opinion was that Waldhausen's K–theory did not give one any useful information here. Thomason's inspiration was to realise that the functor i is just an idempotent completion, and that this is quite enough to give a great deal of information in K–theory.

In this survey, we begin by giving a brief account of Waldhausen's [22] general localisation theorem in K–theory. Then we explain the way Thomason [19] applies Waldhausen's general theorem to get a long exact sequence for the K–theory of schemes. Next we recall my own paper [9], generalising Thomason's Key Lemma. Finally there is an account of recent work [11] by Ranicki and myself.

When I found the proof of the Key Lemma given in [9], I did not have in mind any application of the more general result. I advertised the article mostly for the fact that the proof of the more general statement turns out to be much simpler than Thomason's. Thomason's Key Lemma is a statement about extending vector bundles (or more precisely complexes of vector bundles) from an open subscheme U to all of X. My version of the Key Lemma is a formal statement about triangulated categories, devoid of any obvious reference to algebraic geometry. The remarkable fact is that the "abstract nonsense" statement is much easier to prove.

In November 2000, Ranicki asked me if Thomason's theorem could be applied to the universal (Cohn) localisation of a non-commutative ring R. Thomason's statements are in algebraic geometry, and all the rings are commutative. But it turns out that my generalisation of Thomason's Key Lemma does apply. We end this article with a brief resumé of the recent paper [11], in which Ranicki and the author study this.

1   Preliminaries, based on Waldhausen's work

Thomason's theorem is an application of Waldhausen's localisation theorem. For this reason, we begin with a brief review of Waldhausen's work. Waldhausen's foundational article, on which this is based, is [22]. In Section 1 of [19], Thomason specialises some of Waldhausen's more general theorems, to the situation of interest in his article.

Let $\mathcal{C}$ be a category with cofibrations and weak equivalences. Out of $\mathcal{C}$ Waldhausen constructs a spectrum, denoted $K(\mathcal{C})$. In Thomason's [19], the category $\mathcal{C}$ is assumed to be a full subcategory of the category of chain complexes over some abelian category, the cofibrations are maps of complexes which are split monomorphisms in each degree, and the weak equivalences contain the quasi–isomorphisms. We will call such categories *permissible Waldhausen categories*. In this article, we may assume that all categories with cofibrations and weak equivalences are permissible Waldhausen categories.

Remark 1.1. Thomason's term for them is *complicial biWaldhausen categories*.

Given a permissible Waldhausen category $\mathcal{C}$, one can form its derived category; just invert the weak equivalences. We denote this derived category by $D(\mathcal{C})$. We have two major theorems here, both of which are special cases of more general theorems of Waldhausen. The first theorem may be found in Thomason's [19, Theorem 1.9.8]:

Theorem 1.2. (Waldhausen's Approximation Theorem). *Let* $F :$ $\mathcal{C} \longrightarrow \mathcal{D}$ *be an exact functor of permissible Waldhausen categories (categories of chain complexes, as above). Suppose that the induced map of derived categories*

$$D(F) : D(\mathcal{C}) \longrightarrow D(\mathcal{D})$$

is an equivalence of categories. Then the induced map of spectra

$$K(F) : K(\mathcal{C}) \longrightarrow K(\mathcal{D})$$

is a homotopy equivalence.

In this sense, Waldhausen's K–theory is almost an invariant of the derived categories. To construct it one needs to have a great deal more structure. One must begin with a permissible category with cofibrations and weak equivalences. But the Approximation Theorem asserts that the dependence on the added structure is not strong.

Next we state Waldhausen's Localisation Theorem. The statement we give is an easy consequence of Theorem 1.2, coupled with Waldhausen's [22, 1.6.4] or Thomason's [19, 1.8.2]:

Theorem 1.3. (Waldhausen's Localisation Theorem). *Let* **R**, **S** *and* **T** *be permissible Waldhausen categories. Suppose*

$$\mathbf{R} \longrightarrow \mathbf{S} \longrightarrow \mathbf{T}$$

are exact functors of permissible Waldhausen categories. Suppose further that the induced triangulated functors of derived categories

$$D(\mathbf{R}) \longrightarrow D(\mathbf{S}) \longrightarrow D(\mathbf{T})$$

compose to zero, that the functor $D(\mathbf{R}) \longrightarrow D(\mathbf{S})$ is fully faithful, and that the natural map

$$\frac{D(\mathbf{S})}{D(\mathbf{R})} \longrightarrow D(\mathbf{T})$$

is an equivalence of categories. Then the sequence of spectra

$$K(\mathbf{R}) \longrightarrow K(\mathbf{S}) \longrightarrow K(\mathbf{T})$$

is a homotopy fibration.

To obtain a homotopy fibration using Waldhausen's localisation theorem, we need to produce three permissible Waldhausen categories, and a sequence

$$\mathbf{R} \longrightarrow \mathbf{S} \longrightarrow \mathbf{T}$$

so that

$$\frac{D(\mathbf{S})}{D(\mathbf{R})} \longrightarrow D(\mathbf{T})$$

is an equivalence of categories. In particular, we want to find triangulated categories $\mathcal{R}^c = D(\mathbf{R})$, $\mathcal{S}^c = D(\mathbf{S})$ and $\mathcal{T}^c = D(\mathbf{T})$ so that $\mathcal{T}^c = \mathcal{S}^c/\mathcal{R}^c$. Of course, it is not enough to just find the triangulated categories $\mathcal{R}^c$, $\mathcal{S}^c$ and $\mathcal{T}^c$; to apply the localisation theorem, we must also find the permissible Waldhausen categories **R**, **S** and **T**, and the exact functors

$$\mathbf{R} \longrightarrow \mathbf{S} \longrightarrow \mathbf{T}.$$

In Theorem 1.2, we learned that the K–theory is largely independent of the choices of **R**, **S** and **T**. This being a survey article, we will allow ourselves some latitude. Thomason is careful to check that the choices of permissible Waldhausen categories can be made; we will consider this a technical point, and explain only how to produce $\mathcal{R}^c = D(\mathbf{R})$, $\mathcal{S}^c = D(\mathbf{S})$ and $\mathcal{T}^c = D(\mathbf{T})$. We will also commit the notational sin of writing $K(\mathcal{R}^c)$ for $K(\mathbf{R})$, where $\mathcal{R}^c = D(\mathbf{R})$, and similarly $K(\mathcal{S}^c)$ for $K(\mathbf{S})$, and $K(\mathcal{T}^c)$ for $K(\mathbf{T})$.

2   Thomason's localisation theorem

Now it is time to state Thomason's localisation theorem.

Theorem 2.1. (Thomason's localisation theorem). *Let X be a quasi–projective, noetherian scheme, let U be an open subset, and let $Z = X - U$ be the complement. Let $D^b(\mathrm{Vect}/X)$ be the derived category of bounded complexes of vector bundles on X, and let $D^b(\mathrm{Vect}/U)$ be the derived category of bounded complexes of vector bundles on U. Let $D^b_Z(\mathrm{Vect}/X)$ be the derived category of bounded complexes of vector bundles on X, which are supported on Z; that is, they become acyclic when restricted to $U = X - Z$.*

Then there is a localisation exact sequence

$$K_{i+1}(D^b(\mathrm{Vect}/X))$$

$$\downarrow$$

$$K_{i+1}(D^b(\mathrm{Vect}/U)) \longrightarrow K_i(D^b_Z(\mathrm{Vect}/X)) \longrightarrow K_i(D^b(\mathrm{Vect}/X))$$

$$\downarrow$$

$$K_i(D^b(\mathrm{Vect}/U))$$

which is exact as long as $i \geq 0$. No assertion is being made about the surjectivity of $K_0(D^b(\mathrm{Vect}/X)) \longrightarrow K_0(D^b(\mathrm{Vect}/U))$.

Note: we know that $K(D^b(\mathrm{Vect}/X)) = K(X)$ and $K(D^b(\mathrm{Vect}/U)) = K(U)$. That is, Waldhausen's K–theory, of the derived category of bounded complexes of vector bundles, agrees with Quillen's K–theory of the exact category of vector bundles. Waldhausen set up the machinery and explicitly proved special cases of the above. The general statement may be found in Gillet's [5, 6.2]. More discussion may be found in Thomason's [19, 1.11.7]. Using the isomorphisms $K(D^b(\mathrm{Vect}/X)) = K(X)$ and $K(D^b(\mathrm{Vect}/U)) = K(U)$, the exact sequence of Theorem 2.1 becomes the localisation exact sequence for singular varieties, as discussed at the beginning of the Introduction.

The proof of Thomason's localisation theorem follows from the following lemma

Lemma 2.2. (Thomason's Key Lemma). *Suppose $D^b_Z(\mathrm{Vect}/X)$, $D^b(\mathrm{Vect}/X)$ and $D^b(\mathrm{Vect}/U)$ are as in Theorem 2.1. Then the natural map*

$$i : \frac{D^b(\mathrm{Vect}/X)}{D^b_Z(\mathrm{Vect}/X)} \longrightarrow D^b(\mathrm{Vect}/U)$$

is nearly an equivalence of categories. It is a fully faithful functor. While not every object of $D^b(\text{Vect}/U)$ lies in the image, up to splitting idempotents it does. That is, given any object $t \in D^b(\text{Vect}/U)$, there exists an object $t' \in D^b(\text{Vect}/U)$, with $t \oplus t'$ isomorphic to an object $i(s)$ in the image of $\frac{D^b(\text{Vect}/X)}{D^b_Z(\text{Vect}/X)}$.

By Waldhausen's localisation theorem (Theorem 1.3), there is a homotopy fibration

$$K(D^b_Z(\text{Vect}/X)) \longrightarrow K(D^b(\text{Vect}/X)) \longrightarrow K\left(\frac{D^b(\text{Vect}/X)}{D^b_Z(\text{Vect}/X)}\right).$$

Next one has Grayson's Cofinality Theorem [7]. In this context, it says the following. We know that

$$i: \frac{D^b(\text{Vect}/X)}{D^b_Z(\text{Vect}/X)} \longrightarrow D^b(\text{Vect}/U)$$

is fully faithful, and every object of $D^b(\text{Vect}/U)$ is a direct summand of an object in the image of i. Hence the map

$$K(i): K\left(\frac{D^b(\text{Vect}/X)}{D^b_Z(\text{Vect}/X)}\right) \longrightarrow K(D^b(\text{Vect}/U))$$

is an isomorphism in π_i for all $i \geq 1$, and is an injection on π_0. Using this, Theorem 2.1 easily follows from Lemma 2.2.

Remark 2.3. The key to everything is Thomason's Key Lemma (Lemma 2.2). It is perhaps enlightening to realise that before Thomason's [19], all that was known was that the functor

$$i: \frac{D^b(\text{Vect}/X)}{D^b_Z(\text{Vect}/X)} \longrightarrow D^b(\text{Vect}/U)$$

is not onto; the counterexample in the footnote on page 61 shows that the image is not all of $D^b(\text{Vect}/U)$. Before Thomason's work, one knew the counterexample; and one therefore knew that the functor i could not always be an equivalence. It is probably for this reason that nobody thought there was much chance of being able to apply Waldhausen's localisation theorem. Thomason's true inspiration was to realise that the counterexample is a red herring. Sure, the functor i is not an equivalence. But it very nearly is, and the very nearly is quite good enough for a localisation exact sequence in K–theory.

Let me state, more concretely, what Thomason's Key Lemma means for chain complexes of vector bundles on X. The Key Lemma is, at some level, a statement about extending bounded complexes of vector bundles from U to X, and also about extending morphisms of such complexes from U to X. The following combines Propositions 5.2.2, 5.2.3 and 5.2.4 of [19].

Lemma 2.4. (Thomason's Key Lemma, without mention of the word "category"). *Let X be a quasi–projective, noetherian scheme, U an open subset, $Z = X - U$ the complement. Then we have*

(i) *If t is a bounded chain complex of vector bundles on U, then there exists a bounded chain complex of vector bundles t' on U, with $t \oplus t'$ quasi–isomorphic to the restriction of a bounded chain complex of vector bundles on X.*

(ii) *Suppose s and s' are bounded chain complexes of vector bundles on X, and suppose $f : s|_U \longrightarrow s'|_U$ is a chain map of chain complexes of vector bundles, given over $U \subset X$. Then there exists another bounded chain complex of vector bundles $\overline{s}$ on X, and a diagram of morphisms of vector bundles on X*

$$\overline{s} \xrightarrow{\ \overline{f}\ } s'$$
$$p \downarrow$$
$$s$$

so that, when we restrict to $U \subset X$, the map $p|_U$ is a quasi–isomorphism, and

$$f \circ \{p|_U\} \cong \overline{f}|_U,$$

that is, the maps $f \circ \{p|_U\}$ and $\overline{f}|_U$ are homotopic.

(iii) *Suppose s and s' are chain complexes of vector bundles on X, and suppose $f : s \longrightarrow s'$ is a chain map. Suppose that the restriction $f|_U$ factors through an acyclic complex of vector bundles on U. Then there is a map f' homotopic to f, so that f' factors through a chain complex of vector bundles on X with an acyclic restriction to U.*

Remark 2.5. The proof of the Key Lemma given by Thomason has since been simplified, and the statement generalised. But this detracts nothing from Thomason's achievement. He was the one who realised that, notwithstanding the counterexample in the footnote on page 61, there was a positive theorem to be proved.

In the introduction to [19], Thomason tells us that the inspiration for the Key Lemma came from the ghost of a deceased friend, who kept haunting his dreams. In fact, Thomason made the friend a coauthor of the paper. Clearly, it is not for me to make judgements about the sources of other peoples' inspirations. There is very little doubt that the theorem really was an inspired improvement over what was known before.

Remark 2.6. As was mentioned in the Introduction, Thomason treats the case where X is not necessarily quasi–projective or noetherian. The more general theorem requires some modification. First, we must assume both X and U to be quasi–compact and quasi–separated. Next, we need to replace the derived categories of vector bundles by the derived categories of perfect complexes. That is, we replace

$$
\begin{array}{ccc}
D^b_Z(\text{Vect}/X) & \text{by} & D^{\text{perf}}_Z(X) \\
D^b(\text{Vect}/X) & \text{by} & D^{\text{perf}}(X) \\
D^b(\text{Vect}/U) & \text{by} & D^{\text{perf}}(U).
\end{array}
$$

There is still a functor

$$
i : \frac{D^{\text{perf}}(X)}{D^{\text{perf}}_Z(X)} \longrightarrow D^{\text{perf}}(U)
$$

The functor i is still fully faithful, and every object of $D^{\text{perf}}(U)$ is a direct summand of an object in the image of i. It is no longer clear whether $K(D^{\text{perf}}(X))$ and $K(D^{\text{perf}}(U))$ agree with Quillen's $K(X)$ and $K(U)$, respectively. Thomason makes the point that if they do not, then $K(D^{\text{perf}}(X))$ and $K(D^{\text{perf}}(U))$ should be taken as the "good" definition for $K(X)$ and $K(U)$.

Thomason also develops a non–connective version of these spectra, allowing one to continue the long exact sequence of K–groups. The reader is referred to Thomason's own account of his work, in the survey article [18]. Unlike this survey, Thomason lays stress on the generality of his results.

3   A generalisation

The proof of Thomason's Key Lemma has since been substantially simplified, and the statement generalised. The first generalisation was due to Dongyuan Yao [26], a student of Thomason's. Yao proved a more general statement than Thomason, but his proof followed the outlines of Thomason's. It was not a simplification; rather, it was a *tour de force,* pushing the methods to their limits. Since Yao's work, I have given two simpler proofs of much more general theorems, by completely different methods.

The idea of my generalisation of the Key Lemma is to look at a larger diagram. So far, we have been looking at functors of triangulated categories

$$D_Z^b(\text{Vect}/X) \longrightarrow D^b(\text{Vect}/X) \longrightarrow D^b(\text{Vect}/U).$$

Now we wish to embed this in a larger diagram

$$
\begin{array}{ccccc}
D_Z^b(\text{Vect}/X) & \longrightarrow & D^b(\text{Vect}/X) & \longrightarrow & D^b(\text{Vect}/U) \\
\downarrow & & \downarrow & & \downarrow \\
D_Z(\text{qc}/X) & \longrightarrow & D(\text{qc}/X) & \longrightarrow & D(\text{qc}/U)
\end{array}
$$

The categories $D(\text{qc}/X)$ and $D(\text{qc}/U)$ are the unbounded derived categories of quasi-coherent sheaves on X and on U, respectively. The category $D_Z(\text{qc}/X)$ is the full subcategory of all objects in $D(\text{qc}/X)$ which become acyclic when restricted to U.

It is very easy to see that

$$\frac{D(\text{qc}/X)}{D_Z(\text{qc}/X)} \;=\; D(\text{qc}/U).$$

In the world of quasi–coherent sheaves, there is no obstruction to extending from U to X. There is even a functor which does the extension, the pushforward by the inclusion $U \longrightarrow X$. The problem is to understand the relation between the top and the bottom rows. Let us now give a couple of definitions.

Definition 3.1. *Let $\mathcal{T}$ be a triangulated category, containing all small coproducts of its objects. An object $c \in \mathcal{T}$ is called* compact *if every map from c to any coproduct factors through a finite part of the coproduct. That is, any map*

$$c \longrightarrow \coprod_{\lambda \in \Lambda} t_\lambda$$

factors as

$$c \longrightarrow \coprod_{i=1}^{n} t_{\lambda_i} \longrightarrow \coprod_{\lambda \in \Lambda} t_\lambda.$$

Definition 3.2. *The full subcategory $\mathcal{T}^c \subset \mathcal{T}$ has for its objects all the compact objects of $\mathcal{T}$.*

The relation between the top and bottom rows of the diagram

$$D^b_Z(\mathrm{Vect}/X) \longrightarrow D^b(\mathrm{Vect}/X) \longrightarrow D^b(\mathrm{Vect}/U)$$
$$\downarrow \qquad\qquad\qquad \downarrow \qquad\qquad\qquad \downarrow$$
$$D_Z(\mathrm{qc}/X) \longrightarrow D(\mathrm{qc}/X) \longrightarrow D(\mathrm{qc}/U)$$

turns out to be that the top row is just the compact objects in the bottom row. The generalisation of Thomason's localisation theorem, which may be found in my article [9], asserts the following.

Theorem 3.3. *Let $\mathcal{S}$ be a triangulated category containing all small co-products. Let $\mathcal{R} \subset \mathcal{S}$ be a full triangulated subcategory, closed under the formation of the coproducts in $\mathcal{S}$ of any set of its objects. Form the category $\mathcal{T} = \mathcal{S}/\mathcal{R}$. We always have a diagram*

$$\mathcal{R}^c \qquad\qquad \mathcal{S}^c \qquad\qquad \mathcal{T}^c$$
$$\downarrow \qquad\qquad \downarrow \qquad\qquad \downarrow$$
$$\mathcal{R} \longrightarrow \mathcal{S} \longrightarrow \mathcal{T}$$

Assume further that there exist

(i) *A set of objects $S \subset \mathcal{S}^c$ so that, for any $s \in \mathcal{S}$, we have*

$$\{\forall c \in S, \operatorname{Hom}(c, s) = 0\} \Longrightarrow \{s = 0\}.$$

(ii) *A set of objects $R \subset \mathcal{R} \cap \mathcal{S}^c$ so that, for any $r \in \mathcal{R}$, we have*

$$\{\forall c \in R, \operatorname{Hom}(c, r) = 0\} \Longrightarrow \{r = 0\}.$$

Then the natural map $\mathcal{R} \longrightarrow \mathcal{S}$ takes compact objects to compact objects, and so does the natural map $\mathcal{S} \longrightarrow \mathcal{T}$. In other words, our diagram may be completed to

$$\mathcal{R}^c \longrightarrow \mathcal{S}^c \longrightarrow \mathcal{T}^c$$
$$\downarrow \qquad\qquad \downarrow \qquad\qquad \downarrow$$
$$\mathcal{R} \longrightarrow \mathcal{S} \longrightarrow \mathcal{T}$$

Of course, the composite $\mathcal{R}^c \longrightarrow \mathcal{S}^c \longrightarrow \mathcal{T}^c$ must vanish, since it is just the restriction to $\mathcal{R}^c$ of a vanishing functor on $\mathcal{R}$. We therefore have a factorisation of $\mathcal{S}^c \longrightarrow \mathcal{T}^c$ as

$$\mathcal{S}^c \longrightarrow \mathcal{S}^c/\mathcal{R}^c \xrightarrow{\ i\ } \mathcal{T}^c.$$

The functor $i : \mathcal{S}^c/\mathcal{R}^c \longrightarrow \mathcal{T}^c$ is fully faithful, and every object of $\mathcal{T}^c$ is a direct summand of an object in the image of i.

Remark 3.4. In other words, Thomason's Key Lemma is a special case of a very general theorem about triangulated categories. In a recent book [10], I generalise the theorem even further, to deal with the large cardinal case. There are now three proofs of Thomason's Key Lemma. The original proof Thomason gave in [18], the proof presented in my old paper [9], and the proof in the book [10]. My two proofs, in [9] and [10], are quite different from each other.

4   The generalisation to universal localisation

In November 2000, I happened to be passing through Edinburgh. Ranicki asked me whether it is possible to apply Thomason's localisation theorem to the universal localisation of non–commutative rings. My first question was "What is the universal localisation of non–commutative rings"? Fortunately the readers of this volume are undoubtedly infinitely more knowledgeable about this than I was in November 2000.

The answer to Ranicki's question turns out to be "Yes". At least, as long as we are willing to apply my generalisation of Thomason's Key Lemma. In the remainder of this article I sketch the joint work which Ranicki and I have done on this problem.

Suppose we are given a non-commutative ring A, and a set σ of maps $\{s_i : P_i \longrightarrow Q_i\}$ of finitely generated, projective left A–modules. From Cohn [1] and Schofield [16] we know that there exists a universal homomorphism $A \longrightarrow \sigma^{-1}A$. We remind the reader what this means. Consider the category of all ring homomorphisms $A \longrightarrow B$ so that, for all i, the map

$$B \otimes_A P_i \xrightarrow{\ 1 \otimes_A s_i\ } B \otimes_A Q_i$$

is an isomorphism. The homomorphism $A \longrightarrow \sigma^{-1}A$ is, by definition, the initial object in this category. Now we wish to apply the generalised Key Lemma (Theorem 3.3) to this situation.

The generalised Key Lemma begins with triangulated categories $\mathcal{R} \subset \mathcal{S}$. Both categories are assumed to contain the coproducts of any set of their objects, and the inclusion is assumed to preserve the coproducts. If we are going to apply the Key Lemma, we need to begin by intelligently choosing the categories $\mathcal{R} \subset \mathcal{S}$. Moreover, there is not a lot to work with. The only reasonable choice for $\mathcal{S}$ is $\mathcal{S} = D(A)$, the unbounded derived category of all left A–modules. Next we define $\mathcal{R}$.

Definition 4.1. *We are given a set of maps $\sigma = \{s_i : P_i \longrightarrow Q_i\}$. We can view these as objects in $S = D(A)$ just by turning them into complexes*

$$\cdots \longrightarrow 0 \longrightarrow P_i \xrightarrow{s_i} Q_i \longrightarrow 0 \longrightarrow \cdots$$

The category $\mathcal{R} \subset S$ is defined to be the smallest triangulated subcategory of $S = D(A)$, which contains σ and is closed in S under the formation of arbitrary coproducts of its objects.

Remark 4.2. It is easy to show that the categories $\mathcal{R} \subset S$ above satisfy the hypotheses (i) and (ii) of Theorem 3.3. For the set $S \subset S^c$ one may take $S = \{\Sigma^n A, n \in \mathbb{Z}\}$. For $R \subset \mathcal{R} \cap S^c$ we can take the set σ of all complexes

$$\cdots \longrightarrow 0 \longrightarrow P_i \xrightarrow{s_i} Q_i \longrightarrow 0 \longrightarrow \cdots$$

This also makes it clear why we need to assume P_i and Q_i finitely generated and projective; it is to guarantee that the complexes lie in S^c.

The conclusions of the generalised Key Lemma therefore hold. The interest becomes in seeing what, if anything, the diagram

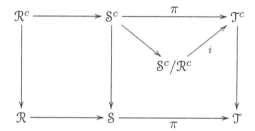

tells us about the ring $\sigma^{-1}A$ and its K–theory. What follows is a list of some of the results Ranicki and I obtained in [11].

In the diagram above we have a functor $\pi : S \longrightarrow \mathcal{T}$. As the diagram shows, we allow ourselves to name the restriction of π to $S^c \subset S$ by the same symbol π. There is an object $A \in S = D(A)$, namely the complex which is A in degree 0 and vanishes in all other degrees. The functor π takes A to an object πA in $\mathcal{T}$. The functor π takes morphisms $A \longrightarrow A$ to morphisms $\pi A \longrightarrow \pi A$. Being an additive functor, it must respect the addition and composition of such maps. Hence it gives a ring homomorphism

$$\pi : \mathrm{End}_S(A, A) \longrightarrow \mathrm{End}_{\mathcal{T}}(\pi A, \pi A).$$

But $\mathrm{End}_S(A, A) = A^{\mathrm{op}}$, since the homomorphisms of A as a left A–module are just right multiplication by A. The first result we have is

Theorem 4.3. *The ring homomorphism above*

$$\pi : A^{\mathrm{op}} = \mathrm{End}_{\mathcal{S}}(A, A) \longrightarrow \mathrm{End}_{\mathcal{T}}(\pi A, \pi A)$$

satisfies the universal property defining $\sigma^{-1}A$. *That is,*

$$\sigma^{-1}A = \left\{ \mathrm{End}_{\mathcal{T}}(\pi A, \pi A) \right\}^{\mathrm{op}}.$$

Proof. This is Theorem 3.21 coupled with Propositions 2.7 and 2.8 of [11]. □

The next question to ask is whether the triangulated category $\mathcal{T}$ is equivalent to $D(\sigma^{-1}A)$, with an equivalence taking $\sigma^{-1}A \in D(\sigma^{-1}A)$ to $\pi A \in \mathcal{T}$. Fortunately, we know necessary and sufficient conditions for this. The next theorem is a result of 'tilting theory', and may be found in the work of Rickard.

Theorem 4.4. *For there to exist a natural isomorphism* $D(\sigma^{-1}A) \longrightarrow \mathcal{T}$ *taking* $\sigma^{-1}A \in D(\sigma^{-1}A)$ *to* $\pi A \in \mathcal{T}$, *it is necessary and sufficient to have*

$$\mathrm{Hom}_{\mathcal{T}}(\pi A, \Sigma^n \pi A) = \begin{cases} \sigma^{-1}A & \text{if } n = 0 \\ 0 & \text{if } n \neq 0 \end{cases}$$

Proof. The object A is a compact generator for $\mathcal{S} = D(A)$. Theorem 3.3 tells us that the object πA must be a compact object in $\mathcal{T}$. The fact that A generates $\mathcal{S}$ immediately gives that πA generates $\mathcal{T} = \mathcal{S}/\mathcal{R}$. But now the statement is a very easy consequence of tilting theory; see Rickard's [15]. □

In the light of Rickard's theorem, it becomes interesting to compute when the groups $\mathrm{Hom}_{\mathcal{T}}(\pi A, \Sigma^n \pi A)$ vanish. In this direction, we have

Proposition 4.5. *For any choice of a ring A and a set* $\sigma = \{s_i : P_i \longrightarrow Q_i\}$ *of maps of finitely generated, projective left R–modules, we have*

$$\{n > 0\} \Longrightarrow \{\mathrm{Hom}_{\mathcal{T}}(\pi A, \Sigma^n \pi A) = 0\}.$$

We also know the following:

$$\{\forall n < 0, \ \mathrm{Hom}_{\mathcal{T}}(\pi A, \Sigma^n \pi A) = 0\} \Longleftrightarrow \{\forall n > 0, \ \mathrm{Tor}_n^A(\sigma^{-1}A, \sigma^{-1}A) = 0\}.$$

Proof. Theorem 10.8 of [11]. □

Comment 4.6. The proofs of Theorem 4.3 and Proposition 4.5 turn out to amount to computations of groups of the form $\mathrm{Hom}_{\mathcal{J}}(\pi A, \Sigma^n \pi A)$. This is obvious for Proposition 4.5, and less obvious but true for Theorem 4.3. The proofs of these theorems hinge on the fact that the functor $\pi : \mathcal{S} \longrightarrow \mathcal{J}$ has a right adjoint G. Therefore,

$$
\begin{aligned}
\mathrm{Hom}_{\mathcal{J}}(\pi A, \Sigma^n \pi A) \;\; &= \;\; \mathrm{Hom}_{\mathcal{S}}(A, \Sigma^n G \pi A) \quad \text{by adjunction} \\
&= \;\; H^n(G\pi A) \qquad\qquad \text{because, for any } X \in \mathcal{S} = D(A),\\
&\qquad\qquad\qquad\qquad\qquad \mathrm{Hom}_{\mathcal{S}}(A, \Sigma^n X) = H^n(X).
\end{aligned}
$$

The proof therefore turns into a study of the object $G\pi A$ and its cohomology. In fact, many of the statements cited in [11] are phrased in terms of $H^n(G\pi A)$, rather than $\mathrm{Hom}_{\mathcal{J}}(\pi A, \Sigma^n \pi A)$.

Caution 4.7. In Proposition 4.5, the reader should note that we are *not* asserting that, for $n \geq 0$, there is an isomorphism

$$
\mathrm{Hom}_{\mathcal{J}}(\pi A, \Sigma^{-n} \pi A) \simeq \mathrm{Tor}_n^A(\sigma^{-1}A, \sigma^{-1}A).
$$

Rather, the two are related by a spectral sequence, which permits one to prove Proposition 4.5. The most precise statement is that, up to a shift by 1, the first non-vanishing $\mathrm{Hom}_{\mathcal{J}}(\pi A, \Sigma^{-n}\pi A)$ agrees with the first non-vanishing $\mathrm{Tor}_n^A(\sigma^{-1}A, \sigma^{-1}A)$. The precise statement runs as follows.

(i) $\mathrm{Tor}_1^A(\sigma^{-1}A, \sigma^{-1}A) = 0$.

(ii) If, for all $1 \leq i \leq n$ we have

$$
\mathrm{Hom}_{\mathcal{J}}(\pi A, \Sigma^{-i}\pi A) = 0,
$$

then for all $1 \leq i \leq n$ we also have $\mathrm{Tor}_{i+1}^A(\sigma^{-1}A, \sigma^{-1}A) = 0$. Furthermore

$$
\mathrm{Tor}_{n+2}^A(\sigma^{-1}A, \sigma^{-1}A) \;\; \cong \;\; \mathrm{Hom}_{\mathcal{J}}(\pi A, \Sigma^{-n-1}\pi A).
$$

The proof of (i), that is the vanishing of $\mathrm{Tor}_1^A(\sigma^{-1}A, \sigma^{-1}A)$, may be found in Corollary 3.27 of [11]. For the proof of (ii) see Corollary 3.31 loc.cit.

Definition 4.8. *We say that a universal localisation* $A \longrightarrow \sigma^{-1}A$ *is* stably flat *if, for all* $n \geq 1$, $\mathrm{Tor}_n^A(\sigma^{-1}A, \sigma^{-1}A) = 0$.

Remark 4.9. Combining Rickard's Theorem 4.4 and Proposition 4.5, we have a complete understanding of what Thomason's Key Lemma yields in the stably flat case. If $A \longrightarrow \sigma^{-1}A$ is stably flat, then $\mathcal{J} = D(\sigma^{-1}A)$, which makes $\mathcal{T}^c = D^c(\sigma^{-1}A)$. (For any ring B, $D^c(B)$ will be our notation for the compact objects $\{D(B)\}^c$). Our pretty diagram becomes

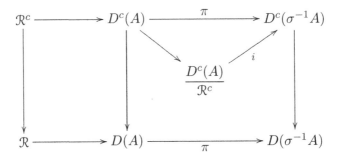

We know that, for any ring B, Waldhausen's K–theory of $D^c(B)$ agrees with Quillen's K–theory of B. We have a homotopy fibration

$$K(\mathcal{R}^c) \longrightarrow K(A) \longrightarrow K\left(\frac{D^c(A)}{\mathcal{R}^c}\right).$$

Grayson's cofinality theorem, coupled with the fact that the functor i is only an idempotent completion, tells us that the map

$$K\left(\frac{D^c(A)}{\mathcal{R}^c}\right) \xrightarrow{K(i)} K(\sigma^{-1}A)$$

is an isomorphism in π_i for $i \geq 1$, and an injection on π_0.

The only thing we might wish to have is a better understanding of $\mathcal{R}^c$ and its K–theory. For this, the following result is useful:

Theorem 4.10. *If our set* $\sigma = \{s_i : P_i \longrightarrow Q_i\}$ *consists only of monomorphisms, then we define an exact category* $\mathcal{E}$. *In our paper [11], $\mathcal{E}$ is called the category of* (A, σ)*–modules. The objects of $\mathcal{E}$ are the A–modules M of projective dimension ≤ 1, such that*

(i) $\{\sigma^{-1}A\} \otimes_A M = 0$, *and*

(ii) $\mathrm{Tor}_1^A(\sigma^{-1}A, M) = 0$.

We assert that $\mathcal{R}^c = D^b(\mathcal{E})$, *and hence the Waldhausen K–theory $K(\mathcal{R}^c)$ equals the Quillen K–theory $K(\mathcal{E})$.*

Proof. Lemma 11.9 in [11]. Note that Theorem 4.10 does not assume that $A \longrightarrow \sigma^{-1}A$ is stably flat.   □

It remains to discuss what we know in the case where $A \longrightarrow \sigma^{-1}A$ is not stably flat. In this case, we know that $\mathcal{T}$ is not equal to $D(\sigma^{-1}A)$. But it turns out that one always has

Lemma 4.11. *The natural functor* $\mathbb{S} = D(A) \longrightarrow D(\sigma^{-1}A)$, *taking* $X \in D(A)$ *to* $\{\sigma^{-1}A\} \otimes_A^L X$, *factors uniquely through* $\pi : D(A) = \mathbb{S} \longrightarrow \mathcal{T}$. *That is, we have a factorisation*

$$D(A) \xrightarrow{\ \pi\ } \mathcal{T} \xrightarrow{\ T\ } D(\sigma^{-1}A).$$

Furthermore, the functor $T : \mathcal{T} \longrightarrow D(\sigma^{-1}A)$ *takes compact objects in* $\mathcal{T}$ *to compact objects in* $D(\sigma^{-1}A)$.

Proof. Proposition 5.1 in [11]. $\qquad\qquad\qquad\qquad\qquad\qquad\qquad$ □

This allows us to extend the commutative diagram of Thomason's Key Lemma, to obtain a better version

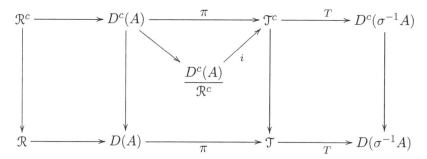

One way to state Remark 4.9 is that, assuming $A \longrightarrow \sigma^{-1}A$ is stably flat, we have that the functor T above is an isomorphism. Without any stable flatness hypothesis, we always have

Proposition 4.12. *The map* $T : \mathcal{T}^c \longrightarrow D^c(\sigma^{-1}A)$ *always induces isomorphisms on* K_0 *and* K_1.

Proof. For K_0 this is Theorem 8.5 of [11], for K_1 it is Corollary 9.2 loc.cit. □

Remark 4.13. Proposition 4.12 gives us, without any stable flatness hypothesis, the exact sequence in low dimensional K–theory

$$K_1(\mathcal{R}^c) \longrightarrow K_1(A) \longrightarrow K_1(\sigma^{-1}A) \longrightarrow K_0(\mathcal{R}^c) \longrightarrow K_0(A) \longrightarrow K_0(\sigma^{-1}A).$$

Remark 4.14. It seems only right to mention that much work has been done on this by others, with many beautiful results by, for example, Grayson [6], Vogel [20] and [21], Schofield [16], Farber and Vogel [3], Farber and Ranicki [2], Ranicki [13], Ranicki and Weiss [14] and Weibel and Yao [23]. In the introduction to [11] we give a review of what the other authors obtained, and of the extent to which our results give an improvement. Let me not repeat this here. The point of this survey was to explain how our approach was descended from the work of Thomason.

References

[1] Paul M. Cohn, *Free rings and their relations*, Academic Press, London, 1971, London Mathematical Society Monographs, No. 2.

[2] Michael Farber and Andrew Ranicki, *The Morse-Novikov theory of circle-valued functions and noncommutative localization*, e-print http://arXiv.math.DG/9812122, Tr. Mat. Inst. Steklova **225** (1999), no. Solitony Geom. Topol. na Perekrest., 381–388.

[3] _____ and Pierre Vogel, *The Cohn localization of the free group ring*, Math. Proc. Cambridge Philos. Soc. **111** (1992), no. 3, 433–443.

[4] Stephen M. Gersten, *The localization theorem for projective modules*, Comm. Algebra **2** (1974), 307–350.

[5] Henri Gillet, *Riemann-Roch theorems for higher algebraic K-theory*, Adv. in Math. **40** (1981), no. 3, 203–289.

[6] Daniel R. Grayson, *K-theory and localization of noncommutative rings*, J. Pure Appl. Algebra **18** (1980), no. 2, 125–127.

[7] _____, *Exact sequences in algebraic K-theory*, Ill. J. Math. **31** (1987), 598–617.

[8] Marc Levine, *Localization on singular varieties*, Inv. Math. **91** (1988), 423–464.

[9] Amnon Neeman, *The connection between the K-theory localisation theorem of Thomason, Trobaugh and Yao, and the smashing subcategories of Bousfield and Ravenel*, Ann. Sci. École Normale Supérieure **25** (1992), 547–566.

[10] _____, *Triangulated Categories*, Annals of Mathematics Studies, vol. 148, Princeton University Press, Princeton, NJ, 2001.

[11] _____ and Andrew Ranicki, *Noncommutative localization and chain complexes I. Algebraic K- and L-theory*, e-print http://arXiv.math.RA/0109118, Geometry and Topology **8** (2004), 1385–1425.

[12] Daniel Quillen, *Higher algebraic K-theory. I*, Algebraic K-theory, I: Higher K-theories (Proc. Conf., Battelle Memorial Inst., Seattle, Wash., 1972), Lecture Notes in Math., vol. 341, Springer verlag, 1973, pp. 85–147.

[13] Andrew Ranicki, *Exact sequences in the algebraic theory of surgery*, Princeton University Press, Princeton, N.J., 1981.

[14] _____ and Michael Weiss, *Chain complexes and assembly*, Math. Z. **204** (1990), no. 2, 157–185.

[15] Jeremy Rickard, *Morita theory for derived categories*, J. London Math. Soc. **39** (1989), 436–456.

[16] Aidan H. Schofield, *Representation of rings over skew fields*, Cambridge University Press, Cambridge, 1985.

[17] Jean-Pierre Serre, *Faisceaux algébriques cohérents*, Ann. of Math. (2) **61** (1955), 197–278.

[18] Robert W. Thomason and Thomas F. Trobaugh, *Le théorème de localisation en K-théorie algébrique*, C. R. Acad. Sci. Paris Sér. I Math. **307** (1988), no. 16, 829–831.

[19] _____, *Higher algebraic K-theory of schemes and of derived categories*, The Grothendieck Festschrift (a collection of papers to honor Grothendieck's 60'th birthday), vol. 3, Birkhäuser, 1990, pp. 247–435.

[20] Pierre Vogel, *Localization in algebraic L-theory*, Topology Symposium, Siegen 1979 (Proc. Sympos., Univ. Siegen, Siegen, 1979), Springer, Berlin, 1980, pp. 482–495.

[21] _____, *On the obstruction group in homology surgery*, Inst. Hautes Études Sci. Publ. Math. (1982), no. 55, 165–206.

[22] Friedhelm Waldhausen, *Algebraic K-theory of spaces*, Algebraic and geometric topology (New Brunswick, N.J., 1983), Springer Verlag, Berlin, 1985, pp. 318–419.

[23] Charles Weibel and Dongyuan Yao, *Localization for the K-theory of noncommutative rings*, Algebraic K-theory, commutative algebra, and algebraic geometry (Santa Margherita Ligure, 1989), Amer. Math. Soc., Providence, RI, 1992, pp. 219–230.

[24] Charles A. Weibel, *Negative K-theory of varieties with isolated singularities*, J. Pure Appl. Algebra **34** (1984), no. 2-3, 331–342.

[25] _____, *A Brown-Gersten spectral sequence for the K-theory of varieties with isolated singularities*, Adv. in Math. **73** (1989), no. 2, 192–203.

[26] Dongyuan Yao, *Higher algebraic K-theory of admissible abelian categories and localization theorems*, J. Pure Appl. Algebra **77** (1992), no. 3, 263–339.

Centre for Mathematics and its Applications
Mathematical Sciences Institute
John Dedman Building
The Australian National University
Canberra, ACT 0200
AUSTRALIA

e-mail: Amnon.Neeman@anu.edu.au

Noncommutative localization in topology

Andrew Ranicki

Introduction

The topological applications of the Cohn noncommutative localization considered in this paper deal with spaces (especially manifolds) with infinite fundamental group, and involve localizations of infinite group rings and related triangular matrix rings. Algebraists have usually considered noncommutative localization of rather better behaved rings, so the topological applications require new algebraic techniques.

Part 1 is a brief survey of the applications of noncommutative localization to topology: finitely dominated spaces, codimension 1 and 2 embeddings (knots and links), homology surgery theory, open book decompositions and circle-valued Morse theory. These applications involve chain complexes and the algebraic K- and L-theory of the noncommutative localization of group rings.

Part 2 is a report on work on chain complexes over generalized free products and the related algebraic K- and L-theory, from the point of view of noncommutative localization of triangular matrix rings. Following Bergman and Schofield, a generalized free product of rings can be constructed as a noncommutative localization of a triangular matrix ring. The novelty here is the explicit connection to the algebraic topology of manifolds with a generalized free product structure realized by a codimension 1 submanifold, leading to noncommutative localization proofs of the results of Waldhausen and Cappell on the algebraic K- and L-theory of generalized free products. In a sense, this is more in the nature of an application of topology to noncommutative localization! But this algebra has in turn topological applications, since in dimensions $\geqslant 5$ the surgery classification of manifolds within a homotopy type reduces to algebra.

Part 1. A survey of applications

We start by recalling the universal noncommutative localization of P.M.Cohn
[6, 7]. Let A be a ring, and let $\Sigma = \{s : P \to Q\}$ be a set of morphisms
of f.g. projective A-modules. A ring morphism $A \to R$ is Σ-*inverting*
if for every $s \in \Sigma$ the induced morphism of f.g. projective R-modules
$1 \otimes s : R \otimes_A P \to R \otimes_A Q$ is an isomorphism. The noncommutative localization $A \to \Sigma^{-1}A$ is Σ-inverting, and has the universal property that any Σ-
inverting ring morphism $A \to R$ has a unique factorization $A \to \Sigma^{-1}A \to R$.
The applications to topology involve homology with coefficients in a noncommutative localization $\Sigma^{-1}A$.

Homology with coefficients is defined as follows. Let X be a connected
topological space with universal cover $\widetilde{X}$, and let the fundamental group
$\pi_1(X)$ act on the left of $\widetilde{X}$, so that the (singular) chain complex $S(\widetilde{X})$ is a free
left $\mathbb{Z}[\pi_1(X)]$-module complex. Given a morphism of rings $\mathcal{F} : \mathbb{Z}[\pi_1(X)] \to \Lambda$
define the Λ-*coefficient homology of* X to be the Λ-modules

$$H_*(X; \Lambda) = H_*(\Lambda \otimes_{\mathbb{Z}[\pi_1(X)]} S(\widetilde{X})) \ .$$

If X is a CW complex then $S(\widetilde{X})$ is chain equivalent to the cellular free
$\mathbb{Z}[\pi_1(X)]$-module chain complex $C(\widetilde{X})$ with one generator in degree r for
each r-cell of X, and

$$H_*(X; \Lambda) = H_*(\Lambda \otimes_{\mathbb{Z}[\pi_1(X)]} C(\widetilde{X})) \ .$$

1.1   Finite domination

A topological space X is *finitely dominated* if there exist a finite CW complex
K, maps $f : X \to K$, $g : K \to X$ and a homotopy $gf \simeq 1 : X \to X$.
The singular chain complex $S(\widetilde{X})$ of the universal cover $\widetilde{X}$ of X is chain
equivalent to a finite f.g. projective $\mathbb{Z}[\pi_1(X)]$-module chain complex P. Wall
[42] defined the *finiteness obstruction* of X to be the reduced projective class

$$[X] = \sum_{r=0}^{\infty}(-1)^r[P_r] \in \widetilde{K}_0(\mathbb{Z}[\pi_1(X)]) \ ,$$

and proved that $[X] = 0$ if and only if X is homotopy equivalent to a finite
CW complex.

In the applications of the finiteness obstruction to manifold topology $X = \overline{M}$ is an infinite cyclic cover of a compact manifold M – see Chapter 17 of Hughes and Ranicki [18] for the geometric wrapping up procedure which shows that in dimension $\geqslant 5$ every tame manifold end has a neighbourhood which is a finitely dominated infinite cyclic cover $\overline{M}$ of a compact manifold M. Let $f : M \to S^1$ be a classifying map, with $\mathbb{Z}$-equivariant lift $\overline{f} : \overline{M} = f^*\mathbb{R} \to \mathbb{R}$. The non-compact manifold $\overline{M}^+ = \overline{f}^{-1}[0, \infty)$ is finitely dominated, with boundary $\partial\overline{M}^+ = \overline{f}^{-1}(0)$ (assuming f is transverse regular at $0 \in S^1$) and one tame end, and $\pi_1(\overline{M}^+) = \pi_1(\overline{M})$. The finiteness obstruction $[\overline{M}^+] \in \widetilde{K}_0(\mathbb{Z}[\pi_1(\overline{M})])$ is the end obstruction of Siebenmann [38], such that $[\overline{M}^+] = 0$ if and only if the tame end can be closed, i.e. compactified by a manifold with boundary.

Given a ring A let Ω be the set of square matrices $\omega \in M_r(A[z, z^{-1}])$ over the Laurent polynomial extension $A[z, z^{-1}]$ such that the A-module

$$P = \mathrm{coker}(\omega : A[z, z^{-1}]^r \to A[z, z^{-1}]^r)$$

is f.g. projective. The noncommutative Fredholm localization $\Omega^{-1}A[z, z^{-1}]$ has the universal property that a finite f.g. free $A[z, z^{-1}]$-module chain complex C is A-module chain equivalent to a finite f.g. projective A-module chain complex if and only if $H_*(\Omega^{-1}C) = 0$ (Ranicki [29, Proposition 13.9]), with $\Omega^{-1}C = \Omega^{-1}A[z, z^{-1}] \otimes_{A[z,z^{-1}]} C$.

Let M be a connected finite CW complex with a connected infinite cyclic cover $\overline{M}$. The fundamental group $\pi_1(M)$ fits into an extension

$$\{1\} \to \pi_1(\overline{M}) \to \pi_1(M) \to \mathbb{Z} \to \{1\}$$

and $\mathbb{Z}[\pi_1(M)]$ is a twisted Laurent polynomial extension

$$\mathbb{Z}[\pi_1(M)] = \mathbb{Z}[\pi_1(\overline{M})]_\alpha[z, z^{-1}]$$

with $\alpha : \pi_1(\overline{M}) \to \pi_1(\overline{M}); g \mapsto z^{-1}gz$ the monodromy automorphism. For the sake of simplicity only the untwisted case $\alpha = 1$ will be considered here, so that $\pi_1(M) = \pi_1(\overline{M}) \times \mathbb{Z}$. The infinite cyclic cover $\overline{M}$ is finitely dominated if and only if $H_*(M; \Omega^{-1}\mathbb{Z}[\pi_1(M)]) = 0$, with $A = \mathbb{Z}[\pi_1(\overline{M})]$ and $\mathbb{Z}[\pi_1(M)] = A[z, z^{-1}]$. The Farrell-Siebenmann obstruction $\Phi(M) \in \mathrm{Wh}(\pi_1(M))$ of an n-dimensional manifold M with finitely dominated infinite cyclic cover $\overline{M}$ is such that $\Phi(M) = 0$ if (and for $n \geqslant 6$ only if) M is a fibre bundle over S^1 – see [29, Proposition 15.16] for the expression of $\Phi(M)$ in terms of the $\Omega^{-1}\mathbb{Z}[\pi_1(M)]$-coefficient Reidemeister-Whitehead torsion

$$\tau(M; \Omega^{-1}\mathbb{Z}[\pi_1(M)]) = \tau(\Omega^{-1}C(\widetilde{M})) \in K_1(\Omega^{-1}\mathbb{Z}[\pi_1(M)]) .$$

1.2 Codimension 1 splitting

Surgery theory asks whether a homotopy equivalence of manifolds is homotopic (or h-cobordant) to a homeomorphism – in general, the answer is no. There are obstructions in the topological K-theory of vector bundles, in the algebraic K-theory of modules and in the algebraic L-theory of quadratic forms. The algebraic K-theory obstruction lives in the Whitehead group $Wh(\pi)$ of the fundamental group π. The L-theory obstruction lives in one of the surgery groups $L_*(\mathbb{Z}[\pi])$ of Wall [43], and is defined when the topological and algebraic K-theory obstructions vanish. The groups $L_*(\Lambda)$ are defined for any ring with involution Λ to be the generalized Witt groups of nonsingular quadratic forms over Λ. For manifolds of dimension $\geqslant 5$ the vanishing of the algebraic obstructions is both a necessary and sufficient condition for deforming a homotopy equivalence to a homeomorphism. See Ranicki [28] for the reduction of the Browder-Novikov-Sullivan-Wall surgery theory to algebra.

A homotopy equivalence of m-dimensional manifolds $f : M' \to M$ *splits* along a submanifold $N^n \subset M^m$ if f is homotopic to a map (also denoted by f) such that $N' = f^{-1}(N) \subset M'$ is also a submanifold, and the restriction $f| : N' \to N$ is also a homotopy equivalence. For codimension $m - n \geqslant 3$ the splitting obstruction is just the ordinary surgery obstruction $\sigma_*(f|) \in L_n(\mathbb{Z}[\pi_1(N)])$. For codimension $m - n = 1, 2$ the splitting obstructions involve the interplay of the knotting properties of codimension $(m - n)$ submanifolds and Mayer-Vietoris-type decompositions of the algebraic K- and L-groups of $\mathbb{Z}[\pi_1(M)]$ in terms of the groups of $\mathbb{Z}[\pi_1(N)]$, $\mathbb{Z}[\pi_1(M\backslash N)]$.

In the case $m - n = 1$ $\pi_1(M)$ is a generalized free product, i.e. either an amalgamated free product or an HNN extension, by the Seifert-van Kampen theorem. Codimension 1 splitting theorems and the algebraic K- and L-theory of generalized free products are a major ingredient of high-dimensional manifold topology, featuring in the work of Stallings, Browder, Novikov, Wall, Siebenmann, Farrell, Hsiang, Shaneson, Casson, Waldhausen, Cappell, ..., and the author. Noncommutative localization provides a systematic development of this algebra, using the intuition afforded by the topological applications – see Part 2 below for a more detailed discussion.

1.3 Homology surgery theory

For a morphism of rings with involution $\mathcal{F} : \mathbb{Z}[\pi] \to \Lambda$ Cappell and Shaneson [4] considered the problem of whether a Λ-coefficient homology equivalence of manifolds with fundamental group π is H-cobordant to a homeomorphism. Again, the answer is no in general, with obstructions in the topological K-theory of vector bundles and in the homology surgery groups $\Gamma_*(\mathcal{F})$, which are generalized Witt groups of Λ-nonsingular quadratic forms over $\mathbb{Z}[\pi]$. Vogel [39], [40] identified the Λ-coefficient homology surgery groups with the ordinary L-groups of the localization $\Sigma^{-1}\mathbb{Z}[\pi]$ of $\mathbb{Z}[\pi]$ inverting the set Σ of Λ-invertible square matrices over $\mathbb{Z}[\pi]$

$$\Gamma_*(\mathcal{F}) \;=\; L_*(\Sigma^{-1}\mathbb{Z}[\pi]) \;,$$

and identified the relative L-groups $L_*(\mathbb{Z}[\pi] \to \Sigma^{-1}\mathbb{Z}[\pi])$ in the localization exact sequence

$$\cdots \to L_n(\mathbb{Z}[\pi]) \to L_n(\Sigma^{-1}\mathbb{Z}[\pi]) \to L_n(\mathbb{Z}[\pi] \to \Sigma^{-1}\mathbb{Z}[\pi]) \to L_{n-1}(\mathbb{Z}[\pi]) \to \ldots$$

with generalized Witt groups $L_*(\mathbb{Z}[\pi], \Sigma)$ of nonsingular $\Sigma^{-1}\mathbb{Z}[\pi]/\mathbb{Z}[\pi])$-valued quadratic linking forms on Σ-torsion $\mathbb{Z}[\pi]$-modules of homological dimension 1.

1.4   Codimension 2 embeddings

Suppose given a codimension 2 embedding $N^n \subset M^{n+2}$ such as a knot or link. By Alexander duality the $\mathbb{Z}[\pi_1(M)]$-modules

$$H_*(M \backslash N; \mathbb{Z}[\pi_1(M)]) \;\cong\; H^{n+2-*}(M, N; \mathbb{Z}[\pi_1(M)]) \;\; (* \neq 0, n+2)$$

are determined by the homotopy class of the inclusion $N \subset M$. See Ranicki [29] for a general account of high-dimensional codimension 2 embedding theory, including some of the applications of the noncommutative localization $\Sigma^{-1}A$ inverting the set Σ of matrices over $A = \mathbb{Z}[\pi_1(M \backslash N)]$ which become invertible over $\mathbb{Z}[\pi_1(M)]$. The $\Sigma^{-1}A$-modules $H_*(M \backslash N; \Sigma^{-1}A)$ and their Poincaré duality properties reflect more subtle invariants of $N \subset M$ such as knotting and linking.

1.5   Open books

An $(n+2)$-dimensional manifold M^{n+2} is an *open book* if there exists a codimension 2 submanifold $N^n \subset M^{n+2}$ such that the complement $M \backslash N$ is a fibre bundle over S^1. Every odd-dimensional manifold is an open book. Quinn [25] showed that for $k \geqslant 2$ a $(2k+2)$-dimensional manifold M is an open book if and only if an asymmetric form over $\mathbb{Z}[\pi_1(M)]$ associated to M represents 0 in the Witt group. This obstruction was identified in Ranicki [29] with an element in the L-group $L_{2k+2}(\Omega^{-1}\mathbb{Z}[\pi_1(M)][z, z^{-1}])$ of the Fredholm localization of $\mathbb{Z}[\pi_1(M)][z, z^{-1}]$ (cf. section 1.1 above).

1.6   Boundary link cobordism

An n-dimensional μ-component boundary link is a codimension 2 embedding

$$N^n = \bigsqcup_\mu S^n \subset M^{n+2} = S^{n+2}$$

with a μ-component Seifert surface, in which case the fundamental group of the complement $X = M \backslash N$ has a compatible surjection $\pi_1(X) \to F_\mu$ onto the free group on μ generators. Duval [11] used the work of Cappell and Shaneson [5] and Vogel [40] to identify the cobordism group of n-dimensional μ-component boundary links for $n \geqslant 2$ with the relative L-group $L_{n+3}(\mathbb{Z}[F_\mu], \Sigma)$ in the localization exact sequence

$$\cdots \to L_{n+3}(\mathbb{Z}[F_\mu]) \to L_{n+3}(\Sigma^{-1}\mathbb{Z}[F_\mu]) \to L_{n+3}(\mathbb{Z}[F_\mu], \Sigma) \to L_{n+2}(\mathbb{Z}[F_\mu]) \to \cdots$$

with Σ the set of $\mathbb{Z}$-invertible square matrices over $\mathbb{Z}[F_\mu]$. The even-dimensional boundary link cobordism groups are $L_{2*+1}(\mathbb{Z}[F_\mu], \Sigma) = 0$. The cobordism class of a $(2k-1)$-dimensional boundary link $\sqcup_\mu S^{2k-1} \subset S^{2k+1}$ $(k \geqslant 2)$ was identified with the Witt class in $L_{2k+2}(\mathbb{Z}[F_\mu], \Sigma)$ of a $\Sigma^{-1}\mathbb{Z}[F_\mu]/\mathbb{Z}[F_\mu]$-valued nonsingular $(-1)^{k+1}$-quadratic linking form on $H_k(X; \mathbb{Z}[F_\mu])$, generalizing the Blanchfield pairing on the homology of the infinite cyclic cover of a knot $S^{2k-1} \subset S^{2k+1}$. The localization $\Sigma^{-1}\mathbb{Z}[F_\mu]$ was identified by Dicks and Sontag [10] and Farber and Vogel [16] with a ring of rational functions in μ noncommuting variables. The high odd-dimensional boundary link cobordism groups $L_{2*+2}(\mathbb{Z}[F_\mu], \Sigma)$ have been computed by Sheiham [35], [37]. Part I of Ranicki and Sheiham [32] deals with the algebraic K-theory of $A[F_\mu]$ and $\Sigma^{-1}A[F_\mu]$ for any ring A; Part II will deal with the algebraic L-theory of $A[F_\mu]$ and $\Sigma^{-1}A[F_\mu]$ for any ring with involution A.

1.7   Circle-valued Morse theory

Novikov [23] initiated the study of the critical points of Morse functions $f : M \to S^1$ on compact manifolds M using the 'Novikov ring'

$$\mathbb{Z}((z)) = \mathbb{Z}[[z]][z^{-1}] \,.$$

The 'Novikov complex' $C(M, f)$ over $\mathbb{Z}((z))$ has one generator for each critical point of f, and the differentials count the gradient flow lines of the $\mathbb{Z}$-equivariant real-valued Morse function $\overline{f} : \overline{M} \to \mathbb{R}$ on the non-compact infinite cyclic cover $\overline{M} = f^*\mathbb{R}$. The 'Novikov homology'

$$H_*(C(M, f)) = H_*(\overline{M}; \mathbb{Z}((z)))$$

provides lower bounds on the number of critical points of Morse functions in the homotopy class of f, generalizing the inequalities of the classical Morse theory of real-valued functions $M \to \mathbb{R}$. Suppose the Morse function $f : M \to S^1$ has monodromy $\alpha = 1 : \pi_1(\overline{M}) \to \pi_1(\overline{M})$ (for the sake of simplicity), so that $\pi_1(M) = \pi_1(\overline{M}) \times \mathbb{Z}$. Let Σ be the set of square matrices over $\mathbb{Z}[\pi_1(\overline{M})][z]$ which become invertible over $\mathbb{Z}[\pi_1(\overline{M})]$ under the augmentation $z \mapsto 0$. There is a natural morphism from the localization to the completion

$$\Sigma^{-1}\mathbb{Z}[\pi_1(M)] \to \mathbb{Z}\widehat{[\pi_1(M)]} = \mathbb{Z}[\pi_1(\overline{M})][[z]][z^{-1}]$$

which is an injection if $\pi_1(M)$ is abelian or F_μ (Dicks and Sontag [10], Farber and Vogel [16]), but may not be an injection in general (Sheiham [34]). See Pajitnov [24], Farber and Ranicki [15], Ranicki [30], and Cornea and Ranicki [9] for the construction and properties of Novikov complexes of f over $\mathbb{Z}\widehat{[\pi_1(M)]}$ and $\Sigma^{-1}\mathbb{Z}[\pi_1(M)]$. Naturally, noncommutative localization also features in the more general Morse theory of closed 1-forms – see Novikov [23] and Farber [13],[14].

1.8   3- and 4-dimensional manifolds

See Garoufalidis and Kricker [17], Quinn [26] for applications of noncommutative localization in the topology of 3- and 4-dimensional manifolds.

1.9   Homotopy theory

Noncommutative localization also features in homotopy theory – see the paper by Dwyer [12] in these proceedings. The homotopy theoretic localization of Vogel has applications to links (cf. Le Dimet [19], Cochran [8], Levine, Mio and Orr [20]). However, these applications are beyond the scope of this survey.

Part 2. The algebraic K- and L-theory of generalized free products via noncommutative localization

A generalized free product of groups (or rings) is either an amalgamated free product or an HNN extension. The expressions of Schofield [33] of generalized free products as noncommutative localizations of triangular matrix rings combine with the localization exact sequences of Neeman and Ranicki [21] to provide more systematic proofs of the Mayer-Vietoris decompositions of Waldhausen [41] and Cappell [3] of the algebraic K- and L-theory of generalized free products. The topological motivation for these proofs comes from a noncommutative localization interpretation of the Seifert-van Kampen and Mayer-Vietoris theorems. If $(M, N \subseteq M)$ is a two-sided pair of connected CW complexes the fundamental group $\pi_1(M)$ is a generalized free product: an amalgamated free product if N separates M, and an HNN extension otherwise. The morphisms $\pi_1(N) \to \pi_1(M \backslash N)$ determine a triangular $k \times k$ matrix ring A with universal localization the full $k \times k$ matrix ring $\Sigma^{-1}A = M_k(\mathbb{Z}[\pi_1(M)])$ ($k = 3$ in the separating case, $k = 2$ in the non-separating case), such that the corresponding presentation of the $\mathbb{Z}[\pi_1(M)]$-module chain complex $C(\widetilde{M})$ of the universal cover $\widetilde{M}$ is the assembly of an A-module chain complex constructed from the chain complexes $C(\widetilde{N}), C(\widetilde{M \backslash N})$ of the universal covers $\widetilde{N}, \widetilde{M \backslash N}$ of $N, M \backslash N$. The two cases will be considered separately, in sections 2.3, 2.4.

2.1   The algebraic K-theory of a noncommutative localization

Given an injective noncommutative localization $A \to \Sigma^{-1}A$ let $H(A, \Sigma)$ be the exact category of homological dimension 1 A-modules T which admit a f.g. projective A-module resolution

$$0 \longrightarrow P \overset{s}{\longrightarrow} Q \longrightarrow T \longrightarrow 0$$

such that $1 \otimes s : \Sigma^{-1}P \to \Sigma^{-1}Q$ is an $\Sigma^{-1}A$-module isomorphism. The algebraic K-theory localization exact sequence of Schofield [33, Theorem 4.12]

$$K_1(A) \to K_1(\Sigma^{-1}A) \to K_1(A, \Sigma) \to K_0(A) \to K_0(\Sigma^{-1}A)$$

was obtained for any injective noncommutative localization $A \to \Sigma^{-1}A$, with $K_1(A, \Sigma) = K_0(H(A, \Sigma))$. Neeman and Ranicki [21] proved that if

$A \to \Sigma^{-1}A$ is injective and 'stably flat'

$$\mathrm{Tor}_i^A(\Sigma^{-1}A, \Sigma^{-1}A) = 0 \ (i \geqslant 1)$$

then

(i) $\Sigma^{-1}A$ has the chain complex lifting property : every finite f.g. free $\Sigma^{-1}A$-module chain complex C is chain equivalent to $\Sigma^{-1}B$ for a finite f.g. projective A-module chain complex B,

(ii) the localization exact sequence extends to the higher K-groups

$$\cdots \to K_n(A) \to K_n(\Sigma^{-1}A) \to K_n(A, \Sigma) \to K_{n-1}(A) \to \cdots \to K_0(\Sigma^{-1}A)$$

with $K_n(A, \Sigma) = K_{n-1}(H(A, \Sigma))$.

See Neeman, Ranicki and Schofield [22] for explicit examples of injective noncommutative localizations $A \to \Sigma^{-1}A$ which are not stably flat.

2.2 Matrix rings

The amalgamated free product of rings and the HNN construction are special cases of the following type of noncommutative localization of triangular matrix rings.

Given rings A_1, A_2 and an (A_1, A_2)-bimodule B define the triangular 2×2 matrix ring

$$A = \begin{pmatrix} A_1 & B \\ 0 & A_2 \end{pmatrix} .$$

An A-module can be written as

$$M = \begin{pmatrix} M_1 \\ M_2 \end{pmatrix}$$

with M_1 an A_1-module, M_2 an A_2-module, together with an A_1-module morphism $B \otimes_{A_2} M_2 \to M_1$. The injection

$$A_1 \times A_2 \to A \ ; \ (a_1, a_2) \mapsto \begin{pmatrix} a_1 & 0 \\ 0 & a_2 \end{pmatrix}$$

induces isomorphisms of algebraic K-groups

$$K_*(A_1) \oplus K_*(A_2) \cong K_*(A)$$

(Berrick and Keating [2]). The columns of A are f.g. projective A-modules

$$P_1 = \begin{pmatrix} A_1 \\ 0 \end{pmatrix} , \ P_2 = \begin{pmatrix} B \\ A_2 \end{pmatrix}$$

such that

$$P_1 \oplus P_2 = A , \ \operatorname{Hom}_A(P_i, P_i) = A_i \ (i = 1, 2) ,$$
$$\operatorname{Hom}_A(P_1, P_2) = B , \ \operatorname{Hom}_A(P_2, P_1) = 0 .$$

The noncommutative localization of A inverting a non-empty subset $\Sigma \subseteq \operatorname{Hom}_A(P_1, P_2) = B$ is the 2×2 matrix ring

$$\Sigma^{-1} A = M_2(C) = \begin{pmatrix} C & C \\ C & C \end{pmatrix}$$

with C the endomorphism ring of the induced f.g. projective $\Sigma^{-1} A$-module $\Sigma^{-1} P_1 \cong \Sigma^{-1} P_2$. The Morita equivalence

$$\{\Sigma^{-1} A\text{-modules}\} \to \{C\text{-modules}\} ; \ L \mapsto (C \ C) \otimes_{\Sigma^{-1} A} L$$

induces isomorphisms in algebraic K-theory

$$K_*(M_2(C)) \cong K_*(C) .$$

The composite of the functor

$$\{A\text{-modules}\} \to \{\Sigma^{-1} A\text{-modules}\} ; \ M \mapsto \Sigma^{-1} M = \Sigma^{-1} A \otimes_A M$$

and the Morita equivalence is the *assembly* functor

$$\{A\text{-modules}\} \to \{C\text{-modules}\} ;$$

$$M = \begin{pmatrix} M_1 \\ M_2 \end{pmatrix} \mapsto (C \ C) \otimes_A M$$
$$= \operatorname{coker}(C \otimes_{A_1} B \otimes_{A_2} M_2 \to C \otimes_{A_1} M_1 \oplus C \otimes_{A_2} M_2)$$

inducing the morphisms

$$K_*(A) = K_*(A_1) \oplus K_*(A_2) \to K_*(\Sigma^{-1} A) = K_*(C)$$

in the algebraic K-theory localization exact sequence. If B and C are flat A_1-modules and C is a flat A_2-module then the A-module $\begin{pmatrix} C \\ C \end{pmatrix}$ has a 1-dimensional flat A-module resolution

$$0 \to \begin{pmatrix} B \\ 0 \end{pmatrix} \otimes_{A_2} C \to \begin{pmatrix} A_1 \\ 0 \end{pmatrix} \otimes_{A_1} C \oplus \begin{pmatrix} B \\ A_2 \end{pmatrix} \otimes_{A_2} C \to \begin{pmatrix} C \\ C \end{pmatrix} \to 0$$

so that $\Sigma^{-1} A = \begin{pmatrix} C \\ C \end{pmatrix} \oplus \begin{pmatrix} C \\ C \end{pmatrix}$ is stably flat.

There are evident generalizations to $k \times k$ matrix rings for any $k \geqslant 2$.

2.3 *HNN* extensions

The *HNN* extension $R *_{\alpha,\beta} \{z\}$ is defined for any ring morphisms $\alpha, \beta :$ $S \to R$, with

$$\alpha(s)z = z\beta(s) \in R *_{\alpha,\beta} \{z\} \quad (s \in S) .$$

Define the triangular 2×2 matrix ring

$$A = \begin{pmatrix} R & R_\alpha \oplus R_\beta \\ 0 & S \end{pmatrix}$$

with R_α the (R, S)-bimodule R with S acting on R via α, and similarly for R_β. Let $\Sigma = \{\sigma_1, \sigma_2\} \subset \mathrm{Hom}_A(P_1, P_2)$, with

$$\sigma_1 = \begin{pmatrix} (1,0) \\ 0 \end{pmatrix} , \quad \sigma_2 = \begin{pmatrix} (0,1) \\ 0 \end{pmatrix} : P_1 = \begin{pmatrix} R \\ 0 \end{pmatrix} \to P_2 = \begin{pmatrix} R_\alpha \oplus R_\beta \\ S \end{pmatrix} .$$

The A-modules P_1, P_2 are f.g. projective since $P_1 \oplus P_2 = A$. Theorem 13.1 of [33] identifies

$$\Sigma^{-1}A = M_2(R *_{\alpha,\beta} \{z\}) .$$

Example Let $(M, N \subseteq M)$ be a non-separating pair of connected *CW* complexes such that N is two-sided in M (i.e. has a neighbourhood $N \times [0, 1] \subseteq M$) with $M \backslash N = M_1$ connected

$$M = M_1 \cup_{N \times \{0,1\}} N \times [0, 1]$$

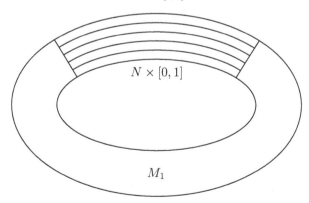

By the Seifert-van Kampen theorem, the fundamental group $\pi_1(M)$ is the *HNN* extension determined by the morphisms $\alpha, \beta : \pi_1(N) \to \pi_1(M_1)$ induced by the inclusions $N \times \{0\} \to M_1$, $N \times \{1\} \to M_1$

$$\pi_1(M) = \pi_1(M_1) *_{\alpha,\beta} \{z\} ,$$

so that
$$\mathbb{Z}[\pi_1(M)] = \mathbb{Z}[\pi_1(M_1)] *_{\alpha,\beta} \{z\} .$$

As above, define a triangular 2×2 matrix ring
$$A = \begin{pmatrix} \mathbb{Z}[\pi_1(N)] & \mathbb{Z}[\pi_1(M_1)]_\alpha \oplus \mathbb{Z}[\pi_1(M_1)]_\beta \\ 0 & \mathbb{Z}[\pi_1(M)] \end{pmatrix}$$

with noncommutative localization
$$\Sigma^{-1}A = M_2(\mathbb{Z}[\pi_1(M_1)] *_{\alpha,\beta} \{z\}) = M_2(\mathbb{Z}[\pi_1(M)]) .$$

Assume that $\pi_1(N) \to \pi_1(M)$ is injective, so that the morphisms α, β are injective, and the universal cover $\widetilde{M}$ is a union
$$\widetilde{M} = \bigcup_{g \in [\pi_1(M):\pi_1(M_1)]} g\widetilde{M_1}$$

of translates of the universal cover $\widetilde{M_1}$ of M_1, and
$$g_1\widetilde{M_1} \cap g_2\widetilde{M_1} = \begin{cases} h\widetilde{N} & \text{if } g_1 \cap g_2 z = h \in [\pi_1(M) : \pi_1(N)] \\ g_1\widetilde{M_1} & \text{if } g_1 = g_2 \\ \emptyset & \text{if } g_1 \neq g_2 \text{ and } g_1 \cap g_2 z = \emptyset \end{cases}$$

with $h\widetilde{N}$ the translates of the universal cover $\widetilde{N}$ of N. In the diagram it is assumed that α, β are isomorphisms

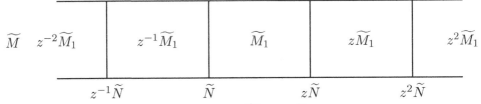

The cellular f.g. free chain complexes $C(\widetilde{M_1})$, $C(\widetilde{N})$ are related by $\mathbb{Z}[\pi_1(M_1)]$-module chain maps
$$i_\alpha : \mathbb{Z}[\pi_1(M_1)]_\alpha \otimes_{\mathbb{Z}[\pi_1(N)]} C(\widetilde{N}) \to C(\widetilde{M_1}) ,$$
$$i_\beta : \mathbb{Z}[\pi_1(M_1)]_\beta \otimes_{\mathbb{Z}[\pi_1(N)]} C(\widetilde{N}) \to C(\widetilde{M_1})$$

defining a f.g. projective A-module chain complex $\begin{pmatrix} C(\widetilde{M_1}) \\ C(\widetilde{N}) \end{pmatrix}$ with assembly the cellular f.g. free $\mathbb{Z}[\pi_1(M)]$-module chain complex of $\widetilde{M}$
$$\text{coker}\left(i_\alpha - z i_\beta : \mathbb{Z}[\pi_1(M)]_\alpha \otimes_{\mathbb{Z}[\pi_1(N)]} C(\widetilde{N}) \to \mathbb{Z}[\pi_1(M)] \otimes_{\mathbb{Z}[\pi_1(M_1)]} C(\widetilde{M_1}) \right)$$
$$= C(\widetilde{M})$$

by the Mayer-Vietoris theorem. $\square$

Let $R *_{\alpha,\beta} \{z\}$ be an HNN extension of rings in which the morphisms $\alpha, \beta : S \to R$ are both injections of (S, S)-bimodule direct summands, and R_α, R_β are flat S-modules. (This is the case in the above example if $\pi_1(N) \to \pi_1(M)$ is injective). Then the natural ring morphisms

$$R \to R *_{\alpha,\beta} \{z\} \;, \; S \to R *_{\alpha,\beta} \{z\} \;,$$

$$A = \begin{pmatrix} R & R_\alpha \oplus R_\beta \\ 0 & S \end{pmatrix} \to \Sigma^{-1} A = M_2(R *_{\alpha,\beta} \{z\})$$

are injective, and $\Sigma^{-1} A$ is a stably flat universal localization, with

$$H(A, \Sigma) = \mathrm{Nil}(R, S, \alpha, \beta)$$

the nilpotent category of Waldhausen [41]. The chain complex lifting property of $\Sigma^{-1} A$ gives a noncommutative localization proof of the existence of Mayer-Vietoris presentations for finite f.g. free $R *_{\alpha,\beta} \{z\}$-module chain complexes C

$$0 \longrightarrow R *_{\alpha,\beta} \{z\} \otimes_S E \xrightarrow{i_\alpha - z i_\beta} R *_{\alpha,\beta} \{z\} \otimes_R D \longrightarrow C \longrightarrow 0$$

with D (resp. E) a finite f.g. free R- (resp. S-) module chain complex ([41], Ranicki [31]). The algebraic K-theory localization exact sequence of [21]

$$\cdots \to K_{n+1}(A, \Sigma) = K_n(S) \oplus K_n(S) \oplus \widetilde{\mathrm{Nil}}_n(R, S, \alpha, \beta)$$

$$\xrightarrow{\begin{pmatrix} \alpha & \beta & 0 \\ 1 & 1 & 0 \end{pmatrix}} K_n(A) = K_n(R) \oplus K_n(S)$$

$$\to K_n(\Sigma^{-1} A) = K_n(R *_{\alpha,\beta} \{z\}) \to \cdots$$

is just the stabilization by $1 : K_*(S) \to K_*(S)$ of the Mayer-Vietoris exact sequence of [41]

$$\cdots \longrightarrow K_n(S) \oplus \widetilde{\mathrm{Nil}}_n(R, \alpha, \beta) \xrightarrow{(\alpha - \beta) \oplus 0} K_n(R) \longrightarrow K_n(R *_{\alpha,\beta} \{z\}) \longrightarrow \cdots$$

In particular, for $\alpha = \beta = 1 : S = R \to R$ the HNN extension is just the Laurent polynomial extension

$$R *_{\alpha,\beta} \{z\} = R[z, z^{-1}]$$

and the Mayer-Vietoris exact sequence splits to give the original splitting of Bass, Heller and Swan [1]

$$K_1(R[z, z^{-1}]) \;=\; K_1(R) \oplus K_0(R) \oplus \widetilde{\mathrm{Nil}}_0(R) \oplus \widetilde{\mathrm{Nil}}_0(R)$$

as well as its extension to the Quillen higher K-groups K_*.

2.4   Amalgamated free products

The amalgamated free product $R_1 *_S R_2$ is defined for any ring morphisms $i_1 : S \to R_1$, $i_2 : S \to R_2$, with

$$r_1 i_1(s) * r_2 \;=\; r_1 * i_2(s) r_2 \in R_1 *_S R_2 \quad (r_1 \in R_1, r_2 \in R_2, s \in S) \;.$$

Define the triangular 3×3 matrix ring

$$A \;=\; \begin{pmatrix} R_1 & 0 & R_1 \\ 0 & R_2 & R_2 \\ 0 & 0 & S \end{pmatrix}$$

and the A-module morphisms

$$\sigma_1 \;=\; \begin{pmatrix} 1 \\ 0 \\ 0 \end{pmatrix} \;:\; P_1 \;=\; \begin{pmatrix} R_1 \\ 0 \\ 0 \end{pmatrix} \to P_3 \;=\; \begin{pmatrix} R_1 \\ R_2 \\ S \end{pmatrix} \;,$$

$$\sigma_2 \;=\; \begin{pmatrix} 0 \\ 1 \\ 0 \end{pmatrix} \;:\; P_2 \;=\; \begin{pmatrix} 0 \\ R_2 \\ 0 \end{pmatrix} \to P_3 \;=\; \begin{pmatrix} R_1 \\ R_2 \\ S \end{pmatrix} \;.$$

The A-modules P_1, P_2, P_3 are f.g. projective since $P_1 \oplus P_2 \oplus P_3 = A$. The noncommutative localization of A inverting $\Sigma = \{\sigma_1, \sigma_2\}$ is the full 3×3 matrix ring

$$\Sigma^{-1} A \;=\; M_3(R_1 *_S R_2)$$

(a modification of Theorem 4.10 of [33]).

Example Let $(M, N \subseteq M)$ be a separating pair of CW complexes such that N has a neighbourhood $N \times [0, 1] \subseteq M$ and

$$M \;=\; M_1 \cup_{N \times \{0\}} N \times [0, 1] \cup_{N \times \{1\}} M_2$$

with M_1, M_2, N connected.

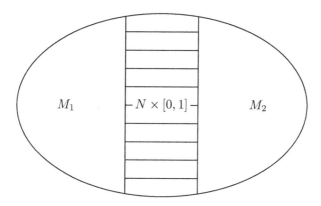

By the Seifert-van Kampen theorem, the fundamental group of M is the amalgamated free product

$$\pi_1(M) = \pi_1(M_1) *_{\pi_1(N)} \pi_1(M_2) \,,$$

so that

$$\mathbb{Z}[\pi_1(M)] = \mathbb{Z}[\pi_1(M_1)] *_{\mathbb{Z}[\pi_1(N)]} \mathbb{Z}[\pi_1(M_2)] \,.$$

As above, define a triangular matrix ring

$$A = \begin{pmatrix} \mathbb{Z}[\pi_1(M_1)] & 0 & \mathbb{Z}[\pi_1(M_1)] \\ 0 & \mathbb{Z}[\pi_1(M_2)] & \mathbb{Z}[\pi_1(M_2)] \\ 0 & 0 & \mathbb{Z}[\pi_1(N)] \end{pmatrix}$$

with noncommutative localization

$$\Sigma^{-1}A = M_3(\mathbb{Z}[\pi_1(M_1)] *_{\mathbb{Z}[\pi_1(N)]} \mathbb{Z}[\pi_1(M_2)]) = M_3(\mathbb{Z}[\pi_1(M)]) \,.$$

Assume that $\pi_1(N) \to \pi_1(M)$ is injective, so that the morphisms

$$i_j \colon \pi_1(N) \to \pi_1(M_j) \,, \ \pi_1(M_j) \to \pi_1(M) \ (j = 1, 2)$$

are all injective, and the universal cover $\widetilde{M}$ of M is a union

$$\widetilde{M} = \bigcup_{g_1 \in [\pi_1(M):\pi_1(M_1)]} g_1 \widetilde{M}_1 \cup \bigcup_{h \in [\pi_1(M):\pi_1(N)]} h\widetilde{N} \bigcup_{g_2 \in [\pi_1(M):\pi_1(M_2)]} g_2 \widetilde{M}_2$$

of $[\pi_1(M) : \pi_1(M_1)]$ translates of the universal cover $\widetilde{M}_1$ of M_1 and $[\pi_1(M) : \pi_1(M_2)]$ translates of the universal cover $\widetilde{M}_2$ of M_2, with intersection the

$[\pi_1(M) : \pi_1(N)]$ translates of the universal cover $\widetilde{N}$ of N.

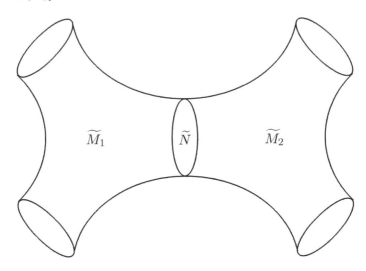

The cellular f.g. free chain complexes $C(\widetilde{M_j})$, $C(\widetilde{N})$ are related by the $\mathbb{Z}[\pi_1(M_j)]$-module chain maps

$$i_j \; : \; \mathbb{Z}[\pi_1(M_j)] \otimes_{\mathbb{Z}[\pi_1(N)]} C(\widetilde{N}) \to C(\widetilde{M_j}) \quad (j = 1, 2)$$

induced by the inclusions $i_j : N \to M_j$, defining a f.g. projective A-module chain complex $\begin{pmatrix} C(\widetilde{M_1}) \\ C(\widetilde{M_2}) \\ C(\widetilde{N}) \end{pmatrix}$ with assembly the cellular f.g. free $\mathbb{Z}[\pi_1(M)]$-module chain complex of $\widetilde{M}$

$$\mathrm{coker}\Bigg(\begin{pmatrix} 1 \otimes i_1 \\ 1 \otimes i_2 \end{pmatrix} : \mathbb{Z}[\pi_1(M)] \otimes_{\mathbb{Z}[\pi_1(N)]} C(\widetilde{N}) \to$$
$$\mathbb{Z}[\pi_1(M)] \otimes_{\mathbb{Z}[\pi_1(M_1)]} C(\widetilde{M_1}) \oplus \mathbb{Z}[\pi_1(M)] \otimes_{\mathbb{Z}[\pi_1(M_1)]} C(\widetilde{M_2}) \Bigg)$$
$$= C(\widetilde{M})$$

by the Mayer-Vietoris theorem.   □

Let $R_1 *_S R_2$ be an amalgamated free product of rings in which the morphisms $i_1 : S \to R_1$, $i_2 : S \to R_2$ are both injections of (S, S)-bimodule direct summands, and R_1, R_2 are flat S-modules. (This is the case in the above example if $\pi_1(N) \to \pi_1(M)$ is injective). Then the natural ring morphisms

$$R_1 \to R_1 *_S R_2 \ , \ R_2 \to R_1 *_S R_2 \ , \ S \to R_1 *_S R_2 \ ,$$

$$A = \begin{pmatrix} R_1 & 0 & R_1 \\ 0 & R_2 & R_2 \\ 0 & 0 & S \end{pmatrix} \to \Sigma^{-1}A = M_3(R_1 *_S R_2)$$

are injective, and $\Sigma^{-1}A$ is a stably flat noncommutative localization, with

$$H(A, \Sigma) = \mathrm{Nil}(R_1, R_2, S)$$

the nilpotent category of Waldhausen [41]. The chain complex lifting property of $\Sigma^{-1}A$ gives a noncommutative localization proof of the existence of Mayer-Vietoris presentations for finite f.g. free $R_1 *_S R_2$-module chain complexes C

$$0 \longrightarrow R_1 *_S R_2 \otimes_S E \longrightarrow R_1 *_S R_2 \otimes_{R_1} D_1 \oplus R_1 *_S R_2 \otimes_{R_2} D_2 \longrightarrow C \longrightarrow 0$$

with D_i (resp. E) a finite f.g. free R_i- (resp. S-) module chain complex ([41], Ranicki [31]). The algebraic K-theory localization exact sequence of [21]

$$\cdots \to K_{n+1}(A, \Sigma) = K_n(S) \oplus K_n(S) \oplus \widetilde{\mathrm{Nil}}_n(R_1, R_2, S)$$

$$\xrightarrow{\begin{pmatrix} i_1 & 0 & 0 \\ 0 & i_2 & 0 \\ 1 & 1 & 0 \end{pmatrix}} K_n(A) = K_n(R_1) \oplus K_n(R_2) \oplus K_n(S)$$

$$\to K_n(\Sigma^{-1}A) = K_n(R_1 *_S R_2) \to \cdots$$

is just the stabilization by $1 : K_*(S) \to K_*(S)$ of the Mayer-Vietoris exact sequence of [41]

$$\cdots \longrightarrow K_n(S) \oplus \widetilde{\mathrm{Nil}}_n(R_1, R_2, S)$$

$$\xrightarrow{\begin{pmatrix} i_1 & 0 \\ i_2 & 0 \end{pmatrix}} K_n(R_1) \oplus K_n(R_2) \longrightarrow K_n(R_1 *_S R_2) \longrightarrow \cdots$$

2.5   The algebraic L-theory of a noncommutative localization

See Chapter 3 of Ranicki [27] for the algebraic L-theory of a commutative or Ore localization.

The algebraic L-theory of a ring A depends on an involution, that is a function $^- : A \to A; a \mapsto \bar{a}$ such that

$$\overline{a+b} = \bar{a}+\bar{b} , \ \overline{ab} = \bar{b}\bar{a} , \ \bar{\bar{a}} = a , \ \bar{1} = 1 \ (a,b \in A) .$$

Vogel [40] extended the quadratic L-theory localization exact sequence of [27]

$$\cdots \to L_n^I(A) \to L_n(\Sigma^{-1}A) \to L_n(A,\Sigma) \to L_{n-1}^I(A) \to \cdots$$

to the noncommutative case, with $A \to \Sigma^{-1}A$ injective, $L_n^I(A)$ the projective L-groups of A decorated by

$$\begin{aligned}
I &= \ker(K_0(A) \to K_0(\Sigma^{-1}A)) \\
&= \operatorname{im}(K_1(A,\Sigma) \to K_0(A)) \subseteq K_0(A) ,
\end{aligned}$$

$L_n(\Sigma^{-1}A)$ the free L-groups of $\Sigma^{-1}A$ and $L_n(A,\Sigma) = L_{n-1}(H(A,\Sigma))$. (See [21] for the symmetric L-theory localization exact sequence in the stably flat case). At first sight, it does not appear possible to apply this sequence to the triangular matrix rings of sections 2.2, 2.3, 2.4. How does one define an involution on a triangular matrix ring

$$A = \begin{pmatrix} A_1 & B \\ 0 & A_2 \end{pmatrix} \ ?$$

The trick is to observe that if A_1, A_2 are rings with involution, and (B,β) is a nonsingular symmetric form over A_1 such that B is an (A_1, A_2)-bimodule then A has a *chain duality* in the sense of Definition 1.1 of Ranicki [28], sending an A-module $M = \begin{pmatrix} M_1 \\ M_2 \end{pmatrix}$ to the 1-dimensional A-module chain complex

$$TM \ : \ TM_1 = \begin{pmatrix} M_1^* \\ 0 \end{pmatrix} \to TM_0 = \begin{pmatrix} B \otimes_{A_2} M_2^* \\ M_2^* \end{pmatrix} .$$

The quadratic L-groups of A with respect to this chain duality are just the relative L-groups in the exact sequence

$$\cdots \to L_n(A) \to L_n(A_2) \xrightarrow{(B,\beta)\otimes_{A_2} ^-} L_n(A_1) \to L_{n-1}(A) \to \cdots .$$

In particular, for generalized free products of rings with involution the triangular matrix rings A of section 2.3, 2.4 have such chain dualities, and in the injective case the torsion L-groups $L_*(A, \Sigma) = L_{*-1}(H(A, \Sigma))$ in the localization exact sequence

$$\cdots \to L_n^I(A) \to L_n(\Sigma^{-1}A) \to L_n(A, \Sigma) \to L_{n-1}^I(A) \to \cdots$$

are just the unitary nilpotent L-groups UNil_* of Cappell [3].

References

[1] H. Bass, A. Heller and R. Swan, *The Whitehead group of a polynomial extension*, Publ. Math. I.H.E.S. 22, 61–80 (1964)

[2] A. J. Berrick and M. E. Keating, *The K-theory of triangular matrix rings*, Contemp. Maths. **55**, 69–74, A.M.S. (1986)

[3] S. Cappell, *Unitary nilpotent groups and hermitian K-theory*, Bull. A.M.S. 80, 1117–1122 (1974)

[4] ———— and J. Shaneson, *The codimension two placement problem, and homology equivalent manifolds*, Ann. of Maths. 99, 277–348 (1974)

[5] ———— and ———— , *Link cobordism*, Comm. Math. Helv. 55, 20–49 (1980)

[6] P. M. Cohn, *Free rings and their relations*, Academic Press (1971)

[7] ———— , *Free ideal rings and localization in general rings*, CUP (2005)

[8] T. D. Cochran, *Localization and finiteness in link concordance*. Proceedings of the 1987 Georgia Topology Conference (Athens, GA, 1987). Topology Appl. 32, 121–133 (1989)

[9] O. Cornea and A. Ranicki, *Rigidity and glueing for the Morse and Novikov complexes*, http://arXiv.math.AT/0107221, J. Eur. Math. Soc. 5, 343–394 (2003)

[10] W. Dicks and E. Sontag, *Sylvester domains*, J. Pure Appl. Algebra 13, 243–275 (1978)

[11] J. Duval, *Forme de Blanchfield et cobordisme d'entrelacs bords*, Comm. Math. Helv. 61, 617–635 (1986)

[12] W. G. Dwyer, *Noncommutative localization in homotopy theory*, in this volume, pp. 23–39.

[13] M. Farber, *Morse-Novikov critical point theory, Cohn localization and Dirichlet units*, Commun. Contemp. Math 1, 467–495 (1999)

[14] _____ , *Topology of closed one-forms*, Mathematical Surveys and Monographs 108, A.M.S. (2004)

[15] _____ and A. Ranicki, *The Morse-Novikov theory of circle-valued functions and noncommutative localization*, http://arXiv.math.DG/9812122, Proc. 1998 Moscow Conference for S.P.Novikov's 60th Birthday, Proc. Steklov Inst. 225, 381–388 (1999)

[16] _____ and P. Vogel, *The Cohn localization of the free group ring*, Math. Proc. Camb. Phil. Soc. 111, 433–443 (1992)

[17] S. Garoufalidis and A. Kricker, *A rational noncommutative invariant of boundary links*, http://arXiv.math.GT/0105028 Geometry and Topology 8, 115–204 (2004)

[18] B. Hughes and A. Ranicki, *Ends of complexes*, Cambridge Tracts in Mathematics 123, CUP (1996)

[19] J-Y. Le Dimet, *Cobordisme d'enlacements de disques*, Mém. Soc. Math. France (N.S.) No. 32, (1988)

[20] J. Levine, W. Mio and K. E. Orr, *Links with vanishing homotopy invariant.* Comm. Pure Appl. Math. 46, 213–220 (1993)

[21] A. Neeman and A. Ranicki, *Noncommutative localization and chain complexes I. Algebraic K- and L-theory*, http://arXiv.math.RA/0109118, Geometry and Topology 8, 1385–1421 (2004)

[22] _____ , _____ and A. Schofield, *Representations of algebras as universal localizations*, http://arXiv.math.RA/0205034, Math. Proc. Camb. Phil. Soc. 136, 105–117 (2004)

[23] S. P. Novikov, *The hamiltonian formalism and a multi-valued analogue of Morse theory*, Russian Math. Surveys 37:5, 1–56 (1982)

[24] A. Pajitnov, *Incidence coefficients in the Novikov complex for Morse forms: rationality and exponential growth properties*,

http://arXiv.math.DG-GA/9604004, St. Petersburg Math. J. 9, 969–1006 (1998)

[25] F. Quinn, *Open book decompositions, and the bordism of automorphisms*, Topology 18, 55–73 (1979)

[26] ———, *Dual decompositions of 4-manifolds II. Linear link invariants*, http://arXiv.math.GT/0109148 (2001)

[27] A. Ranicki, *Exact sequences in the algebraic theory of surgery*, Mathematical Notes 26, Princeton (1981)

[28] ———, *Algebraic L-theory and topological manifolds*, Cambridge Tracts in Mathematics 102, CUP (1992)

[29] ———, *High dimensional knot theory*, Springer Mathematical Monograph, Springer (1998)

[30] ———, *The algebraic construction of the Novikov complex of a circle-valued Morse function*, http://arXiv.math.AT/9903090, Math. Ann. 322, 745–785 (2002)

[31] ———, *Algebraic and combinatorial codimension 1 transversality*, http://arXiv.math.AT/0308111, Geometry and Topology Monographs 7, Proc. Casson Fest, 145–180 (2004)

[32] ——— and D. Sheiham, *Blanchfield and Seifert algebra in high dimensional boundary link theory I. Algebraic K-theory*, arXiv:math.AT/0508405

[33] A. Schofield, *Representations of rings over skew fields*, LMS Lecture Notes 92, Cambridge (1985)

[34] D. Sheiham, *Noncommutative characteristic polynomials and Cohn localization*, J. London Math. Soc. 64, 13–28 (2001)

[35] ———, *Invariants of boundary link cobordism*, Edinburgh Ph. D. thesis (2001) http://arXiv.math.AT/0110249, A.M.S. Memoir 165 (2003)

[36] ———, *Whitehead groups of localizations and the endomorphism class group*, http://arXiv.math.KT/0209311, J. Algebra 270, 261–280 (2003)

[37] ———, *Invariants of boundary link cobordism II. The Blanchfield-Duval form*, in this volume, pp. 143–219, http://arXiv.math.AT/0404229.

[38] L. Siebenmann, *The obstruction to finding the boundary of an open manifold of dimension greater than five*, Princeton Ph.D. thesis (1965). http://www.maths.ed.ac.uk/~aar/surgery/sieben.pdf

[39] P. Vogel, *On the obstruction group in homology surgery*, Publ. Math. I.H.E.S. 55, 165–206 (1982)

[40] ———— , *Localisation non commutative de formes quadratiques*, Springer Lecture Notes 967, 376–389 (1982)

[41] F. Waldhausen, *Algebraic K-theory of generalized free products*, Ann. of Maths. 108, 135–256 (1978)

[42] C.T.C. Wall, *Finiteness conditions for CW complexes*, Ann. of Maths. 81, 55–69 (1965)

[43] ———— , *Surgery on compact manifolds*, Academic Press (1970), 2nd edition A.M.S. (1999)

School of Mathematics
University of Edinburgh
James Clerk Maxwell Building
King's Buildings
Mayfield Road
Edinburgh EH9 3JZ
SCOTLAND, UK

e-mail a.ranicki@ed.ac.uk

L^2-Betti numbers, Isomorphism Conjectures and Noncommutative Localization

Holger Reich

Abstract

In this paper we want to discuss how the question about the rationality of L^2-Betti numbers is related to the Isomorphism Conjecture in algebraic K-theory and why in this context noncommutative localization appears as an important tool.

L^2-*Betti numbers* are invariants of spaces which are defined analogously to the ordinary Betti-numbers but they take information about the fundamental group into account and are a priori real valued.

The *Isomorphism Conjecture in algebraic K-theory* predicts that $K_0(\mathbb{C}\Gamma)$, the Grothendieck group of finitely generated projective $\mathbb{C}\Gamma$-modules, should be computable from the K-theory of the complex group rings of finite subgroups of Γ.

Given a commutative ring one can always invert the set of all non-zerodivisors. Elements in the resulting ring have a nice description in terms of fractions. For noncommutative rings like group rings this may no longer be the case and other concepts for a *noncommutative localization* can be more suitable for specific problems.

The question whether L^2-Betti numbers are always rational numbers was asked by Atiyah in [1]. The question turns out to be a question about modules over the group ring of the fundamental group Γ. In [33] Linnell was able to answer the question affirmatively if Γ belongs to a certain class of

Keywords Novikov-Shubin Invariants, Noncommutative Power Series
Research supported by the SFB "Geometrische Strukturen in der Mathematik" in Münster, Germany

groups which contains free groups and is stable under extensions by elementary amenable groups (one also needs a bound on the orders of finite subgroups). In fact Linnell proves the stronger result that there exists a semisimple subring in $\mathcal{U}\Gamma$, the algebra of operators affiliated to the group von Neumann algebra, which contains the complex group ring.

The main purpose of this short survey is to give a conceptual framework for Linnell's result, to explain how the question about the rationality of L^2-Betti numbers relates to the Isomorphism conjecture, and why this may involve studying noncommutative localizations of group rings. (The impatient reader should right away take a look at Proposition 3.4, Theorem 6.3 and Addendum 6.4.)

Since probably not every reader is familiar with all three circles of ideas – L^2-Betti numbers – Isomorphism Conjectures – Noncommutative Localization – the paper contains introductions to all of these.

After a brief introduction to group von Neumann algebras and the notion of Γ-dimension we proceed to explain the algebra $\mathcal{U}\Gamma$ of operators affiliated to a group von Neumann algebra and introduce L^2-Betti numbers in a very algebraic fashion. (Once $\mathcal{U}\Gamma$ has been defined there is no more need for Hilbert-spaces.) Section 3 explains the Atiyah Conjecture and contains in particular Proposition 3.4 which is a kind of strategy for its proof. That Proposition says that if one can factorize the inclusion $\mathbb{C}\Gamma \subset \mathcal{U}\Gamma$ over a ring $\mathcal{S}\Gamma$ with good ring-theoretical properties in such way that a certain K-theoretic condition is satisfied, then the Atiyah conjecture follows. In Section 5 we present a number of candidates for the ring $\mathcal{S}\Gamma$. To do this we first review a number of concepts from the theory of noncommutative localization in Section 4. One of the candidates is the universal localization of $\mathbb{C}\Gamma$ with respect to all matrices that become invertible over $\mathcal{U}\Gamma$. Section 6 contains Linnell's result. We would like to emphasize that the intermediate rings Linnell constructs can also be viewed as universal localizations, see Addendum 6.4 (U). In Section 7 we discuss the Isomorphism Conjecture which seems to be closely related to the K-theoretical condition mentioned above. In the last Section we discuss to what extent the functor $- \otimes_{\mathbb{C}\Gamma} \mathcal{U}\Gamma$, which plays an important role when one studies L^2-Betti numbers, is exact. The only new result in this paper is the following.

Theorem. *Let $\mathcal{D}\Gamma$ denote the division closure of $\mathbb{C}\Gamma$ inside $\mathcal{U}\Gamma$. The flat dimension of $\mathcal{D}\Gamma$ over $\mathbb{C}\Gamma$ is smaller than 1 for groups in Linnell's class which have a bound on the orders of finite subgroups.*

For the division closure see Definition 4.8, for Linnell's class of groups see Definition 6.1. The result is proven as Theorem 8.6 below. As an immediate corollary one obtains.

Corollary. *If the infinite group* Γ *belongs to Linnell's class* $\mathcal{C}$ *and has a bound on the orders of finite subgroups then the* L^2-*Euler characteristic (and hence the ordinary one whenever defined) satisfies*

$$\chi^{(2)}(\Gamma) \leq 0.$$

We also would like to mention that the above theorem leads to interesting non-trivial examples of stably flat universal localizations which appear in [45].

A reader who is interested in more information about L^2-Betti numbers and the Atiyah Conjecture should consult the book [39]. Almost all topics discussed here are also treated there in detail. More information and further results about the Atiyah Conjecture can be found in [34], [24], [55] and [56].

Acknowledgement

I would like to thank Wolfgang Lück who as my thesis advisor introduced me to the Atiyah Conjecture and the Isomorphism Conjecture. I would also like to thank Thomas Schick. In discussions with him the idea that the Corollary above should be true evolved. Furthermore I would like to thank Andrew Ranicki and the ICMS in Edinburgh for organizing the lively and interesting Workshop on Noncommutative Localization.

1 The von Neumann Dimension

In this section we want to introduce group von Neumann algebras and explain a notion of dimension for finitely generated projective modules over such algebras.

For a (discrete) group Γ we denote by $\mathbb{C}\Gamma$ the complex group ring and by $l^2\Gamma$ the complex Hilbert space with orthonormal basis Γ. Each group element operates from the left on $l^2\Gamma$. Linearly extending this action we obtain an inclusion

$$\mathbb{C}\Gamma \to \mathcal{B}(l^2\Gamma)$$

into the algebra $\mathcal{B}(l^2\Gamma)$ of bounded linear operators on the Hilbert space $l^2\Gamma$. The group von Neumann algebra $\mathcal{N}\Gamma$ is defined as the closure of $\mathbb{C}\Gamma$ inside $\mathcal{B}(l^2\Gamma)$ with respect to the weak (or strong, it doesn't matter) operator topology. This algebra is closed under taking the adjoint, i.e. $a^* \in \mathcal{N}\Gamma$ for every operator $a \in \mathcal{N}\Gamma$.

Digression 1.1. A von Neumann algebra is by definition a ∗-closed subalgebra of the algebra of bounded linear operators on some Hilbert-space which is closed with respect to the strong (or weak, it doesn't matter) operator topology. Similarly a C^*-algebra can be defined as a ∗-closed subalgebra of the algebra of bounded operators on some Hilbert space which is closed with respect to the topology given by the operator-norm. Every von Neumann algebra is in particular a C^*-algebra. In the situation described above the operator-norm closure of $\mathbb{C}\Gamma$ inside $\mathcal{B}(l^2\Gamma)$ defines the so called reduced C^*-algebra $C_r^*\Gamma$ and we have a natural inclusion of $C_r^*\Gamma$ in $\mathcal{N}\Gamma$.

The bicommutant theorem of von Neumann (see for example Theorem 5.3.1 in [28]) is a first hint that the definition of $\mathcal{N}\Gamma$ is very natural also from a purely algebraic point of view (at least if we agree to consider $\mathcal{B}(l^2\Gamma)$ as something natural). It says that the von Neumann algebra is the double commutant of $\mathbb{C}\Gamma$, i.e.

$$\mathcal{N}\Gamma = \mathbb{C}\Gamma'',$$

where for a subset $A \subset \mathcal{B}(l^2\Gamma)$ we write $A' = \{b \in \mathcal{B}(l^2\Gamma)|ba = ab \text{ for all } a \in A\}$ for the commutant of A in $\mathcal{B}(l^2\Gamma)$.

The group von Neumann algebra comes equipped with a natural trace. This trace is given as follows:

$$\text{tr}_\Gamma : \mathcal{N}\Gamma \quad \to \quad \mathbb{C}$$
$$a \quad \mapsto \quad \langle a(e), e \rangle.$$

Here $\langle -, - \rangle$ denotes the inner product in $l^2\Gamma$ and e is the unit element of the group considered as a vector in $l^2\Gamma$. Applied to an element $a = \sum a_g g$ in the group ring $\mathbb{C}\Gamma$ the trace yields the coefficient of the identity element a_e. Of course we have the trace property $\text{tr}_\Gamma(ab) = \text{tr}_\Gamma(ba)$. Once we have such a trace there is a standard procedure to assign a complex number to each finitely generated projective $\mathcal{N}\Gamma$-module: if $p = p^2 = (p_{ij}) \in M_n(\mathcal{N}\Gamma)$ is an idempotent matrix over $\mathcal{N}\Gamma$ which represents P, i.e. such that $P \cong \text{im}(p : \mathcal{N}\Gamma^n \to \mathcal{N}\Gamma^n)$ then we set

$$\dim_\Gamma P = \Sigma_{i=1}^n \text{tr}_\Gamma(p_{ii}) \tag{1}$$

and call it the Γ-dimension of P. We have the following standard facts.

Proposition 1.2. *The Γ-dimension has the following properties.*

    (i) $\dim_\Gamma P$ is a nonnegative real number.

    (ii) $\dim_\Gamma P$ depends only on the isomorphism class of P.

(iii) **Normalization.** *We have* $\dim_\Gamma \mathcal{N}\Gamma = 1$.

(iv) **Additivity.** *If* $0 \to L \to M \to N \to 0$ *is a short exact sequence of finitely generated projective modules then*

$$\dim_\Gamma M = \dim_\Gamma L + \dim_\Gamma N.$$

(v) **Faithfulness.** $\dim_\Gamma P = 0$ *if and only if* $P = 0$.

Proof. (i) follows since one can always arrange that the idempotent $p = p^2$ in (1) is a projection, i.e. $p = p^2 = p^*$ (see for example Proposition 4.6.2 on p.23 in [5]). (v) follows from the fact that the trace is faithful, i.e. $\operatorname{tr}(a^*a) = 0$ implies $a = 0$. (ii)-(iv) are straightforward. □

Let $K_0(\mathcal{N}\Gamma)$ denote the Grothendieck-group of finitely generated projective $\mathcal{N}\Gamma$-modules then because of (i)-(iv) above we obtain a homomorphism

$$K_0(\mathcal{N}\Gamma) \xrightarrow{\;\dim_\Gamma\;} \mathbb{R}.$$

We recall some terminology, compare page 5 in [57].

Definition 1.3. A projective rank function ρ on a ring R is a homomorphism $\rho : K_0(R) \to \mathbb{R}$ satisfying $\rho([R^1]) = 1$ and $\rho([P]) \geq 0$ for every finitely generated projective R-module P. It is called faithful if moreover $\rho([P]) = 0$ implies $P = 0$.

In this terminology we can summarize the content of the proposition above by saying that $\dim_\Gamma : K_0(\mathcal{N}\Gamma) \to \mathbb{R}$ is a faithful projective rank function. Other natural examples of faithful projective rank functions occur as follows: Suppose the ring R is embedded in a simple artinian ring $M_n(D)$, where D is a skew field. Then $P \mapsto \frac{1}{n} \dim_D P \otimes_R M_n(D)$ defines a faithful projective rank function on R. We would like to emphasize the following additional properties of the Γ-dimension for $\mathcal{N}\Gamma$-modules which are not true for arbitrary projective rank functions. They give further justification for the use of the word "dimension" in this context.

Proposition 1.4. *The Γ-dimension satisfies:*

(v) **Monotony.** *The $\mathcal{N}\Gamma$-dimension is monotone, i.e. $P \subset Q$ implies that $\dim_\Gamma P \leq \dim_\Gamma Q$.*

(vi) **Cofinality.** *If $P = \bigcup_{i \in I} P_i$ is a directed union of submodules then*

$$\dim_\Gamma P = \sup_{i \in I} \dim_\Gamma P_i.$$

Of course cofinality implies monotony. To convince the reader that these properties are not automatic for projective rank functions we would like to treat an example.

Example 1.5. Let Γ be a free group on two generators x and y. By work of Cohn [11] we know that $\mathbb{C}\Gamma$ is a free ideal ring. In particular every finitely generated projective module is free and taking its rank yields an isomorphism

$$K_0(\mathbb{C}\Gamma) \xrightarrow{\cong} \mathbb{Z}.$$

This is a faithful projective rank function with values in $\mathbb{Z}$. However there is an exact sequence

$$0 \longrightarrow \mathbb{C}\Gamma^2 \xrightarrow{(x-1,y-1)} \mathbb{C}\Gamma \longrightarrow \mathbb{C} \longrightarrow 0$$

which shows that the rank function is not monotone. (Geometrically the above resolution of $\mathbb{C}$ is obtained as the cellular chain complex with complex coefficients of the universal cover $E\Gamma$ of the model for the classifying space $B\Gamma$ given by the wedge of two circles.)

In fact one can always compose $\dim_\Gamma$ with the natural map $K_0(\mathbb{C}\Gamma) \to K_0(\mathcal{N}\Gamma)$. In this way we obtain naturally a faithful projective rank function on $\mathbb{C}\Gamma$ for every group Γ. One rediscovers the example above in the case where Γ is the free group on two generators.

2   The Algebra of Operators affiliated to $\mathcal{N}\Gamma$.

The category of finitely generated projective $\mathcal{N}\Gamma$-modules has one drawback: it is not abelian. In particular if we start out with a complex of finitely generated projective $\mathcal{N}\Gamma$-modules then the homology modules are not necessarily finitely generated projective and hence the $\mathcal{N}\Gamma$-dimension as explained above is a priori not available. But this is exactly what we would like to do in order to define L^2-Betti numbers, i.e. we want to consider

$$C_*^{cell}(\widetilde{X}) \otimes_{\mathbb{Z}\Gamma} \mathcal{N}\Gamma$$

the cellular chain-complex of the universal covering of a CW-complex X of finite type tensored up to $\mathcal{N}\Gamma$ and assign a dimension to the homology modules.

There are several ways to get around this problem. The traditional way to deal with it is to work with certain Hilbert spaces with an isometric Γ-operation instead of modules, e.g. with $l^2\Gamma^n$ instead of $\mathcal{N}\Gamma^n$. These Hilbert spaces have a Γ-dimension and one (re-)defines the homology as the kernel of the differentials modulo the *closure* of their images. This is then again a Hilbert space with an isometric Γ-action and has a well defined Γ-dimension. A different approach is taken in [36]: *finitely presented* $\mathcal{N}\Gamma$-modules do form an abelian category (because $\mathcal{N}\Gamma$ is a semihereditary ring) and the $\mathcal{N}\Gamma$-dimension can be extended to these modules in such a way that the properties (i)-(vi) still hold. (In fact in [37] the Γ-dimension is even extended to arbitrary $\mathcal{N}\Gamma$-modules.)

A third possible approach is to introduce the algebra $\mathcal{U}\Gamma$ of operators affiliated to $\mathcal{N}\Gamma$. This algebra has better ring-theoretic properties and indeed finitely generated projective $\mathcal{U}\Gamma$-modules do form an abelian category. Moreover the notion of Γ-dimension extends to that algebra. We want to explain this approach in some detail in this section.

Recall that an unbounded operator $a : \mathrm{dom}(a) \to H$ on a Hilbert space H is a linear map which is defined on a linear subspace $\mathrm{dom}(a) \subset H$ called the domain of a. It is called densely defined if $\mathrm{dom}(a)$ is a dense subspace of H and it is called closed if its graph considered as a subspace of $H \oplus H$ is closed. Each bounded operator is closed and densely defined. For unbounded operators a and b the symbol $a \subset b$ means that restricted to the possibly smaller domain of a the two operators coincide. The following definition goes back to [44].

Definition 2.1 (Affiliated Operators). A closed and densely defined (possibly unbounded) operator $a : \mathrm{dom}(a) \to l^2\Gamma$ is affiliated to $\mathcal{N}\Gamma$ if $ba \subset ab$ for all $b \in \mathcal{N}\Gamma'$. The set

$$\mathcal{U}\Gamma = \{a : \mathrm{dom}(a) \to l^2\Gamma \mid a \text{ is } \begin{matrix} \text{closed,} \\ \text{densely defined} \\ \text{and affiliated to } \mathcal{N}\Gamma \end{matrix} \}$$

is called the algebra of operators affiliated to $\mathcal{N}\Gamma$.

Remark 2.2. Each group element $\gamma \in \Gamma$ acts by right multiplication on $l^2\Gamma$. This defines an element $r_\gamma \in \mathcal{N}\Gamma'$ (we had Γ acting from the left when we defined $\mathcal{N}\Gamma$). In order to prove that a closed densely defined operator a

is affiliated it suffices to check that its domain $\mathrm{dom}(a)$ is Γ-invariant and that for all vectors $v \in \mathrm{dom}(a)$ we have $r_\gamma a(v) = a r_\gamma(v)$ for all $\gamma \in \Gamma$. In this sense the affiliated operators are precisely the Γ-equivariant unbounded operators.

Observe that the naive composition of two unbounded operators c and d yields an operator dc which is only defined on $c^{-1}(\mathrm{dom}(d))$. Similarly addition is only defined on the intersection of the domains. It is hence not obvious that $\mathcal{U}\Gamma$ is an algebra.

Proposition 2.3. *The set $\mathcal{U}\Gamma$ becomes a $\mathbb{C}$-algebra if we define addition and a product as the closure of the naive addition respectively composition of operators.*

Proof. This is proven in Chapter XVI in [44]. A proof is reproduced in Appendix I in [53] and also in Chapter 8 of [39]. □

The subalgebra of all bounded operators in $\mathcal{U}\Gamma$ is $\mathcal{N}\Gamma$. In contrast to $\mathcal{N}\Gamma$ there seems to be no useful topology on $\mathcal{U}\Gamma$. So we left the realm of C^*-algebras and C^*-algebraic methods. The reason $\mathcal{U}\Gamma$ is nevertheless very useful is that we have gained good ringtheoretic properties. Let us recall the definition of von Neumann regularity.

Definition 2.4. A ring R is called von Neumann regular if one of the following equivalent conditions is satisfied.

(i) Every R-module M is flat, i.e. for every module M the functor $- \otimes_R M$ is exact.

(ii) Every finitely presented R-module is already finitely generated projective.

(iii) The category of finitely generated projective R-modules is abelian.

(iv) For all $x \in R$ there exists a $y \in R$ such that $xyx = x$.

Proof. For (i) $\Leftrightarrow$ (iv) see for example Theorem 4.2.9 in [64]. (i) $\Rightarrow$ (ii) follows since every finitely presented flat R-module is projective, see Theorem 3.2.7 in [64]. Since the tensor product is compatible with colimits, directed colimits are exact and every module is a directed colimit of finitely presented modules we obtain (ii) $\Rightarrow$ (i). For (ii) $\Rightarrow$ (iii) one needs to check that cokernels and kernels between finitely generated projectives are again finitely generated projective. But a cokernel is essentially a finitely presented module. The argument for the kernel and (iii) $\Rightarrow$ (ii) are elementary. □

Note that in particular fields, skew fields, simple artinian rings and semi-simple rings are von Neumann regular (every module is projective over such rings). The first condition says that von Neumann regular rings form a very natural class of rings from a homological algebra point of view: they constitute precisely the rings of weak homological dimension 0. The last condition, which seems less conceptional to modern eyes, was von Neumann's original definition [61] and has the advantage that one can explicitly verify it in the case we are interested in. More information about von Neumann regular rings can be found in [21].

Proposition 2.5. *The algebra $\mathcal{U}\Gamma$ is a von Neumann regular ring.*

Proof. Using the polar decomposition and functional calculus one can explicitly construct a y as it is required in the characterization 2.4 (iii) of von Neumann regularity given above. Compare Proposition 2.1 (v) in [54]. □

In order to define L^2-Betti numbers it remains to establish a notion of dimension for finitely generated projective $\mathcal{U}\Gamma$-modules.

Proposition 2.6. *We have the following facts about the inclusion $\mathcal{N}\Gamma \subset \mathcal{U}\Gamma$.*

(i) *The natural map $K_0(\mathcal{N}\Gamma) \to K_0(\mathcal{U}\Gamma)$ is an isomorphism. In particular there is a Γ-dimension for finitely generated projective $\mathcal{U}\Gamma$-modules which we simply define via the following diagram:*

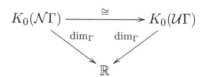

(ii) *The ring $\mathcal{U}\Gamma$ is the Ore-localization (compare Proposition 4.3) of $\mathcal{N}\Gamma$ with respect to the multiplicative subset of all non-zerodivisors. In particular $- \otimes_{\mathcal{N}\Gamma} \mathcal{U}\Gamma$ is an exact functor.*

Proof. See [54] Proposition 6.1 (i) and Proposition 2.1 (iii). □

If we now start with a finitely presented (as opposed to finitely generated projective) $\mathbb{C}\Gamma$-module M then because of 2.4 (ii) we know that $M \otimes_{\mathbb{C}\Gamma} \mathcal{U}\Gamma$ is a finitely generated projective $\mathcal{U}\Gamma$-module and it makes sense to consider its Γ-dimension.

Remark 2.7. The assignment $M \mapsto \dim_\Gamma(M \otimes_{\mathbb{C}\Gamma} \mathcal{U}\Gamma)$ is a Sylvester module rank function for finitely presented $\mathbb{C}\Gamma$-modules in the sense of Chapter 7 in [57].

We are now prepared to give a definition of L^2-Betti numbers using the Γ-dimension for $\mathcal{U}\Gamma$-modules. Let X be a CW-complex of finite type, i.e. there are only finitely many cells in each dimension. Let $\widetilde{X}$ denote the universal covering. It carries a natural CW-structure and a cellular free $\Gamma = \pi_1(X)$-action. There is one Γ-orbit of cells in $\widetilde{X}$ for each cell in X and in particular the cellular chain complex $C_*^{cell}(\widetilde{X})$ is a complex of finitely generated free $\mathbb{Z}\Gamma$-modules.

Definition 2.8. For a CW-complex X of finite type we define its L^2-Betti numbers as
$$b_p^{(2)}(X) = \dim_{\mathcal{U}\Gamma} H_p(C_*^{cell}(\widetilde{X}) \otimes_{\mathbb{Z}\Gamma} \mathcal{U}\Gamma).$$

Note that by 2.4 (iii) the homology modules are finitely generated projective $\mathcal{U}\Gamma$-modules and hence have a well defined $\mathcal{U}\Gamma$-dimension.

Remark 2.9. As already mentioned it is possible to extend the notion of Γ-dimension to arbitrary $\mathcal{N}\Gamma$-modules in such a way that one still has "additivity" and "cofinality" [37]. Of course one has to allow the value ∞, and in cases where this value occurs one has to interpret "additivity" and "cofinality" suitably. In [54] it is shown that analogously there is a Γ-dimension for arbitrary $\mathcal{U}\Gamma$-modules which is compatible with the one for $\mathcal{N}\Gamma$-modules in the sense that for an $\mathcal{N}\Gamma$-module M we have

$$\dim_{\mathcal{U}\Gamma} M \otimes_{\mathcal{N}\Gamma} \mathcal{U}\Gamma = \dim_{\mathcal{N}\Gamma} M. \tag{2}$$

Both notions of extended dimension can be used to define L^2-Betti numbers for arbitrary spaces by working with the singular instead of the cellular chain complex. From 2.6 (ii) we conclude that for a complex C_* of $\mathcal{N}\Gamma$-modules we have

$$H_*(C_* \otimes_{\mathcal{N}\Gamma} \mathcal{U}\Gamma) = H_*(C_*) \otimes_{\mathcal{N}\Gamma} \mathcal{U}\Gamma.$$

If we combine this with (2) we see that the two possible definitions of L^2-Betti numbers coincide. In the following we will not deal with L^2-Betti numbers in this generality. We restrict our attention to CW-complexes of finite type and hence to finitely generated projective $\mathcal{U}\Gamma$-modules.

In order to illustrate the notions defined so far we would like to go through two easy examples.

Example 2.10. Suppose Γ is a finite group of order $\#\Gamma$. In this case all the functional analysis is irrelevant. We have $\mathbb{C}\Gamma = \mathcal{N}\Gamma = \mathcal{U}\Gamma$ and $l^2\Gamma = \mathbb{C}\Gamma$. A finitely generated projective module P is just a finite dimensional complex Γ-representation. One can check that

$$\dim_\Gamma P = \frac{1}{\#\Gamma} \dim_\mathbb{C} P.$$

Example 2.11. Suppose $\Gamma = C$ is the infinite cyclic group written multiplicatively with generator $z \in C$. In this case (using Fourier transformation) the Hilbert-space $l^2\Gamma$ can be identified with $L^2(S^1)$, the square integrable functions on the unit circle equipped with the standard normalized measure $\mu = \frac{1}{2\pi}dz$. Under this correspondence the group element z corresponds to the function $z \mapsto z$, where we think of S^1 as embedded in the complex plane. The algebras $\mathbb{C}\Gamma$, $\mathcal{N}\Gamma$ and $\mathcal{U}\Gamma$ can be identified as follows:

$$
\begin{array}{rcll}
\mathbb{C}\Gamma & \leftrightarrow & \mathbb{C}[z^{\pm 1}] & \text{Laurent-polynomials considered} \\
& & & \text{as functions on } S^1 \\[2mm]
\mathcal{N}\Gamma & \leftrightarrow & L^\infty(S^1) & \text{essentially bounded} \\
& & & \text{functions on } S^1 \\[2mm]
\mathcal{U}\Gamma & \leftrightarrow & L(S^1) & \text{measurable functions} \\
& & & \text{on } S^1
\end{array}
$$

The action on $L^2(S^1)$ in each case is simply given by multiplication of functions. The trace on $\mathcal{N}\Gamma$ becomes the integral $f \mapsto \int_{S^1} f d\mu$. For a measurable subset $A \subset S^1$ let χ_A denote its characteristic function. Then $p = \chi_A$ is a projection and $P_A = pL^\infty(S^1)$ is a typical finitely generated projective $L^\infty(S^1)$-module. We have

$$\dim_\Gamma P_A = \mathrm{tr}_\Gamma(p) = \int_{S^1} \chi_A d\mu = \mu(A).$$

In particular we see that every nonnegative real number can occur as the Γ-dimension of a finitely generated projective $\mathcal{N}\Gamma$- or $\mathcal{U}\Gamma$-module. The module $L^\infty(S^1)/(z-1)L^\infty(S^1)$ is an example of a module which becomes trivial (and hence projective) over $L(S^1)$, because $(z-1)$ becomes invertible. In fact one can show that there is an isomorphism

$$K_0(L^\infty(S^1)) \cong K_0(L(S^1)) \xrightarrow{\;\cong\;} L^\infty(S^1; \mathbb{Z}),$$

where $L^\infty(S^1; \mathbb{Z})$ denotes the space of integer valued measurable bounded functions on S^1, compare Proposition 6.1 (iv) in [54]. Every such function can be written in a unique way as a finite sum $f = \sum_{n=-\infty}^{\infty} n \cdot \chi_{A_n}$ with $A_n = f^{-1}(\{n\}) \subset S^1$ and corresponds to $\sum_{n=-\infty}^{\infty} n[P_{A_n}]$ under the above isomorphism.

Once we have the notion of L^2-Betti numbers it is natural to define

$$\chi^{(2)}(X) = \sum (-1)^i b_i^{(2)}(X).$$

A standard argument shows that for a finite CW-complex this L^2-Euler characteristic coincides with the ordinary Euler-characteristic. But in fact since L^2-Betti numbers tend to vanish more often than the ordinary Betti-numbers the L^2-Euler characteristic is often defined in cases where the ordinary one is not. We also define L^2-Betti numbers and the L^2-Euler characteristic of a group as

$$b_p^{(2)}(\Gamma) = b_p^{(2)}(B\Gamma) \text{ and } \chi^{(2)}(\Gamma) = \chi^{(2)}(B\Gamma).$$

As an example of an application we would like to mention the following result which is due to Cheeger and Gromov [8].

Theorem 2.12. *Let Γ be a group which contains an infinite amenable normal subgroup, then*

$$b_p^{(2)}(\Gamma) = 0 \text{ for all } p, \text{ and hence } \chi^{(2)}(\Gamma) = 0.$$

3   The Atiyah Conjecture

The question arises which real numbers do actually occur as values of L^2-Betti numbers. This question was asked by Atiyah in [1] where he first introduced the notion of L^2-Betti numbers. (The definition of L^2-Betti numbers at that time only applied to manifolds and was given in terms of the Laplace operator on the universal covering.) It turns out that the question about the values can be phrased as a question about the passage from finitely presented $\mathbb{Z}\Gamma$- or $\mathbb{Q}\Gamma$-modules to $\mathcal{U}\Gamma$-modules.

Proposition 3.1. *Let Λ be an additive subgroup of $\mathbb{R}$ which contains $\mathbb{Z}$. Let Γ be a finitely presented group. The following two statements are equivalent.*

(i) For all CW-complexes X of finite type with fundamental group Γ and all $p \geq 0$ we have

$$b_p^{(2)}(X) \in \Lambda.$$

(ii) For all finitely presented $\mathbb{Z}\Gamma$-modules M we have

$$\dim_{\mathcal{U}\Gamma}(M \otimes_{\mathbb{Z}\Gamma} \mathcal{U}\Gamma) \in \Lambda.$$

Proof. Using the additivity of the dimension and the fact that the finitely generated free modules of the complex $C_*^{cell}(\widetilde{X}) \otimes_{\mathbb{Z}\Gamma} \mathcal{U}\Gamma$ have integer dimensions (ii) $\Rightarrow$ (i) is straightforward. For the reverse direction one needs to construct a CW-complex X with fundamental group Γ such that the presentation matrix of M appears as the, say 5-th differential in $C_*^{cell}(\widetilde{X})$ whereas the 4-th differential is zero. This is possible by standard techniques. For details see Lemma 10.5 in [39]. $\square$

More generally one can induce up finitely presented modules over $R\Gamma$ for every coefficient ring R with $\mathbb{Z} \subset R \subset \mathbb{C}$ and ask about the values of the corresponding Γ-dimensions. Let $S \subset R$ be a multiplicatively closed subset. Since each finitely presented $(S^{-1}R)\Gamma$-module is induced from a finitely presented $R\Gamma$-module (clear denominators in a presentation matrix) we can without loss of generality assume that R is a field. In the following we will work for simplicity with the maximal choice $R = \mathbb{C}$.

Let us describe a candidate for Λ. We denote by $\frac{1}{\#\mathcal{F}in\Gamma}\mathbb{Z}$ the additive subgroup of $\mathbb{R}$ which is generated by the set of numbers

$$\{\frac{1}{|H|} \mid H \text{ a finite subgroup of } \Gamma\} \,.$$

If there is a bound on the orders of finite subgroups then $\frac{1}{\#\mathcal{F}in\Gamma}\mathbb{Z} = \frac{1}{l}\mathbb{Z}$ where l is the least common multiple of the orders of finite subgroups. If Γ is torsionfree then $\frac{1}{\#\mathcal{F}in\Gamma}\mathbb{Z} = \mathbb{Z}$.

The following Conjecture turned out to be too optimistic in general (compare Remark 3.8 below). But it still has a chance of being true if one additionally assumes a bound on the orders of finite subgroups.

Conjecture 3.2 (Strong Atiyah Conjecture). *Let M be a finitely presented $\mathbb{C}\Gamma$-module then*

$$\dim_{\mathcal{U}\Gamma} M \otimes_{\mathbb{C}\Gamma} \mathcal{U}\Gamma \in \frac{1}{\#\mathcal{F}in\Gamma}\mathbb{Z}.$$

We will see below in 5.3 that this Conjecture implies the Zero-Divisor Conjecture.

Remark 3.3. As explained above the conjecture makes sense with any field $\mathbb{F}$ such that $\mathbb{Q} \subset \mathbb{F} \subset \mathbb{C}$ as coefficients for the group ring. With $\mathbb{F} = \mathbb{Q}$ the conjecture is equivalent to the corresponding conjecture about the values of L^2-Betti numbers. The conjecture with $\mathbb{F} = \mathbb{C}$ clearly implies the conjecture formulated with smaller fields.

To get a first idea let us discuss the Conjecture in the easy case where Γ is the infinite cyclic group. We have already seen in Example 2.11 that in this case the inclusion $\mathbb{C}\Gamma \subset \mathcal{U}\Gamma$ can be identified with $\mathbb{C}[z^{\pm 1}] \subset L(S^1)$, the Laurent polynomials considered as functions on S^1 inside the algebra of all measurable functions on S^1. Clearly $\mathbb{C}\Gamma$ corresponds to $\mathbb{C}[z^{\pm 1}]$. The crucial observation now is that in this case we find a field in between $\mathbb{C}\Gamma$ and $\mathcal{U}\Gamma$. Let $\mathbb{C}(z)$ denote the field of fractions of the polynomial ring $\mathbb{C}[z]$ then we have

$$\mathbb{C}[z^{\pm 1}] \subset \mathbb{C}(z) \subset L(S^1).$$

Now let M be a finitely presented $\mathbb{C}[z^{\pm 1}]$-module then $M \otimes_{\mathbb{C}[z^{\pm 1}]} \mathbb{C}(z)$ is a finitely generated free $\mathbb{C}(z)$-module because $\mathbb{C}(z)$ is a field and hence $M \otimes_{\mathbb{C}[z^{\pm 1}]} L(S^1)$ is a finitely generated free $L(S^1)$-module. In particular its Γ-dimension is an integer as predicted by Conjecture 3.2.

Note that $\mathbb{C}(z)$ is not contained in the group von Neumann algebra $L^\infty(S^1)$ because a rational function like for example $z \mapsto \frac{1}{z-1}$ which has a pole on S^1 can not be essentially bounded. It hence was crucial for this proof that we had the algebra of affiliated operators $\mathcal{U}\Gamma$, here $L(S^1)$, available.

The following generalizes these simple ideas.

Proposition 3.4. *Suppose the inclusion map $\mathbb{C}\Gamma \to \mathcal{U}\Gamma$ factorizes over a ring $\mathcal{S}\Gamma$ such that the following two conditions are fulfilled.*

(K) The composite map

$$\mathrm{colim}_{H \in \mathcal{F}in\Gamma} K_0(\mathbb{C}H) \longrightarrow K_0(\mathbb{C}\Gamma) \longrightarrow K_0(\mathcal{S}\Gamma)$$

     is surjective.

(R) The ring $\mathcal{S}\Gamma$ is von Neumann regular.

Then Conjecture 3.2 holds for the group Γ.

In the source of the map in (K) the colimit is taken over the finite subgroups of Γ. The structure maps in the colimit are induced by inclusions $K \subset H$ and conjugation maps $c_g : H \to H^g$, $h \mapsto ghg^{-1}$.

We will see below (compare Theorem 6.3) that there is a reasonably large class of groups for which a factorization of the inclusion $\mathbb{C}\Gamma \to \mathcal{U}\Gamma$ as required above is known to exist. In order to prove Proposition 3.4 we need one more fact about Γ-dimensions.

Proposition 3.5. *The Γ-dimension is compatible with induction, i.e. if G is a subgroup of Γ then there is a natural inclusion $\mathcal{U}G \subset \mathcal{U}\Gamma$ and for a finitely generated projective $\mathcal{U}G$-module P we have*

$$\dim_{\mathcal{U}\Gamma} P \otimes_{\mathcal{U}G} \mathcal{U}\Gamma = \dim_{\mathcal{U}G} P.$$

Proof. There exists a natural inclusion $i : \mathcal{U}G \to \mathcal{U}\Gamma$ which extends the inclusion $i : \mathcal{N}G \to \mathcal{N}\Gamma$ because $\mathcal{U}G$ is the Ore localization of $\mathcal{N}G$. The latter inclusion is compatible with the trace, i.e. $\operatorname{tr}_\Gamma(i(a)) = \operatorname{tr}_G(a)$ for $a \in \mathcal{N}\Gamma$, see Lemma 1.24 in [39]. The claim follows from these facts. □

Proof of Proposition 3.4. Let M be a finitely presented $\mathbb{C}\Gamma$-module. Then also $M \otimes_{\mathbb{C}\Gamma} \mathcal{S}\Gamma$ is finitely presented and hence finitely generated projective by 2.4 (ii) because we assume that $\mathcal{S}\Gamma$ is von Neumann regular. In particular $M \otimes_{\mathbb{C}\Gamma} \mathcal{S}\Gamma$ defines a class in $K_0(\mathcal{S}\Gamma)$. Our second assumption implies that this class comes from $\operatorname{colim}_{H \in \mathcal{F}in\Gamma} K_0(\mathbb{C}H)$ via the natural map. It remains to check that the composition

$$\operatorname{colim}_{H \in \mathcal{F}in\Gamma} K_0(\mathbb{C}H) \longrightarrow K_0(\mathbb{C}\Gamma) \longrightarrow K_0(\mathcal{S}\Gamma) \longrightarrow K_0(\mathcal{U}\Gamma) \xrightarrow{\dim_\Gamma} \mathbb{R}$$

lands inside the subgroup $\frac{1}{\#\mathcal{F}in\Gamma}\mathbb{Z}$ of $\mathbb{R}$. But from Example 2.10 together with Proposition 3.5 we conclude that for a finite subgroup H and a finitely generated projective $\mathbb{C}H$-module P we have

$$\dim_{\mathcal{U}\Gamma} P \otimes_{\mathbb{C}H} \mathcal{U}\Gamma = \dim_{\mathbb{C}H} P = \frac{1}{\#H} \dim_{\mathbb{C}} P.$$

□

Remark 3.6. From 2.4 (iv) it follows that the homomorphic image of a von Neumann regular ring is again von Neumann regular. In particular the image of $\mathcal{S}\Gamma$ in $\mathcal{U}\Gamma$ would be von Neumann regular if $\mathcal{S}\Gamma$ is. (But it is not clear that the induced map for K_0 is surjective, compare Question 7.5.)

Note 3.7. Suppose $\mathcal{S}\Gamma$ is a subring of $\mathcal{U}\Gamma$ which contains $\mathbb{C}\Gamma$. If we assume the properties (K) and (R) and additionally we assume that Γ has a bound on the orders of finite subgroups, then $\mathcal{S}\Gamma$ is semisimple.

Proof. The assumptions imply that the projective rank function

$$P \mapsto \dim_\Gamma P \otimes_{\mathcal{S}\Gamma} \mathcal{U}\Gamma$$

for finitely generated $\mathcal{S}\Gamma$ modules takes values in $\frac{1}{l}\mathbb{Z}$, where l is the least common multiple of the orders of finite subgroups. Since each finitely generated projective $\mathcal{S}\Gamma$-module is a subset of a $\mathcal{U}\Gamma$-module it is easy to see that the projective rank function is faithful. In order to prove that a von Neumann regular ring is semisimple it suffices to show that there are no infinite chains of ideals, see page 21 in [21]. Since each ideal is a direct summand of $\mathcal{S}\Gamma$ and each subideal of a given ideal is a direct summand this can be checked using the faithful projective rank function with values in $\frac{1}{l}\mathbb{Z}$. □

Remark 3.8. The lamplighter group is the semidirect product of $\mathbb{Z}$ and $\bigoplus_{-\infty}^{\infty} \mathbb{Z}/2$ where $\mathbb{Z}$ acts via shift on $\bigoplus_{-\infty}^{\infty} \mathbb{Z}/2$. The orders of finite subgroups that occur are precisely all powers of 2. Conjecture 3.2 hence predicts $\mathbb{Z}[\frac{1}{2}]$ as the range for the dimensions. However in [24] a finitely presented $\mathbb{Q}\Gamma$-module is constructed whose Γ-dimension is $\frac{1}{3}$.

4   Noncommutative Localization

Our next aim is to present several candidates for the ring $\mathcal{S}\Gamma$ which appears in Proposition 3.4. In order to do this we first want to fix some language and review a couple of concepts from the theory of localization for noncommutative rings. For more on this subject the reader should consult Chapter II in [58], Chapter 7 in [12] and Chapter 4 in [57].

Ore Localization

Classically the starting point for the localization of rings is the wish that certain *elements* in the ring should become invertible. In mathematical terms we have the following universal property.

Definition 4.1. Let $T \subset R$ be a subset which does not contain any zero-divisors. A ring homomorphism $f : R \to S$ is called T-inverting if $f(t)$ is invertible for all $t \in T$. A T-inverting ring homomorphism $i : R \to R_T$ is called universally T-inverting if it has the following universal property: given any T-inverting ring homomorphism $f : R \to S$ there exists a unique ring homomorphism $\Phi : R_T \to S$ such that

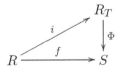

commutes.

A generator and relation construction shows that there always exists a universal T-inverting ring and as usual it is unique up to canonical isomorphism. Given a ring homomorphism $R \to S$ let us agree to write

$$T(R \to S)$$

for the set of elements in R which become invertible in S. If one replaces T by $\overline{T} = T(R \to R_T)$ the universal inverting ring does not change. We can hence always assume that T is multiplicatively closed. A natural maximal choice for T is the set $NZD(R)$ of all non-zerodivisors of R.

If the ring R is commutative it is well known that there is a model for R_T whose elements are "fractions" or more precisely equivalence classes of pairs $(a, t) \in R \times T$. For noncommutative rings the situation is more complicated. It goes back to Ore that under a suitable assumption such a calculus of fractions still exists.

Definition 4.2. A multiplicatively closed subset $T \subset R$ which does not contain zero-divisors or zero itself satisfies the right Ore-condition if for given $(a, s) \in R \times T$ there always exists a $(b, t) \in R \times T$ such that $at = sb$.

It is clear that this condition is necessary if a calculus of right fractions exists because we need to be able to write a given wrong way (left) fraction $s^{-1}a$ as bt^{-1}. It is a bit surprising that this is the only condition.

Proposition 4.3. *Let $T \subset R$ be a multiplicatively closed subset without zero divisors which satisfies the right Ore condition, then there exists a ring RT^{-1} and a universal T-inverting ring-homomorphism $i : R \to RT^{-1}$ such that every element of RT^{-1} can be written as $i(a)i(t)^{-1}$ with $(a, t) \in R \times T$.*

Proof. Elements in RT^{-1} are equivalence classes of pairs $(a, t) \in R \times T$. The pair (a, t) is equivalent to (b, s) if there exist elements u, $v \in R$ such that $au = bv$, $su = tv$ and $su = tv \in S$. For more details see Chapter II in [58]. □

Remark 4.4. Ore-localization is an exact functor, i.e. RT^{-1} is a flat R-module, see page 57 in [58].

Example 4.5. Let Γ be the free group on two generators x and y. The group ring $\mathbb{C}\Gamma$ does not satisfy the Ore condition with respect to the set $NZD(\mathbb{C}\Gamma)$ of all non-zerodivisors. Let $C \subset \Gamma$ be the infinite cyclic subgroup generated by x. Now $x - 1$ is a non-zerodivisor since it becomes invertible in $\mathcal{U}C$ (compare Example 2.11) and therefore in the overring $\mathcal{U}\Gamma$. In fact every non-trivial element in $\mathbb{C}\Gamma$ is a non-zerodivisor since one can embed $\mathbb{C}\Gamma$ in a skew field. The Ore condition would imply the existence of $(b, t) \in \mathbb{C}\Gamma \times NZD(\mathbb{C}\Gamma)$ with $(y - 1)t = (x - 1)b$ alias

$$(x - 1)^{-1}(y - 1) = bt^{-1}.$$

This implies that $(-b, t)^{tr}$ is in the kernel of the map $(x - 1, y - 1) : \mathbb{C}\Gamma^2 \to \mathbb{C}\Gamma$. But this map is injective, compare Example 1.5.

Localizing Matrices

Instead of elements one can try to invert maps. Let Σ be a set of homomorphisms between right R-modules. A ring homomorphism $R \to S$ is called Σ-inverting if for every map $\alpha \in \Sigma$ the induced map $\alpha \otimes_R \mathrm{id}_S$ is an isomorphism.

Definition 4.6. A Σ-inverting ring homomorphism $i : R \to R_\Sigma$ is called universal Σ-inverting if it has the following universal property. Given any Σ-inverting ring homomorphism $f : R \to S$ there exists a unique ring homomorphism $\Psi : R_\Sigma \to S$ such that the following diagram commutes.

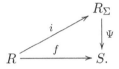

From now on let us assume that Σ is a set of matrices over R. For a ring homomorphism $R \to S$ we will write

$$\Sigma(R \to S)$$

for the set of all matrices over R which become invertible over S. One can always replace a given set of matrices Σ by $\overline{\Sigma} = \Sigma(R \to R_\Sigma)$ without changing the universal Σ-inverting ring homomorphism. There are different constructions which prove the existence of a universal Σ-inverting ring homomorphism. One possibility is a generator and relation construction where one starts with the free ring on a set of symbols $\overline{a}_{i,j}$ where $(a_{i,j})$ runs through the matrices in Σ and imposes the relations which are given in matrix form as $\overline{A}A = A\overline{A} = 1$, compare Theorem 2.1 in [12]. For more information the reader should consult Chapter 7 in [12] and Chapter 4 in [57].

Another construction due to Malcolmson [41] (see also [4]), a kind of calculus of fractions for matrices, allows a certain amount of control over the ring R_Σ.

As an easy example we would like to mention the following: A set of matrices is lower multiplicatively closed if $1 \in \Sigma$ and $a, b \in \Sigma$ implies that

$$\begin{pmatrix} a & 0 \\ c & b \end{pmatrix} \in \Sigma$$

for arbitrary matrices c of suitable size. Observe that $\Sigma(R \to S)$ is always lower multiplicatively closed.

Proposition 4.7 (Cramer's rule). *Let R be a ring and Σ be a lower multiplicatively closed set of matrices over R then every matrix a over R_Σ satisfies an equation of the form*

$$
s \begin{pmatrix} 1 & 0 \\ 0 & a \end{pmatrix} \begin{pmatrix} 1 & x \\ 0 & 1 \end{pmatrix} = b
$$

with $s \in \Sigma$, $x \in M(R_\Sigma)$ and $b \in M(R)$.

Proof. See Theorem 4.3 on page 53 in [57]. □

In particular every matrix a over R_Σ is stably associated over R_Σ to a matrix b over R, i.e. there exist invertible matrices c, $d \in GL(R_\Sigma)$ such that

$$
c \begin{pmatrix} a & 0 \\ 0 & 1_n \end{pmatrix} d^{-1} = \begin{pmatrix} b & 0 \\ 0 & 1_m \end{pmatrix}
$$

with suitable m and n.

Division Closure and Rational Closure

Recall that for a given ring homomorphism $R \to S$ we denoted by $T(R \to S)$ the set of all elements in R which become invertible in S and by $\Sigma(R \to S)$ the set of all matrices over R that become invertible over S. The universal localizations $R_{T(R \to S)}$ and $R_{\Sigma(R \to S)}$ come with a natural map to S. In the case where $R \to S$ is injective one may wonder whether these maps embed the universal localizations into S. The intermediate rings in the following definition serve as potential candidates for such embedded versions of the universal localizations.

Definition 4.8. Let S be a ring.

(i) A subring $R \subset S$ is called division closed in S if $T(R \subset S) = R^\times$, i.e. for every element $r \in R$ which is invertible in S the inverse r^{-1} lies already in R.

(ii) A subring $R \subset S$ is called rationally closed in S if $\Sigma(R \subset S) = GL(R)$, i.e. for every matrix A over R which is invertible over S the entries of the inverse matrix A^{-1} are all in R.

(iii) Given a subring $R \subset S$ the division closure of R in S denoted

$$
\mathcal{D}(R \subset S)
$$

is the smallest division closed subring of S which contains R.

(iv) Given a subring $R \subset S$ the rational closure of R in S denoted by

$$\mathcal{R}(R \subset S)$$

is the smallest rationally closed subring of S containing R.

Note that the intersection of division closed intermediate rings is again division closed and similarly for rationally closed rings. This proves the existence of the division and rational closure. Moreover we really have closure-operations, i.e.

$$\mathcal{D}(\mathcal{D}(R \subset S) \subset S) = \mathcal{D}(R \subset S) \text{ and}$$
$$\mathcal{R}(\mathcal{R}(R \subset S) \subset S) = \mathcal{R}(R \subset S).$$

In [12, Chapter 7, Theorem 1.2] it is shown that the set

$$\{a_{i,j} \in S \mid (a_{i,j}) \text{ invertible over } S, \ (a_{i,j})^{-1} \text{ matrix over } R\} \qquad (3)$$

is a subring of S and that it is rationally closed. Since this ring is contained in $\mathcal{R}(R \subset S)$ the two rings coincide. The following observation is very useful in our context.

Proposition 4.9. *A von Neumann regular ring R is division closed and rationally closed in every overring.*

Proof. Suppose $a \in R$ is not invertible in R, then the corresponding multiplication map $l_a : R \to R$ is not an isomorphism. Therefore the kernel or the cokernel is non-trivial. Both split of as direct summands because the ring is von Neumann regular. The corresponding projection onto the kernel or cokernel is given by left multiplication with a suitable idempotent. This idempotent shows that a must be a zerodivisor and hence can not become invertible in any overring. A matrix ring over a von Neumann regular ring is again von Neumann regular and the same reasoning applied to matrix rings over R yields that R is also rationally closed in every overring. □

In particular note that once we know that the division closure $\mathcal{D}(R \subset S)$ is von Neumann regular then it coincides with the rational closure $\mathcal{R}(R \subset S)$. The following proposition relates the division respectively rational closure to the universal localizations $R_{T(R \subset S)}$ and $R_{\Sigma(R \subset S)}$.

Proposition 4.10. *Let $R \subset S$ be a ring extension.*

(i) *The map $R_{T(R \subset S)} \to S$ given by the universal property factorizes over the division closure.*

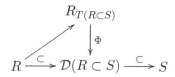

(ii) *If the pair $(R, T(R \subset S))$ satisfies the right Ore condition, then Φ is an isomorphism.*

(iii) *The map $R_{\Sigma(R \subset S)} \to S$ given by the universal property factorizes over the rational closure.*

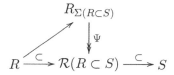

The map Ψ is always surjective.

Proof. (i) This follows from the definitions. (ii) Note that $T(R \subset S)$ always consists of non-zerodivisors. Thus we can choose a ring of right fractions as a model for $R_{T(R \subset S)}$. Every element in $\text{im}\Phi$ is of the form at^{-1} with $t \in T(R \subset S)$. Such an element is invertible in S if and only if $a \in T(R \subset S)$. We see that the image of Φ is division closed and hence Φ is surjective. On the other hand the abstract fraction $at^{-1} \in RT(R \subset T)^{-1}$ is zero if and only if $a = 0$ because $T(R \subset S)$ contains no zerodivisors, so Φ is injective. (iii) Only the last statement is maybe not obvious. By Cohn's description of the rational closure (compare (3)) we need to find a preimage for $a_{i,j}$, where $(a_{i,j})$ is a matrix invertible over S whose inverse lies over R. The generator and relation construction of the universal localization immediately gives such an element. $\qquad\square$

In general it is not true that the map Ψ is injective.

5   Some Candidates for $\mathcal{S}\Gamma$

We are now prepared to describe the candidates for the ring $\mathcal{S}\Gamma$ which appears in Proposition 3.4. We consider the ring extension $\mathbb{C}\Gamma \subset \mathcal{U}\Gamma$ and define

$$\begin{aligned}
\mathcal{D}\Gamma &= \mathcal{D}(\mathbb{C}\Gamma \subset \mathcal{U}\Gamma), \\
\mathcal{R}\Gamma &= \mathcal{R}(\mathbb{C}\Gamma \subset \mathcal{U}\Gamma), \\
\mathbb{C}\Gamma_T &= \mathbb{C}\Gamma_{T(\mathbb{C}\Gamma \subset \mathcal{U}\Gamma)} \text{ and} \\
\mathbb{C}\Gamma_\Sigma &= \mathbb{C}\Gamma_{\Sigma(\mathbb{C}\Gamma \subset \mathcal{U}\Gamma)}.
\end{aligned}$$

These rings are organized in the following diagram

$$(4)$$

A first hint that the rational or division closure may be a good candidate for $\mathcal{S}\Gamma$ is the following result which is implicit in [33]. At the same time its proof illustrates the usefulness of Cramer's rule 4.7.

Proposition 5.1. *If Γ is a torsionfree group then the Strong Atiyah Conjecture 3.2 implies that $\mathcal{R}\Gamma$ is a skew field.*

Proof. For $x \in \mathcal{U}\Gamma$ let $l_x : \mathcal{U}\Gamma \to \mathcal{U}\Gamma$ denote left multiplication with x. From the additivity and faithfulness of the dimension it follows that x is invertible if and only if $\dim \operatorname{im}(l_x) = 1$ or equivalently $\dim \ker(l_x) = 0$ or equivalently $\dim \operatorname{coker}(l_x) = 0$. Now let X be a matrix over $\mathcal{R}\Gamma$ then by 4.10 (iii) we know that we can lift it to a matrix over $\mathbb{C}\Gamma_\Sigma$. Using Cramer's rule 4.7 and projecting down again we see that we can find invertible matrices A and $B \in GL(\mathcal{R}\Gamma)$ such that

$$C = A \begin{pmatrix} 1_n & 0 \\ 0 & X \end{pmatrix} B$$

is a matrix over $\mathbb{C}\Gamma$. In particular if $0 \neq x \in \mathcal{R}\Gamma$ then for $X = (x)$ we know that there exists an n such that

$$\dim \operatorname{im}(l_x) + n = \dim(\operatorname{im} \begin{pmatrix} 1_n & 0 \\ 0 & l_x \end{pmatrix}) = \dim(\operatorname{im} C) \in \mathbb{Z}$$

because we assume for the matrix C over $\mathbb{C}\Gamma$ that the dimension of its image is an integer. It follows that $\dim \operatorname{im}(l_x) = 1$ and hence that x is invertible in $\mathcal{R}\Gamma$. $\square$

Note 5.2. If one of the rings $\mathcal{D}\Gamma$ or $\mathcal{R}\Gamma$ is a skew field then so is the other and the two coincide.

Proof. If $\mathcal{D}\Gamma$ is a skew field then it is also rationally closed, see 4.9. If $\mathcal{R}\Gamma$ is a skew field then $\mathcal{D}\Gamma$ is a division closed subring of a skew field and hence itself a skew field. $\square$

Corollary 5.3. *The Atiyah Conjecture 3.2 implies the Zero-Divisor Conjecture, i.e. the conjecture that the complex group ring of a torsionfree group does not contain any zero-divisors.*

Remark 5.4. One can show that for a torsionfree amenable group the Atiyah Conjecture 3.2 is equivalent to the Zero-Divisor Conjecture, see Lemma 10.16 in [39].

Another natural question is in how far the rings discussed above depend functorially on the group. Since an arbitrary group homomorphism $G \to G'$ does not induce a map from $\mathcal{U}G$ to $\mathcal{U}G'$ we can not expect functoriality but at least we have the following.

Note 5.5. An injective group homomorphism induces maps on the rings $\mathcal{D}\Gamma$, $\mathcal{R}\Gamma$, $\mathbb{C}\Gamma_T$ and $\mathbb{C}\Gamma_\Sigma$. These maps are compatible with the maps in diagram 4 above.

Proof. We already know that the inclusion $\mathbb{C}\Gamma \subset \mathcal{U}\Gamma$ is functorial for injective group homomorphisms. Let G be a subgroup of Γ. Since $\mathcal{U}G$ is von Neumann regular it is division closed and rationally closed in every overring, compare 4.9. Therefore $\mathcal{D}\Gamma \cap \mathcal{U}G$ is division-closed in $\mathcal{U}\Gamma$ and $\mathcal{D}G \subset \mathcal{D}\Gamma \cap \mathcal{U}G \subset \mathcal{D}\Gamma$. Analogously one argues for the rational closure. One immediately checks that $T(\mathbb{C}G \subset \mathcal{U}G) \subset T(\mathbb{C}\Gamma \subset \mathcal{U}\Gamma)$ and $\Sigma(\mathbb{C}G \subset \mathcal{U}G) \subset \Sigma(\mathbb{C}\Gamma \subset \mathcal{U}\Gamma)$. The universal properties imply the statement for $\mathbb{C}\Gamma_T$ and $\mathbb{C}\Gamma_\Sigma$. $\square$

6 Linnell's Result

Before we state Linnell's result we would like to introduce the class of groups it applies to.

Definition 6.1 (Linnell's class of groups). Let $\mathcal{C}$ be the smallest class of groups which has the following properties.

(LC1) Free groups are contained in $\mathcal{C}$.

(LC2) If $1 \to G \to \Gamma \to H \to 1$ is an exact sequence of groups such that G lies in $\mathcal{C}$ and H is finite or finitely generated abelian then Γ lies in $\mathcal{C}$.

(LC3) The class $\mathcal{C}$ is closed under directed unions, i.e. if a group $\Gamma = \bigcup_{i \in I} \Gamma_i$ is a directed union of subgroups Γ_i which lie in $\mathcal{C}$ then Γ also lies in $\mathcal{C}$.

To put this definition into perspective we would like to make a couple of remarks.

Remark 6.2. (i) If one replaces (LC1) above by the requirement that the trivial group belongs to $\mathcal{C}$ one obtains the smaller class of elementary amenable groups. Compare [9] and [29]. Elementary amenable groups are in particular amenable (see [16]) but it is not easy to find amenable groups that are not elementary amenable [23]. A group which contains a non-abelian free subgroup is not amenable.
(ii) One can show that if Γ lies in $\mathcal{C}$ and A is an elementary amenable normal subgroup then Γ/A also belongs to $\mathcal{C}$.
(iii) The class $\mathcal{C}$ is closed under free products.

In [33] Linnell proves Conjecture 3.2 for groups in the class $\mathcal{C}$ which additionally have a bound on the orders of finite subgroups. In fact by carefully investigating the proof given there one can obtain the following statements.

Theorem 6.3. *Suppose the group Γ lies in $\mathcal{C}$ and has a bound on the orders of finite subgroups then*

(K)   The composition

$$\mathrm{colim}_{H \in \mathcal{F}in\Gamma} K_0(\mathbb{C}H) \to K_0(\mathbb{C}\Gamma) \to K_0(\mathcal{D}\Gamma)$$

is surjective.

(R)   The ring $\mathcal{D}\Gamma$ is semi-simple and hence $\mathcal{D}\Gamma = \mathcal{R}\Gamma$ by Proposition 4.9.

As already mentioned this result is essentially contained in [33]. In the above formulation it is proven in [53]. The proof is published in Chapter 10 in [39]. Below we will only make a couple of remarks about the proof.
Since we formulated the theorem with the division closure $\mathcal{D}\Gamma$ the reader may get the impression that this is the best candidate for an intermediate ring $\mathcal{S}\Gamma$ as in Proposition 3.4. But in fact the situation is not so clear. We already stated that $\mathcal{D}\Gamma = \mathcal{R}\Gamma$ when the theorem applies. Moreover one can show the following.

Addendum 6.4.

(U)   In the situation of Theorem 6.3 the natural map $\mathbb{C}\Gamma_\Sigma \to \mathcal{R}\Gamma$ is an isomorphism and hence $\mathcal{D}\Gamma = \mathcal{R}\Gamma \cong \mathbb{C}\Gamma_\Sigma$.

(O) If Γ lies in the smaller class of elementary amenable groups and has a bound on the orders of finite subgroups then $\mathbb{C}\Gamma$ satisfies the right Ore condition with respect to the set $\mathrm{NZD}(\mathbb{C}\Gamma)$ of all non-zerodivisors and this set coincides with $T(\mathbb{C}\Gamma \subset \mathcal{U}\Gamma)$. Hence $\mathbb{C}\Gamma_T$ can be realized as a ring of fractions and the natural map $\mathbb{C}\Gamma_T \to \mathcal{D}\Gamma$ is an isomorphism, compare 4.10 (iii).

The statement (O) about the Ore localization appears already in [31]
We will now make some comments about the proof of Theorem 6.3 and the Addendum 6.4. As one might guess from the description of the class of groups to which the Theorem (and the Addendum) applies the proof proceeds via transfinite induction on the class of groups, i.e. one proves the following statements.

(I) (K), (R) and (U) hold for free groups.

(II) If $1 \to G \to \Gamma \to H \to 1$ is an extension of groups where H is finite or infinite cyclic and (K), (R) and (U) hold for G then they hold for Γ. Similar with (O) replacing (U).

(III) If Γ is the directed union of the subgroups Γ_i and (K), (R) and (U) hold for all Γ_i then they hold for Γ if Γ has a bound on the orders of finite subgroups. Similar with (O) replacing (U).

(I) The Kadison Conjecture says that there are no non-trivial idempotents in the group C^*-algebra of a torsionfree group. Linnell observed that Connes conceptional proof of this Conjecture for the free group on two generators given in [14] (see also [27]) can be used to verify the stronger Conjecture 3.2 in this case. Combined with Proposition 5.1 and Note 5.2 one concludes that $\mathcal{D}\Gamma = \mathcal{R}\Gamma$ is a skew field. This yields (R) and also (K) since K_0 of a skew field is $\mathbb{Z}$. Every finitely generated free group is a subgroup of the free group on two generators and every free group is a directed union of finitely generated free subgroups. This is used to pass to arbitrary free groups. Some rather non-trivial facts about group rings of free groups (see [25] and [30]) are used to verify that $\mathcal{R}\Gamma$ also coincides with the universal localization $\mathbb{C}\Gamma_\Sigma$ and hence to verify (U) in this case. Recall that we saw in Example 4.5 that (O) is false for free groups.

(II) For information about crossed products we refer the reader to [46] and to Digression 6.5 below. If $1 \to G \to \Gamma \to H \to 1$ is an extension of groups then every set-theoretical section μ of the quotient map $\Gamma \to H$ (we can always assume $\mu(e) = e$ and $\mu(g^{-1}) = \mu(g)^{-1}$) allows to describe the group ring $\mathbb{C}\Gamma$ as a crossed product $\mathbb{C}G * H$ of the ring $\mathbb{C}G$ with the group H.

Similarly crossed products $\mathcal{D}G * H$, $\mathcal{R}G * H$ and $\mathbb{C}G_\Sigma * H$ exist and can serve as intermediate steps when one tries to prove the statements (R), (K) and (U) for Γ. For example there are natural inclusions

$$\mathbb{C}G * H \to \mathcal{D}G * H \to \mathcal{D}\Gamma.$$

If H is a finite group and $\mathcal{D}G$ is semisimple then $\mathcal{D}G*H$ is semisimple and coincides with $\mathcal{D}\Gamma$. It is relatively easy to verify that $\mathcal{D}G*H$ is noetherian and semiprime if H is an infinite cyclic group and then Goldie's theorem (a criterion for the existence of an Ore-localization, see Section 9.4 in [13]) together with results from [32] are used to verify that $\mathcal{D}\Gamma$ is the Ore-localization of $\mathcal{D}G * H$ with respect to the set of all non-zerodivisors. This is roughly the line of argument in order to verify that condition (R) survives extensions by finite or infinite cyclic groups. Once we know that $\mathcal{D}\Gamma$ is an Ore localization of $\mathcal{D}G * H$ we can combine this with the assumption (which implies that $\mathbb{C}G_\Sigma * H \to \mathcal{D}G * H$ is an isomorphism) in order to verify (U). Similarly iterating Ore localizations one obtains that (O) is stable under extensions with an infinite cyclic group. Moody's Induction Theorem (see Theorem 7.3 below) plays a crucial role in the argument for (K). Moreover one has to assume that the class of groups which appears in the induction hypothesis is already closed under extensions by finite subgroups. Hence one is forced to start the induction with virtually free groups and in particular one has to prove that (K) holds for such groups. For this purpose results of Waldhausen [62] about generalized free products can be used. Moreover the map induced by $\mathbb{C}\Gamma \to \mathbb{C}\Gamma_\Sigma$ on K_0 needs to be studied, compare Question 7.5. Here it is important to deal with the universal matrix localization.

(III) If Γ is the directed union of the subgroups Γ_i, $i \in I$ then $\mathcal{D}\Gamma$ is the directed union of the subrings $\mathcal{D}\Gamma_i$ and similar for $\mathcal{R}\Gamma$. A directed union of von Neumann regular rings is again von Neumann regular (use Definition 2.4 (iv)) so $\mathcal{D}\Gamma$ is at least von Neumann regular if all the $\mathcal{D}\Gamma_i$ are semisimple. The fact that K-theory is compatible with colimits yields that (K) holds for Γ if it holds for all the Γ_i. Now the assumption on the bound of the orders of finite subgroups implies that $\mathcal{D}\Gamma$ is even semisimple by Note 3.7. That (U) and (O) are stable under directed unions is straightforward.

Digression 6.5. A crossed product $R * G = (S, \mu)$ of the ring R with the group G consists of a ring S which contains R as a subring together with an injective map $\mu : G \to S^\times$ such that the following holds.

(i) The ring S is a free R-module with basis $\mu(G)$.

(ii) For every $g \in G$ the conjugation map $c_{\mu(g)} : S \to S$, $\mu(g)s\mu(g)^{-1}$ can be restricted to R.

(iii) For all $g, g' \in G$ the element $\tau(g, g') = \mu(g)\mu(g')\mu(gg')^{-1}$ lies in $R^{\times}$.

7 The Isomorphism Conjecture in algebraic K-theory

Condition (K) in Proposition 3.4 requires that the composite map

$$\text{colim}_{H \in \mathcal{F}in\Gamma} K_0(\mathbb{C}H) \longrightarrow K_0(\mathbb{C}\Gamma) \longrightarrow K_0(\mathcal{S}\Gamma)$$

is surjective. About the first map in this composition there is the following conjecture.

Conjecture 7.1 (Isomorphism Conjecture - Special Case). *For every group Γ the map*

$$\text{colim}_{H \in \mathcal{F}in\Gamma} K_0(\mathbb{C}H) \longrightarrow K_0(\mathbb{C}\Gamma)$$

is an isomorphism. In particular for a torsionfree group Γ we expect

$$K_0(\mathbb{C}\Gamma) \cong \mathbb{Z}.$$

To form the colimit we understand $\mathcal{F}in\Gamma$ as the category whose objects are the finite subgroups of Γ and whose morphisms are generated by inclusion maps $K \subset H$ and conjugation maps $c_g : H \to H^g$, $h \mapsto ghg^{-1}$ with $g \in \Gamma$. Observe that in the torsionfree case the colimit reduces to $K_0(\mathbb{C}) \cong \mathbb{Z}$.

In fact Conjecture 7.1 would be a consequence of a much more general conjecture which predicts the whole algebraic K-theory of a group ring $R\Gamma$ in terms of the K-theory of the coefficients and homological data about the group. This more general conjecture is known as the Farrell-Jones Isomorphism Conjecture for algebraic K-theory [19]. A precise formulation would require a certain amount of preparation and we refer the reader to [15], [2] and in particular to [40] for more information.

Conjecture 7.1 would have the following consequence.

Consequence 7.2. *For every finitely generated projective $\mathbb{C}\Gamma$-module P we have*

$$\dim_{\Gamma}(P \otimes_{\mathbb{C}\Gamma} \mathcal{U}\Gamma) \in \frac{1}{\#\mathcal{F}in\Gamma}\mathbb{Z}.$$

Proof. Use Example 2.10 and Proposition 3.5. □

So if all finitely presented $\mathbb{C}\Gamma$-modules were also finitely generated projective then the Isomorphism Conjecture would imply the Strong Atiyah Conjecture. But of course this is seldom the case.

Conjecture 7.1 is true for infinite cyclic groups and products of such by the Bass-Heller-Swan formula [3]. Cohn's results in [11] imply the Conjecture for free groups. Work of Waldhausen [62] deals with generalized free products and HNN-extensions. (The reader should consult [52] and [45] for a "noncommutative localization"-perspective on these results.) A version of the following result plays also an important role in the proof of 6.3.

Theorem 7.3 (Moody's induction theorem - Special case). *Let Γ be a polycyclic-by-finite group then the map*

$$\mathrm{colim}_{H \in \mathcal{F}in\Gamma} K_0(\mathbb{C}H) \to K_0(\mathbb{C}\Gamma)$$

is surjective.

Proof. See [42], [43], [10] and Chapter 8 in [46]. $\qquad\square$

What happens if we replace the complex coefficients in Conjecture 7.1 by integral coefficients? Thinking about the situation for finite groups it is at first glance very surprising that for infinite groups there are a lot of cases where there are results about $K_0(\mathbb{Z}\Gamma)$ whereas nothing is known about $K_0(\mathbb{C}\Gamma)$. See for example [19]. The reason is that the elements of algebraic K-groups of the *integral* group ring have a topological interpretation. They occur as obstruction groups in certain topological problems. Many people put a lot of effort into solving these topological problems and each time this is successful one obtains a result about the algebraic K-groups of $\mathbb{Z}\Gamma$.

However with integral coefficients one does not expect an isomorphism as in Conjecture 7.1. In the case where Γ is torsionfree one would still expect $K_0(\mathbb{Z}\Gamma) \cong \mathbb{Z}$, but in general so called Nil-groups and also negative K-groups should enter in a "computation" of $K_0(\mathbb{Z}\Gamma)$. Moreover by a result of Swan (see Theorem 8.1 in [59]) the map $K_0(\mathbb{Z}H) \to K_0(\mathbb{Q}H)$ is almost the trivial map for a finite group H, i.e. the map on reduced K-groups $\tilde{K}_0(\mathbb{Z}H) \to \tilde{K}_0(\mathbb{Q}H)$ is trivial. Summarizing: In general in the square

$$
\begin{array}{ccc}
\mathrm{colim}_{H \in \mathcal{F}in\Gamma} K_0(\mathbb{Z}H) & \longrightarrow & K_0(\mathbb{Z}\Gamma) \\
\downarrow & & \downarrow \\
\mathrm{colim}_{H \in \mathcal{F}in\Gamma} K_0(\mathbb{Q}H) & \longrightarrow & K_0(\mathbb{Q}\Gamma).
\end{array}
$$

neither the upper horizontal arrow nor the vertical arrows are surjective. We see that the comparison to the integral group ring is not very useful for the question we are interested in.

The main techniques to prove results about the K-theory of $\mathbb{Z}\Gamma$ stems from "controlled topology". See [51], [49], [50], [17], [18] and [26]. The set-up has been adapted to a more algebraic setting [48] and this "controlled algebra" (see [48],[6] and [47]) was used successfully to obtain "lower bounds" for the K-theory of group rings with arbitrary coefficients under certain curvature conditions on the group [7].

A result about Conjecture 7.1 which uses this "controlled algebra" is the following result from [2]. Recall that a ring is called (right)-regular if it is right noetherian and every finitely generated right R-module admits finite dimensional projective resolution.

Theorem 7.4. *Let Γ be the fundamental group of a closed riemannian manifold with strictly negative sectional curvature. Let R be a regular ring, e.g. $R = \mathbb{C}$ then*

$$K_0(R) \cong K_0(R\Gamma).$$

Moreover $K_{-n}(R\Gamma) = 0$ and $K_1(R\Gamma) = \Gamma_{ab} \otimes_{\mathbb{Z}} K_0(R) \oplus K_1(R)$, where Γ_{ab} denotes the abelianized group.

The assumption about Γ implies that Γ is torsionfree so the above verifies Conjecture 7.1.

The author is optimistic that in the near future techniques similar to the ones used in [2] will lead to further results about Conjecture 7.1. In view of condition (K) in Proposition 3.4 the following seems to be an important question.

Question 7.5. *Are the maps*

$$
\begin{aligned}
K_0(\mathbb{C}\Gamma) &\to K_0(\mathbb{C}\Gamma_\Sigma), \\
K_0(\mathbb{C}\Gamma) &\to K_0(\mathcal{R}\Gamma) \\
or \quad K_0(\mathbb{C}\Gamma) &\to K_0(\mathcal{D}\Gamma)
\end{aligned}
$$

surjective?

Note that this is true for groups in Linnell's class $\mathcal{C}$ with a bound on the orders of finite subgroups by Theorem 6.3 (K).

8 Exactness Properties

In this section we want to investigate to what extent the functor $- \otimes_{\mathbb{C}\Gamma} \mathcal{U}\Gamma$ and related functors are exact. Recall that this functor is crucial for the definition of L^2-Betti numbers, compare Definition 2.8.

Note 8.1. If Γ is elementary amenable and there is a bound on the orders of finite subgroups then $- \otimes_{\mathbb{C}\Gamma} \mathcal{U}\Gamma$ is exact.

Proof. From Addendum 6.4 (O) we know that for these groups $\mathcal{D}\Gamma$ is an Ore-localization of $\mathbb{C}\Gamma$. In particular in this case $- \otimes_{\mathbb{C}\Gamma} \mathcal{D}\Gamma$ is exact. Since by Theorem 6.3 (R) $\mathcal{D}\Gamma$ is also semisimple (and hence von Neumann regular) we know that every module is flat over $\mathcal{D}\Gamma$. □

The following tells us that we cannot always have exactness.

Note 8.2. Suppose for the infinite group Γ the functor $- \otimes_{\mathbb{C}\Gamma} \mathcal{U}\Gamma$ is exact, then all L^2-Betti numbers and also the Euler-characteristic $\chi^{(2)}(\Gamma)$ of the group Γ vanishes.

Proof. Flatness implies

$$H_p(C_*(E\Gamma) \otimes_{\mathbb{Z}\Gamma} \mathcal{U}\Gamma) = Tor_p^{\mathbb{Z}\Gamma}(\mathbb{Z}, \mathcal{U}\Gamma) = Tor_p^{\mathbb{Z}\Gamma}(\mathbb{Z}, \mathbb{Z}\Gamma) \otimes_{\mathbb{Z}\Gamma} \mathcal{U}\Gamma = 0$$

for $p > 0$. Moreover $b_0^{(2)}(\Gamma) = 0$ for every infinite group (see Theorem 6.54 (8) (b) in [39]). □

In particular we see that for the free group on two generators we cannot have exactness. We saw this phenomenon already in Example 1.5 because exactness of $- \otimes_{\mathbb{C}\Gamma} \mathcal{N}\Gamma$ would contradict the monotony of the dimension. (Recall from Proposition 2.6 (ii) that $- \otimes_{\mathcal{N}\Gamma} \mathcal{U}\Gamma$ is always exact.)
More generally we have.

Note 8.3. If Γ contains a nonabelian free group, then neither $\mathcal{D}\Gamma$ nor $\mathcal{R}\Gamma$, $\mathbb{C}\Gamma_\Sigma$ or $\mathcal{U}\Gamma$ can be flat over $\mathbb{C}\Gamma$.

Proof. Every free group contains a free group on two generators. Let $G \subset \Gamma$ be a free subgroup on two generators. Let $\mathbb{C}G^2 \to \mathbb{C}G$ be the injective homomorphism from Example 1.5. Since $\mathbb{C}\Gamma$ is flat over $\mathbb{C}G$ we obtain an injective map $\mathbb{C}\Gamma^2 \to \mathbb{C}\Gamma$. On the other hand since $\mathcal{D}G$ is a skew-field we know that the non-trivial kernel of the corresponding map $\mathcal{D}G^2 \to \mathcal{D}G$ (which must appear for dimension reasons since $-\otimes_{\mathcal{D}G} \mathcal{U}G$ is exact and the Γ-dimension is faithful) is a one-dimensional free module which splits off $\mathcal{D}G^2$ as a direct summand. The same remains true for every overring of $\mathcal{D}G$. In particular for $\mathcal{D}\Gamma$, $\mathcal{R}\Gamma$ and $\mathcal{U}\Gamma$. But also for $\mathbb{C}\Gamma_\Sigma$ because $\mathcal{D}G = \mathcal{R}G \cong \mathbb{C}G_\Sigma$ by Addendum 6.4 (O) and since there is a natural map $\mathbb{C}G_\Sigma \to \mathbb{C}\Gamma_\Sigma$. □

In this context we would also like to mention the following result from [38].

Theorem 8.4. *If Γ is amenable then $\mathcal{N}\Gamma$ (and hence $\mathcal{U}\Gamma$) is dimension-flat over $\mathbb{C}\Gamma$, i.e. for $p > 0$ and every $\mathbb{C}\Gamma$-module M we have*

$$\dim_{\mathcal{N}\Gamma} Tor_p^{\mathbb{C}\Gamma}(M, \mathcal{N}\Gamma) = \dim_{\mathcal{U}\Gamma} Tor_p^{\mathbb{C}\Gamma}(M, \mathcal{U}\Gamma) = 0.$$

Proof. See [38] or Theorem 6.37 on page 259 in [39] and recall that $\mathcal{U}\Gamma$ is flat over $\mathcal{N}\Gamma$ and the $\mathcal{U}\Gamma$-dimension and the $\mathcal{N}\Gamma$-dimension are compatible. □

Given these facts it is tempting to conjecture that $- \otimes_{\mathbb{C}\Gamma} \mathcal{U}\Gamma$ is exact if and only if Γ is amenable. However in [35] it is shown that the condition about the bound on the orders of finite subgroups in Note 8.1 is necessary.

Example 8.5. Let H be a nontrivial finite group and let $H \wr \mathbb{Z}$ denote the semidirect product $\bigoplus_{-\infty}^{\infty} H \rtimes \mathbb{Z}$, where $\mathbb{Z}$ is acting via shift on the $\bigoplus_{-\infty}^{\infty} H$. Then neither $\mathcal{D}\Gamma$ nor $\mathcal{U}\Gamma$ is flat over $\mathbb{C}\Gamma$ (see Theorem 1 in [35]).

The main purpose of this section is to prove the following result which measures the deviation from exactness for groups in Linnell's class.

Theorem 8.6. *Let Γ be in the class $\mathcal{C}$ with a bound on the orders of finite subgroups, then*

$$Tor_p^{\mathbb{C}\Gamma}(-; \mathcal{D}\Gamma) = 0 \qquad \text{for all } p \geq 2.$$

Note that for these groups $\mathcal{D}\Gamma = \mathcal{R}\Gamma \cong \mathbb{C}\Gamma_\Sigma$ is semisimple and therefore the functor $- \otimes_{\mathcal{D}\Gamma} \mathcal{U}\Gamma$ is exact. The functor $- \otimes_{\mathbb{Z}\Gamma} \mathbb{C}\Gamma$ is always exact. Therefore we obtain the corresponding statements for $Tor_p^{\mathbb{C}\Gamma}(-; \mathcal{U}\Gamma)$, $Tor_p^{\mathbb{Z}\Gamma}(-; \mathcal{D}\Gamma)$ and $Tor_p^{\mathbb{Z}\Gamma}(-; \mathcal{U}\Gamma)$.

As an immediate consequence we obtain interesting examples of stably flat universal localizations.

Corollary 8.7. *If Γ lies in Linnell's class $\mathcal{C}$ and has a bound on the orders of finite subgroups then $\mathcal{D}\Gamma \cong \mathbb{C}\Gamma_\Sigma$ is stably flat over $\mathbb{C}\Gamma$, i.e. we have*

$$Tor_p^{\mathbb{C}\Gamma}(\mathcal{D}\Gamma, \mathcal{D}\Gamma) = 0 \qquad \text{for all } p \geq 1.$$

Proof. We know that $\mathcal{D}\Gamma \cong \mathbb{C}\Gamma_\Sigma$ is a universal localization of $\mathbb{C}\Gamma$ and hence $\mathbb{C}\Gamma \to \mathbb{C}\Gamma_\Sigma$ is an epimorphism in the category of rings, see page 56 in [57]. By Theorem 4.8 b) in [57] we know that $Tor_1^{\mathbb{C}\Gamma}(\mathcal{D}\Gamma, \mathcal{D}\Gamma) = 0$. For $p \geq 2$ the result follows from Theorem 8.6. □

Recent work of Neeman and Ranicki [45] show that for universal localizations which are stably flat there exists a long exact localization sequence which extends Schofield's localization sequence for universal localizations (see Theorem 4.12 in [57]) to the left. In the case of Ore-localizations the corresponding sequence was known for a long time, see [20], [22], [65] and [60]. Observe that because of Note 8.3 we know that whenever Γ contains a free group $\mathbb{C}\Gamma_\Sigma$ cannot be an Ore-localization.

Here is another consequence of Theorem 8.6.

Corollary 8.8. *If the infinite group Γ belongs to $\mathcal{C}$ and has a bound on the orders of finite subgroups, then*

$$\chi^{(2)}(\Gamma) \leq 0.$$

Proof. Since the group is infinite we have $b_0^{(2)}(\Gamma) = 0$. Because of

$$H_p(\Gamma; \mathcal{U}\Gamma) = \operatorname{Tor}_p^{\mathbb{C}\Gamma}(\mathbb{C}; \mathcal{U}\Gamma) = 0 \qquad \text{for all } p \geq 0$$

we know that $b_1^{(2)}(\Gamma)$ is the only L^2-Betti number which could possibly be nonzero. $\qquad\square$

The $L^{(2)}$-Euler characteristic coincides with the usual Euler-characteristic and the rational Euler-Characteristic of [63] whenever these are defined. Before we proceed to the proof of Theorem 8.6 we would also like to mention the following consequences for L^2-homology.

Corollary 8.9 (Universal Coefficient Theorem). *Let Γ be in $\mathcal{C}$ with a bound on the orders of finite subgroups. Then there is a universal coefficient theorem for L^2-homology: Let X be a Γ-space whose isotropy groups are all finite, then there is an exact sequence*

$$0 \to H_n(X; \mathbb{Z}) \otimes_{\mathbb{Z}\Gamma} \mathcal{U}\Gamma \to H_n^\Gamma(X; \mathcal{U}\Gamma) \to \operatorname{Tor}_1(H_{n-1}(X; \mathbb{Z}); \mathcal{U}\Gamma) \to 0.$$

Proof. We freely use the dimension theory for arbitrary $\mathcal{U}\Gamma$-modules, compare Remark 2.9. If X has finite isotropy, then the set of singular simplices also has only finite isotropy groups. If H is a finite subgroup of Γ, then $\mathbb{C}[\Gamma/H] \cong \mathbb{C}\Gamma \otimes_{\mathbb{C}H} \mathbb{C}$ is induced from the projective $\mathbb{C}H$-module $\mathbb{C}$ and therefore projective. We see that the singular chain complex with complex coefficients $C_* = C_*^{sing}(X; \mathbb{C})$ is a complex of projective $\mathbb{C}\Gamma$-modules. The E^2-term of the Künneth spectral sequence (compare Theorem 5.6.4 on page 143 in [64])

$$E_{pq}^2 = \operatorname{Tor}_p^{\mathbb{C}\Gamma}(H_q(C_*); \mathcal{D}\Gamma) \Rightarrow H_{p+q}(C_* \otimes \mathcal{D}\Gamma) = H_{p+q}(X; \mathcal{D}\Gamma)$$

is concentrated in two columns. The spectral sequence collapses, and we get exact sequences

$$0 \to H_n(X;\mathbb{C}) \otimes_{\mathbb{C}\Gamma} \mathcal{D}\Gamma \to H_n(X;\mathcal{D}\Gamma) \to \mathrm{Tor}_1^{\mathbb{C}\Gamma}(H_{n-1}(X;\mathbb{C});\mathcal{D}\Gamma) \to 0.$$

Applying the exact functor $- \otimes_{\mathcal{D}\Gamma} \mathcal{U}\Gamma$ yields the result. $\qquad\square$

The proof of Theorem 8.6 depends on the following Lemma.

Lemma 8.10. *(i) Let $R*G \subset S*G$ be compatible with the crossed product structure. Let M be an $R*G$-module. There is a natural isomorphism of right S-modules*

$$\mathrm{Tor}_p^{R*G}(M;S*G) \cong \mathrm{Tor}_p^R(\mathrm{res}_R^{R*G} M;S)$$

for all $p \geq 0$.

(ii) Suppose $R \subset S$ is a ring extension and $R = \bigcup_{i \in I} R_i$ is the directed union of the subrings R_i. Let M be an R-module. Then there is a natural isomorphism of right S-modules

$$\mathrm{Tor}_p^R(M;S) \cong \mathrm{colim}_{i \in I} \mathrm{Tor}_p^{R_i}(\mathrm{res}_{R_i}^R M;S_i) \otimes_{S_i} S$$

for all $p \geq 0$.

Proof. (i) We start with the case $p = 0$. We denote the crossed product structure map by μ, compare Digression 6.5. Define a map

$$h_M : \mathrm{res}_R^{R*G} M \otimes_R S \to M \otimes_{R*G} S*G$$

by $m \otimes s \mapsto m \otimes s$. Obviously h is a natural transformation from the functor $\mathrm{res}_R^{R*G}(-) \otimes_R S$ to $- \otimes_{R*G} S*G$. If $M = R*G$ the map $h_{R*G}^{-1} :$ $R*G \otimes_{R*G} S*G \cong S*G \to \mathrm{res}_R^{R*G} R*G \otimes_R S$ given by $s\mu(g) \mapsto g \otimes c_g^{-1}(s)$ is a well-defined inverse. Since h is compatible with direct sums we see that h_F is an isomorphism for all free modules F. Now if M is an arbitrary module choose a free resolution $F_* \to M$ of M and apply both functors to

$$F_1 \to F_0 \to M \to 0 \to 0.$$

Both functors are right exact, therefore an application of the five lemma yields the result for $p = 0$. Now let $P_* \to M$ be a projective resolution of M, then

$$
\begin{aligned}
\mathrm{Tor}_p^R(\mathrm{res}_R^{R*G} M;S) &= H_p(\mathrm{res}_R^{R*G} P_* \otimes_R S) \\
&\xrightarrow{\cong} H_p(P_* \otimes_{R*G} S*G) \\
&= \mathrm{Tor}_p^R(M;S*G).
\end{aligned}
$$

(ii) Again we start with the case $p = 0$. The natural surjections $\mathrm{res}_{R_i}^R M \otimes_{R_i} S \to M \otimes_R S$ induce a surjective map

$$h_M : \mathrm{colim}_{i \in I} \mathrm{res}_{R_i}^R M \otimes_{R_i} S \to M \otimes_R S$$

which is natural in M. Suppose the element of the colimit represented by $\sum_k m_k \otimes s_k \in \mathrm{res}_{R_i}^R M \otimes_{R_i} S$ is mapped to zero in $M \otimes_R S$. By construction the tensor product $M \otimes_R S$ is the quotient of the free module on the set $M \times S$ by a relation submodule. But every relation involves only finitely many elements of R, so we can find a $j \in I$ such that $\sum_k m_k \otimes s_k = 0$ already in $\mathrm{res}_{R_j}^R M \otimes_{R_j} S$. We see that h_M is an isomorphism. Now let $P_* \to M$ be a projective resolution. Since the colimit is an exact functor it commutes with homology and we get

$$
\begin{aligned}
\mathrm{colim}_{i \in I} \mathrm{Tor}_p^{R_i}(M; S) &= \mathrm{colim}_{i \in I} H_p(\mathrm{res}_{R_i}^R P_* \otimes_{R_i} S) \\
&= H_p(\mathrm{colim}_{i \in I}(\mathrm{res}_{R_i}^R P_* \otimes_{R_i} S)) \\
&\xrightarrow{\cong} H_p(P_* \otimes_R S) \\
&= \mathrm{Tor}_p^R(M; S). \quad \cdot
\end{aligned}
$$

$\square$

Proof of Theorem 8.6. The proof works via transfinite induction over the group as for the proof of Linnell's Theorem 6.3 itself, compare (I), (II) and (III) on page 127.

(I) The statement for free groups is well known: let Γ be the free group generated by the set S. The cellular chain complex of the universal covering of the obvious 1-dimensional classifying space gives a projective resolution of the trivial module of length one

$$0 \to \bigoplus_S \mathbb{C}\Gamma \to \mathbb{C}\Gamma \to \mathbb{C} \to 0.$$

Now if M is an arbitrary $\mathbb{C}\Gamma$-module we apply $- \otimes_{\mathbb{C}} M$ to the above complex and get a projective resolution of length 1 for M (diagonal action). (Use that for P a projective $\mathbb{C}\Gamma$-module $P \otimes_{\mathbb{C}} M$ with the diagonal respectively the left Γ-action are noncanonically isomorphic $\mathbb{C}\Gamma$-modules.)

(II) The next step is to prove that the statement remains true under extensions by finite groups. So let $1 \to G \to \Gamma \to H \to 1$ be an exact sequence with H finite. We know that $\mathcal{D}\Gamma = \mathcal{D}G * H$, see Lemma 10.59 on page 399 in [39] or Proposition 8.13 in [53]. Let M be a $\mathbb{C}\Gamma$-module, then with

Lemma 8.10 and the induction hypothesis we conclude

$$
\begin{aligned}
\mathrm{Tor}_p^{\mathbb{C}\Gamma}(M;\mathcal{D}\Gamma) &= \mathrm{Tor}_p^{\mathbb{C}G*H}(M;\mathcal{D}G*H) \\
&\cong \mathrm{Tor}_p^{\mathbb{C}G}(\mathrm{res}_{\mathbb{C}G}^{\mathbb{C}G*H}M;\mathcal{D}G) \\
&= 0 \quad \text{for } p > 1.
\end{aligned}
$$

The case H infinite cyclic is only slightly more complicated. This time we know from Lemma 10.69 in [39] or Proposition 8.18 in [53] that $\mathcal{D}\Gamma = (\mathcal{D}G * H)T^{-1}$ is an Ore localization, where $T = T(\mathcal{D}G * H \subset \mathcal{U}\Gamma)$, i.e. the set of all elements in $\mathcal{D}G * H$ which become invertible in $\mathcal{U}\Gamma$. Since Ore localization is an exact functor we get

$$
\begin{aligned}
\mathrm{Tor}_p^{\mathbb{C}\Gamma}(M;\mathcal{D}\Gamma) &= \mathrm{Tor}_p^{\mathbb{C}G*H}(M;(\mathcal{D}G * H)T^{-1}) \\
&\cong \mathrm{Tor}_p^{\mathbb{C}G*H}(M;\mathcal{D}G * H) \otimes_{\mathcal{D}G*H} \mathcal{D}\Gamma
\end{aligned}
$$

and conclude again with Lemma 8.10 that this module vanishes if $p > 1$.
(III) The behaviour under directed unions remains to be checked. Let $\Gamma = \bigcup_{i \in I} \Gamma_i$ be a directed union, then using Definition 2.4 we see that $\bigcup_{i \in I} \mathcal{D}\Gamma_i$ is von Neumann regular and it is easy to check that it coincides with the division closure $\mathcal{D}\Gamma$. Now Lemma 8.10 gives

$$
\begin{aligned}
\mathrm{Tor}_p^{\mathbb{C}\Gamma}(M;\mathcal{D}\Gamma) &\cong \mathrm{colim}_{i \in I}\mathrm{Tor}_p^{\mathbb{C}\Gamma_i}(\mathrm{res}_{\mathbb{C}\Gamma_i}^{\mathbb{C}\Gamma}M;\mathcal{D}\Gamma) \\
&= \mathrm{colim}_{i \in I}\mathrm{Tor}_p^{\mathbb{C}\Gamma_i}(\mathrm{res}_{\mathbb{C}\Gamma_i}^{\mathbb{C}\Gamma}M;\mathcal{D}\Gamma_i) \otimes_{\mathcal{D}\Gamma_i} \mathcal{D}\Gamma \\
&= 0 \quad \text{for } p > 1.
\end{aligned}
$$

$\square$

References

[1] M. F. Atiyah. Elliptic operators, discrete groups and von Neumann algebras. *Astérisque*, 32-33:43–72, 1976.

[2] A. Bartels, F. T. Farrell, L. E. Jones, and H. Reich. On the isomorphism conjecture in algebraic K-theory. *Topology*, 40:157–213, 2004.

[3] H. Bass, A. Heller, and R. G. Swan. The Whitehead group of a polynomial extension. *Inst. Hautes Études Sci. Publ. Math.*, 22:61–79, 1964.

[4] J. A. Beachy. On universal localization at semiprime Goldie ideals. In *Ring theory (Granville, OH, 1992)*, pages 41–57. World Sci. Publishing, River Edge, NJ, 1993.

[5] B. Blackadar. *K-theory for operator algebras*. Springer-Verlag, New York, 1986.

[6] G. Carlsson. Homotopy fixed points in the algebraic *K*-theory of certain infinite discrete groups. In *Advances in homotopy theory (Cortona, 1988)*, volume 139 of *London Math. Soc. Lecture Note Ser.*, pages 5–10. Cambridge Univ. Press, Cambridge, 1989.

[7] _____ and E. K. Pedersen. Controlled algebra and the Novikov conjectures for *K*- and *L*-theory. *Topology*, 34(3):731–758, 1995.

[8] J. Cheeger and M. Gromov. L_2-cohomology and group cohomology. *Topology*, 25(2):189–215, 1986.

[9] C. Chou. Elementary amenable groups. *Illinois J. Math.*, 24(3):396–407, 1980.

[10] G. Cliff and A. Weiss. Moody's induction theorem. *Illinois J. Math.*, 32(3):489–500, 1988.

[11] P. M. Cohn. Free ideal rings. *J. Algebra*, 1:47–69, 1964.

[12] _____ *Free rings and their relations*. Academic Press Inc. [Harcourt Brace Jovanovich Publishers], London, second edition, 1985.

[13] _____ *Algebra. Vol. 3*. John Wiley & Sons Ltd., Chichester, second edition, 1991.

[14] A. Connes. Noncommutative differential geometry. *Inst. Hautes Études Sci. Publ. Math.*, (62):257–360, 1985.

[15] J. F. Davis and W. Lück. Spaces over a category and assembly maps in isomorphism conjectures in *K*- and *L*-theory. *K-Theory*, 15(3):201–252, 1998.

[16] M. M. Day. Amenable semigroups. *Illinois J. Math.*, 1:509–544, 1957.

[17] F. T. Farrell. *Lectures on surgical methods in rigidity*. Published for the Tata Institute of Fundamental Research, Bombay, 1996.

[18] _____ The Borel Conjecture. In *Topology of high-dimensional manifolds*, volume 9(1) of *ICTP Lecture Notes*, pages 225–298. ICTP, Trieste, 2002.

[19] _____ and L. E. Jones. Isomorphism conjectures in algebraic K-theory. *J. Amer. Math. Soc.*, 6(2):249–297, 1993.

[20] S. M. Gersten. The localization theorem for projective modules. *Comm. Algebra*, 2:317–350, 1974.

[21] K. R. Goodearl. *von Neumann regular rings*. Pitman (Advanced Publishing Program), Boston, Mass., 1979.

[22] D. R. Grayson. K-theory and localization of noncommutative rings. *J. Pure Appl. Algebra*, 18(2):125–127, 1980.

[23] R. I. Grigorchuk. An example of a finitely presented amenable group that does not belong to the class EG. *Mat. Sb.*, 189(1):79–100, 1998.

[24] _____ , P. A. Linnell, T. Schick, and A. Żuk. On a question of Atiyah. *C. R. Acad. Sci. Paris Sér. I Math.*, 331(9):663–668, 2000.

[25] I. Hughes. Division rings of fractions for group rings. *Comm. Pure Appl. Math.*, 23:181–188, 1970.

[26] L. Jones. Foliated control theory and its applications. In *Topology of high-dimensional manifolds*, volume 9(2) of *ICTP Lecture Notes*, pages 405–460. ICTP, Trieste, 2002.

[27] P. Julg and A. Valette. K-theoretic amenability for $SL_2(\mathbf{Q}_p)$, and the action on the associated tree. *J. Funct. Anal.*, 58(2):194–215, 1984.

[28] R. V. Kadison and J. R. Ringrose. *Fundamentals of the theory of operator algebras. Vol. I.* Academic Press Inc. [Harcourt Brace Jovanovich Publishers], New York, 1983. Elementary theory.

[29] P. H. Kropholler, P. A. Linnell, and J. A. Moody. Applications of a new K-theoretic theorem to soluble group rings. *Proc. Amer. Math. Soc.*, 104(3):675–684, 1988.

[30] J. Lewin. Fields of fractions for group algebras of free groups. *Trans. Amer. Math. Soc.*, 192:339–346, 1974.

[31] P. A. Linnell. Zero divisors and group von Neumann algebras. *Pacific J. Math.*, 149(2):349–363, 1991.

[32] _____ Zero divisors and $L^2(G)$. *C. R. Acad. Sci. Paris Sér. I Math.*, 315(1):49–53, 1992.

[33] _____ Division rings and group von Neumann algebras. *Forum Math.*, 5(6):561–576, 1993.

[34] _____ Analytic versions of the zero divisor conjecture. In *Geometry and cohomology in group theory (Durham, 1994)*, pages 209–248. Cambridge Univ. Press, Cambridge, 1998.

[35] _____ , W. Lück, and T. Schick. The Ore condition, affiliated operators, and the lamplighter group. Proceedings of the School on "High-dimensional Manifold Topology" in ICTP Trieste, 2001, World Scientific, 315–321, 2003.

[36] W. Lück. Hilbert modules and modules over finite von Neumann algebras and applications to L^2-invariants. *Math. Ann.*, 309(2):247–285, 1997.

[37] _____ Dimension theory of arbitrary modules over finite von Neumann algebras and L^2-Betti numbers. I. Foundations. *J. Reine Angew. Math.*, 495:135–162, 1998.

[38] _____ Dimension theory of arbitrary modules over finite von Neumann algebras and L^2-Betti numbers. II. Applications to Grothendieck groups, L^2-Euler characteristics and Burnside groups. *J. Reine Angew. Math.*, 496:213–236, 1998.

[39] _____ L^2-invariants: theory and applications to geometry and K-theory. Springer-Verlag, Berlin, 2002. Ergebnisse der Mathematik und ihrer Grenzgebiete, 3. Folge, Band 44.

[40] _____ and H. Reich. The Baum-Connes and the Farrell-Jones conjectures in K- and L-theory. eprint KT.0402405, 2004.

[41] P. Malcolmson. Construction of universal matrix localization. Advances in non-commutative ring theory, Pittsburgh, 1981. *Lecture Notes in Mathematics* 951, Springer, 1982.

[42] J. A. Moody. Induction theorems for infinite groups. *Bull. Amer. Math. Soc. (N.S.)*, 17(1):113–116, 1987.

[43] _____ Brauer induction for G_0 of certain infinite groups. *J. Algebra*, 122(1):1–14, 1989.

[44] F. Murray and J. von Neumann. On rings of operators. *Annals of Math.*, 37:116–229, 1936.

[45] A. Neeman and A. Ranicki. Noncommutative localization and chain complexes I. e-print AT.0109118 Geometry and Topology 8:1385–1421, 2004.

[46] D. S. Passman. *Infinite crossed products*. Academic Press Inc., Boston, MA, 1989.

[47] E. K. Pedersen. Controlled algebraic *K*-theory, a survey. In *Geometry and topology: Aarhus (1998)*, volume 258 of *Contemp. Math.*, pages 351–368. Amer. Math. Soc., Providence, RI, 2000.

[48] ——— and C. A. Weibel. *K*-theory homology of spaces. In *Algebraic topology (Arcata, CA, 1986)*, pages 346–361. Springer-Verlag, Berlin, 1989.

[49] F. Quinn. Ends of maps. I. *Ann. of Math. (2)*, 110(2):275–331, 1979.

[50] ——— Ends of maps. II. *Invent. Math.*, 68(3):353–424, 1982.

[51] ——— Applications of topology with control. In *Proceedings of the International Congress of Mathematicians, Vol. 1, 2 (Berkeley, Calif., 1986)*, pages 598–606, Providence, RI, 1987. Amer. Math. Soc.

[52] A. Ranicki. Noncommutative localization in topology. In this volume, pp. 81–102. e-print AT.0303046, 2003.

[53] H. Reich. Group von Neumann algebras and related algebras. Dissertation Universität Göttingen, http://www.math.uni-muenster.de/u/lueck/publ/reich/reich.dvi, 1999.

[54] ——— On the *K*- and *L*-theory of the algebra of operators affiliated to a finite von Neumann algebra. *K-Theory*, 24(4):303–326, 2001.

[55] T. Schick. Integrality of L^2-Betti numbers. *Math. Ann.*, 317(4):727–750, 2000.

[56] ——— Erratum for "Integrality of L^2-Betti numbers". *Math. Ann.*, 322(2):421–422, 2002.

[57] A. H. Schofield. *Representation of rings over skew fields*. Cambridge University Press, Cambridge, 1985.

[58] B. Stenström. *Rings of quotients*. Springer-Verlag, New York, 1975. Die Grundlehren der Mathematischen Wissenschaften, Band 217, An introduction to methods of ring theory.

[59] R. G. Swan. Induced representations and projective modules. *Ann. of Math. (2)*, 71:552–578, 1960.

[60] R. W. Thomason and T. Trobaugh. Higher algebraic K-theory of schemes and of derived categories. In *The Grothendieck Festschrift, Vol. III*, volume 88 of *Progr. Math.*, pages 247–435. Birkhäuser Boston, Boston, MA, 1990.

[61] J. von Neumann. *Continuous geometry*, volume 27 of *Princeton Mathematical Series*. Princeton University Press, Princeton, New Jersey, 1960.

[62] F. Waldhausen. Algebraic K-theory of generalized free products. I, II. *Ann. of Math. (2)*, 108(1):135–256, 1978.

[63] C. T. C. Wall. Rational Euler characteristics. *Proc. Cambridge Philos. Soc.*, 57:182–184, 1961.

[64] C. A. Weibel. *An introduction to homological algebra*. Cambridge University Press, Cambridge, 1994.

[65] ―――― and D. Yao. Localization for the K-theory of noncommutative rings. In *Algebraic K-theory, commutative algebra, and algebraic geometry (Santa Margherita Ligure, 1989)*, volume 126 of *Contemp. Math.*, pages 219–230. Amer. Math. Soc., Providence, RI, 1992.

Fachbereich Mathematik
Universität Münster
Einsteinstr. 62
48149 Münster
GERMANY

e-mail: reichh@math.uni-muenster

Invariants of boundary link cobordism II. The Blanchfield-Duval form.

Desmond Sheiham

Abstract

We use the Blanchfield-Duval form to define complete invariants for the cobordism group $C_{2q-1}(F_\mu)$ of $(2q-1)$-dimensional μ-component boundary links (for $q \geq 2$).

The author solved the same problem in earlier work via Seifert forms. Although Seifert forms are convenient in explicit computations, the Blanchfield-Duval form is more intrinsic and appears naturally in homology surgery theory.

The free cover of the complement of a link is constructed by pasting together infinitely many copies of the complement of a μ-component Seifert surface. We prove that the algebraic analogue of this construction, a functor denoted B, identifies the author's earlier invariants with those defined here. We show that B is equivalent to a universal localization of categories and describe the structure of the modules sent to zero. Taking coefficients in a semi-simple Artinian ring, we deduce that the Witt group of Seifert forms is isomorphic to the Witt group of Blanchfield-Duval forms.

1   Introduction

This paper is the second in a series on cobordism (=concordance) groups of a natural class of high-dimensional links. Chapter 1 of the first work [51] discusses background to the problem at greater length but we summarize here some of the key ideas.

1.1   Background

A knot is an embedding of spheres[1] $S^n \subset S^{n+2}$. The following are generalizations:

- A μ-component *link* is an embedding of μ disjoint spheres

$$L = \overbrace{S^n \sqcup \cdots \sqcup S^n}^{\mu} \subset S^{n+2}.$$

- A *boundary link* is a link whose components bound disjoint $(n+1)$-
  manifolds. The union of these $(n+1)$-manifolds is called a *Seifert
  surface*.

- An F_μ-*link* is a pair (L, θ) where L is a link and θ is a homomorphism from the fundamental group $\pi_1(X)$ of the link complement
  $X = S^{n+2} \backslash L$ onto the free group F_μ on μ (distinguished) generators such that some meridian of the ith link component is sent to the
  ith generator.

Not every link is a boundary link; a link L can be refined to an F_μ-link (L, θ)
if and only if L is a boundary link.

Let us call a homomorphism $\theta : \pi_1(X) \to F_\mu$ permissible if it sends some
meridian of the ith link component to the ith generator. There may be
many permissible homomorphisms for a given boundary link but if θ and
θ' are permissible then $\theta' = \alpha\theta$ where α is some "generator conjugating"
automorphism of F_μ (Cappell and Shaneson [6], Ko [27, p660-663]). Homomorphisms $\pi_1(X) \to F_\mu$ correspond to homotopy classes of maps from
the link complement X to a wedge of μ circles and the permissible homotopy classes correspond, by the Pontrjagin-Thom construction, to cobordism
classes of Seifert surfaces (rel L).

Every knot is a (1-component) boundary link and admits precisely one
permissible homomorphism, namely the abelianization

$$\theta : \pi_1(X) \to \pi_1(X)^{\mathrm{ab}} \cong \mathbb{Z}.$$

Among the three generalizations above it is the theory of F_μ-links which
seems to bear the closest resemblance to knot theory.

---

[1]Manifolds are assumed oriented and embeddings are assumed locally flat. One may
work in the category of smooth, *PL* or topological manifolds according to taste, with the
understanding that S^n is permitted exotic structures if one selects the smooth category.

Although one does not hope for a complete classification of knots or F_μ-links in higher dimensions much is known about their classification up to the equivalence relation known as cobordism (or concordance). Two links L^0 and L^1 are called *cobordant* if there is an embedding

$$LI = (S^n \sqcup \cdots \sqcup S^n) \times [0,1] \subset S^{n+2} \times [0,1]$$

which joins $L^0 \subset S^{n+2} \times \{0\}$ to $L^1 \subset S^{n+2} \times \{1\}$. One requires[2] that $(S^n \sqcup \cdots \sqcup S^n) \times \{i\} \subset S^{n+2} \times \{i\}$ for $i = 0$ and $i = 1$ but no such requirement is made when $0 < i < 1$. Boundary links are said to be *boundary cobordant* if there is a cobordism LI whose components bound disjoint $(n+2)$-manifolds in $S^{n+2} \times [0,1]$. Two F_μ-links (L^0, θ^0) and (L^1, θ^1) are called cobordant if there is a pair

$$(LI \ , \ \Theta : \pi_1(S^{n+2} \times [0,1] \backslash LI) \to F_\mu)$$

such that the restrictions of Θ to $\pi_1(X^0)$ and $\pi_1(X^1)$ coincide with θ^0 and θ^1 (up to inner automorphism).

The cobordism classes of knots form an abelian group $C_n(F_1)$ under (ambient) connected sum but this operation does not extend to links in any obvious way. If one attempts to add links L^0 and L^1 there are many inequivalent choices of connecting arc from the ith component of L^0 to the ith component of L^1.

However when $n \geq 2$ connected sum $[L_1, \theta_1] + [L_2, \theta_2]$ of *cobordism classes* of F_μ-links is well-defined; one can remove the ambiguity in the choice of paths by assuming, perhaps after some surgery, that θ_1 and θ_2 are isomorphisms. The set $C_n(F_\mu)$ of cobordism classes of F_μ-links is therefore an abelian group.

When n is even, $C_n(F_\mu)$ is in fact the trivial group [24, 6, 27, 38]; we sketch a proof in [51, Ch1§4.1]. On the other hand J.Levine obtained a complete system of invariants for odd-dimensional knot cobordism groups $C_{2q-1}(F_1)$ for $q \geq 2$ [31] and showed that each is isomorphic to a countable direct sum

$$C_{2q-1}(F_1) \cong \mathbb{Z}^{\oplus\infty} \oplus \left(\frac{\mathbb{Z}}{2\mathbb{Z}}\right)^{\oplus\infty} \oplus \left(\frac{\mathbb{Z}}{4\mathbb{Z}}\right)^{\oplus\infty}. \tag{1}$$

---

[2] LI is also required to meet $S^{n+2} \times \{0\}$ and $S^{n+2} \times \{1\}$ transversely.

The computation of $C_1(F_1)$ remains open. In [51] the author obtained a complete system of invariants for odd-dimensional F_μ-link cobordism groups $C_{2q-1}(F_\mu)$, $q \geq 2$ (including some secondary invariants defined only if certain primary invariants vanish) and found that

$$C_{2q-1}(F_\mu) \cong \mathbb{Z}^{\oplus\infty} \oplus \left(\frac{\mathbb{Z}}{2\mathbb{Z}}\right)^{\oplus\infty} \oplus \left(\frac{\mathbb{Z}}{4\mathbb{Z}}\right)^{\oplus\infty} \oplus \left(\frac{\mathbb{Z}}{8\mathbb{Z}}\right)^{\oplus\infty} \tag{2}$$

for all $q \geq 2$ and all $\mu \geq 2$.

Both (1) and (2) were deduced from a purely algebraic reformulation of F_μ-link cobordism associated to Seifert surfaces: It was proved by Levine [32] in the knot theory case $\mu = 1$ and by Ko [27] and Mio [38] independently in the general case that $C_{2q-1}(F_\mu)$ is isomorphic to the "Witt group of Seifert forms". In the notation of the present paper, which we explain more carefully in Sections 2.4, 4.1 and 4.3,

$$C_{2q-1}(F_\mu) \cong W^{(-1)^q}(\mathcal{S}ei(\mathbb{Z})) \qquad (q \geq 3). \tag{3}$$

The symbol $\mathcal{S}ei(\mathbb{Z})$ denotes[3] a category of "Seifert modules" designed to contain the homology modules of Seifert surfaces among the objects (see Notation 4.1). In the case $\mu = 1$ an object in $\mathcal{S}ei(\mathbb{Z})$ is a finitely generated free $\mathbb{Z}$-module V together with an endomorphism $V \to V$ which carries information about how a Seifert surface is embedded. If $\mu > 1$ then the definition of Seifert module also includes a direct sum decomposition $V = V_1 \oplus \cdots \oplus V_\mu$ which reflects the connected components of a Seifert surface.

The intersection form in a Seifert surface is an isomorphism $\phi : V \to V^*$ in $\mathcal{S}ei(\mathbb{Z})$ which satisfies $\phi^* = (-1)^q \phi$. Such $(-1)^q$-hermitian forms are the generators of the Witt group $W^{(-1)^q}(\mathcal{S}ei(\mathbb{Z}))$. The relations say that certain "metabolic forms" are identified with zero; see Definitions 2.22 and 2.23 below.

Although Seifert surface methods are convenient in explicit computations, it is preferable to define F_μ-link invariants without making a choice of Seifert surface. In the present paper we focus instead on the covering space $\overline{X} \to X$ of a link complement determined by the homomorphism $\theta : \pi_1(X) \twoheadrightarrow F_\mu$. This approach sits more naturally in homology surgery theory and is more amenable to generalization from boundary links to arbitrary links or other manifold embeddings.

We take as starting point the identification

$$C_{2q-1}(F_\mu) \cong W^{(-1)^{q+1}}(\mathcal{F}lk(\mathbb{Z})) \qquad (q \geq 3) \tag{4}$$

---

[3]The category $\mathcal{S}ei(\mathbb{Z})$ was denoted $(P_\mu\text{–}\mathbb{Z})$-Proj in [51].

where $\mathcal{F}lk(\mathbb{Z})$ is a category designed to contain homology modules of the cover $\overline{X}$ (see Definition 2.1 and Notation 2.2). The objects in $\mathcal{F}lk(\mathbb{Z})$ are certain modules over the group ring $\mathbb{Z}[F_\mu]$ of the free group; they are called F_μ-link modules in the present paper although they are more commonly known as link modules.

The F_μ-equivariant Poincaré duality in $\overline{X}$ leads to a $(-1)^{q+1}$-hermitian form ϕ in the category $\mathcal{F}lk(\mathbb{Z})$. This is the Blanchfield-Duval form of the title, originally introduced by Blanchfield [3] in the knot theory case $\mu = 1$. The identity (4) was proved by Kearton for $\mu = 1$ [23, 22] and by Duval [14] for $\mu \geq 2$. Cappell and Shaneson earlier identified the cobordism group $C_n(F_\mu)$ with a Γ-group, an obstruction group in their homology surgery theory [5, 6]. The identification of this Γ-group with the Witt group $W^{(-1)^{q+1}}(\mathcal{F}lk(\mathbb{Z}))$ was due to Pardon [39, 40], Ranicki [42, §7.9] and Smith [52] for $\mu = 1$ and to Duval [14] for $\mu \geq 2$. More general results of Vogel [54, 55] on homology surgery and universal localization are stated elsewhere in this volume [45, §1.4]. An outline of their application to $C_n(F_\mu)$ is given in [51, Ch1,§4.4,5.3].

1.2 Overview

Universal localization plays two roles in this paper. Firstly the "augmentation localization" of the group ring $\mathbb{Z}[F_\mu]$ of the free group appears in the definition of the Blanchfield-Duval form, our main object of study. Secondly, we prove that the category $\mathcal{F}lk(\mathbb{Z})$ of F_μ-link modules is (equivalent to) a universal localization of the category $\mathcal{S}ei(\mathbb{Z})$ of Seifert modules.

Our first aim is to use (4) to distinguish the elements of $C_{2q-1}(F_\mu)$. We define complete invariants (and secondary invariants if certain primary invariants vanish) by analyzing the Witt groups $W^{(-1)^{q+1}}(\mathcal{F}lk(\mathbb{Q}))$. We proceed in three steps, explained in more detail in Section 3, which run parallel to steps 2, 3 and 4 in chapter 2 of [51]:

1. Obtain a direct sum decomposition of $W^{(-1)^{q+1}}(\mathcal{F}lk(\mathbb{Q}))$ by "devissage". One must prove that $\mathcal{F}lk(\mathbb{Q})$ is an abelian category in which each module has a finite composition series.

2. Use hermitian Morita equivalence to show that each summand of the group $W^{(-1)^{q+1}}(\mathcal{F}lk(\mathbb{Q}))$ is isomorphic to some group $W^1(E)$ where E is a division ring of finite dimension over $\mathbb{Q}$.

3. Recall from the literature invariants of each $W^1(E)$.

In the knot theory case $\mu = 1$ there is one summand of $W^{(-1)^{q+1}}(\mathcal{F}lk(\mathbb{Q}))$ for each maximal ideal $(p) \in \mathbb{Q}[z, z^{-1}]$ which is invariant under the involution $z \mapsto z^{-1}$. The generator p is often called an Alexander polynomial. The division ring E coincides with the quotient field $\mathbb{Q}[z, z^{-1}]/(p)$ and $W^1(E)$ is the Witt group of hermitian forms over E (compare Milnor [37]).

The following theorem and corollary are restated and proved in Section 3; see Theorem 3.2 and Corollary 3.3.

Theorem 1.1. *The invariants (and secondary invariants) defined in Section 3 are sufficient to distinguish the elements of the Witt groups $W^\pm(\mathcal{F}lk(\mathbb{Q}))$ of Blanchfield-Duval forms with coefficients in $\mathbb{Q}$.*

Corollary 1.2. *Let $q > 1$ and suppose ϕ^0 and ϕ^1 are the Blanchfield-Duval forms for the $(2q-1)$-dimensional F_μ-links (L^0, θ^0) and (L^1, θ^1) respectively. These two F_μ-links are cobordant if and only if all the invariants (and possible secondary invariants) of*

$$[\mathbb{Q} \otimes_{\mathbb{Z}} (\phi^0 \oplus -\phi^1)] \in W^{(-1)^{q+1}}(\mathcal{F}lk(\mathbb{Q}))$$

defined in Section 3 are trivial.

Corollary 1.2 follows from (4) and the fact that the canonical map

$$W^{(-1)^{q+1}}(\mathcal{F}lk(\mathbb{Z})) \to W^{(-1)^{q+1}}(\mathcal{F}lk(\mathbb{Q}))$$

is an injection, which we deduce from Theorem 1.3 at the end of Section 3.1. Corollary 1.2 is also a consequence of Theorem 1.4 and Theorem B of [51].

Our second aim is to understand the algebraic relationship between the Seifert forms and the Blanchfield-Duval form of an F_μ-link and prove that the cobordism invariants defined in [51] using Seifert forms are equivalent to those defined in Section 3 via the Blanchfield-Duval form. Example 4.3 gives a sample calculation of the Seifert form invariants in [51].

In the knot theory case $\mu = 1$, the relationship between Seifert and Blanchfield forms has been investigated extensively by Kearton [23], Levine [33, §14], Farber [15, §7.1] and Ranicki ([43, ch32],[44]). For $\mu \geq 1$ K.H.Ko [28] used geometric arguments to obtain a formula for Cappell and Shaneson's homology surgery obstruction in terms of the Seifert form. A formula for the Blanchfield-Duval form in terms of the Seifert form, again based on geometric arguments, can also be found in Cochran and Orr [7, Thm4.2] in the slightly more general context of "homology boundary links".

M.Farber related Seifert and Blanchfield-Duval forms of F_μ-links in a purely algebraic way [16, 17]. Although the present paper is logically independent of his work, we take up a number of his ideas in Sections 4 and 5, providing a systematic treatment in the language of hermitian categories. Whereas Farber takes coefficients in a field or in $\mathbb{Z}$, in these sections we allow the coefficients to lie in an arbitrary associative ring A.

The first step is to show that an F_μ-link module admits a canonical Seifert module structure (cf [16, p193]). An F_μ-link module $M \in \mathcal{F}\mathrm{lk}(A)$ is not in general finitely generated (or projective) as an A-module so we introduce a larger category $\mathcal{S}\mathrm{ei}_\infty(A)$ which contains $\mathcal{S}\mathrm{ei}(A)$ as a full subcategory (see Notation 4.1). We obtain a "forgetful" functor

$$U : \mathcal{F}\mathrm{lk}(A) \to \mathcal{S}\mathrm{ei}_\infty(A).$$

For example, in the case $\mu = 1$ of knot theory, an object in $\mathcal{F}\mathrm{lk}(A)$ is a module M over the ring $A[z, z^{-1}]$ of Laurent polynomials with a presentation

$$0 \to (A[z, z^{-1}])^m \xrightarrow{\sigma} (A[z, z^{-1}])^m \to M \to 0$$

such that $1 - z : M \to M$ is an isomorphism. The Seifert module $U(M)$ is the A-module M together with the endomorphism $(1 - z)^{-1}$.

If $A = k$ is a field, Farber defined, for each $M \in \mathcal{F}\mathrm{lk}(k)$, the "minimal lattice" [16, p194-199] of M, a Seifert submodule of $U(M)$ which is of finite k-dimension. We prefer to work directly with $U(M)$ which is defined regardless of the coefficients and avoids technicalities of Farber's definition. His minimal lattice becomes isomorphic to $U(M)$ after one performs a universal localization of categories which we describe a few paragraphs below.

Given a Seifert surface for an F_μ-link one can construct the free cover by cutting the link complement along the Seifert surface and gluing together infinitely many copies of the resulting manifold in the pattern of the Cayley graph of F_μ. Figure 1 illustrates the geometric construction in the case of a 2-component link.

The algebraic analogue of this geometric construction is a functor

$$B : \mathcal{S}\mathrm{ei}(A) \to \mathcal{F}\mathrm{lk}(A)$$

from Seifert modules to F_μ-link modules (see Definition 5.1). Since U takes values in the larger category $\mathcal{S}\mathrm{ei}_\infty(A)$ we expand the domain of B to $\mathcal{S}\mathrm{ei}_\infty(A)$, by necessity replacing $\mathcal{F}\mathrm{lk}(A)$ by a larger category $\mathcal{F}\mathrm{lk}_\infty(A)$. This process of enlargement stops here for there are functors

$$U : \mathcal{F}\mathrm{lk}_\infty(A) \to \mathcal{S}\mathrm{ei}_\infty(A)$$
$$B : \mathcal{S}\mathrm{ei}_\infty(A) \to \mathcal{F}\mathrm{lk}_\infty(A).$$

Figure 1

We show in Section 5.2 that B is left adjoint to U. Roughly speaking, this means that $B(V)$ is the "free" F_μ-link module generated by the Seifert module V (with respect to the functor U). In other words, B is universal (up to equivalence) among functors from Seifert modules to F_μ-link modules.

Returning our attention to the subcategories $\mathcal{S}\mathrm{ei}(A)$ and $\mathcal{F}\mathrm{lk}(A)$ whose definitions involve a "finitely generated projective" condition we show that B is compatible with the notions of duality in $\mathcal{S}\mathrm{ei}(A)$ and $\mathcal{F}\mathrm{lk}(A)$, extending

B to a "duality-preserving functor" between "hermitian categories"

$$(B, \Phi, -1) : \mathcal{S}ei(A) \rightarrow \mathcal{F}lk(A). \tag{5}$$

(see definitions 2.13 and 2.24 and proposition 5.4). The following theorem concerns the induced homomorphism of Witt groups:

$$B : W^{\pm}(\mathcal{S}ei(A)) \rightarrow W^{\mp}(\mathcal{F}lk(A)). \tag{6}$$

Recall that by Wedderburn's Theorem, a ring A is semi-simple and Artinian if and only if it is a product of matrix rings over division rings.

Theorem 1.3. *If A is a semi-simple Artinian ring then (6) is an isomorphism.*

The map (6) will be considered for more general rings A in subsequent work (joint with A.Ranicki) [46]. It follows from the isomorphisms (3) and (4) above that (6) is an isomorphism when $A = \mathbb{Z}$.

Theorem 1.4. *The duality-preserving functor $(B, \Phi, -1)$ identifies the Seifert form invariants of [51] with the Blanchfield-Duval form invariants of Section 3.*

Theorem 6.5 below is a more precise statement of Theorem 1.4. The invariants of [51] are outlined in Section 4.3. Theorem 1.3 is proved in two stages. The first stage is to establish that, for any ring A, there is an equivalence between $(B, \Phi, -1)$ and a certain universal localization of hermitian categories. Taking up Farber's terminology we call a Seifert module $V \in \mathcal{S}ei_{\infty}(A)$ primitive if $B(V) \cong 0$. We denote by $\mathcal{P}rim_{\infty}(A)$ the category of primitive modules. One may write

$$\mathcal{P}rim_{\infty}(A) = \mathrm{Ker}(\, B : \mathcal{S}ei_{\infty}(A) \rightarrow \mathcal{F}lk_{\infty}(A)\,).$$

The category quotient

$$F : \mathcal{S}ei_{\infty}(A) \rightarrow \mathcal{S}ei_{\infty}(A)/\mathcal{P}rim_{\infty}(A) \tag{7}$$

is universal among functors which make invertible morphisms whose kernel and cokernel are primitive. In particular, primitive modules in $\mathcal{S}ei_{\infty}(A)$ are made isomorphic to 0 in $\mathcal{S}ei_{\infty}(A)/\mathcal{P}rim_{\infty}(A)$.

Since $B : \mathcal{S}ei_{\infty}(A) \rightarrow \mathcal{F}lk_{\infty}(A)$ is left adjoint to U it follows that B exhibits the same universal property as F although only "up to natural isomorphism" (Proposition 5.14). We conclude that B is equivalent to F

and, with a little extra work, establish that $(B, \Phi, -1) : \mathcal{S}ei(A) \to \mathcal{F}lk(A)$ is equivalent to a universal localization of hermitian categories (Theorem 5.17 and Proposition 5.22).

In the knot theory case $\mu = 1$ the category $\mathcal{S}ei_\infty(A)$ coincides with the category of (left) modules over the polynomial ring $A[s]$ in a central indeterminate s. Setting $t = s(1 - s)$, the functor (7) is the central localization $A[s, t^{-1}] \otimes_{A[s]} \_$ from the category of $A[s]$-modules to the category of $A[s, t^{-1}]$-modules; see Farber [15, Thm2.6] and Ranicki [44].

Pere Ara recently gave an independent proof [2, Thm 6.2] using Farber's minimal lattice that if $A = k$ is a field then $\mathcal{F}lk(k)$ is equivalent to a localization of $\mathcal{S}ei(k)$ by a category of primitive modules (for all $\mu \geq 1$)[4].

The second stage in the proof of Theorem 1.3 involves the analysis of primitive modules. A Seifert module V is called "trivially primitive" if the endomorphism with which it is endowed is either zero or the identity. In Proposition 5.28 we show that every primitive module $V \in \mathcal{P}rim_\infty(A)$ is composed of (possibly infinitely many) trivially primitive modules. Restricting attention to $\mathcal{S}ei(A)$ we show that if A is semi-simple and Artinian then every primitive module is composed of finitely many trivially primitive modules in $\mathcal{S}ei(A)$ (cf Farber [17, §3,§7.10]). The proof of Theorem 1.3 is completed in Section 6.1 by establishing that the Witt group of the subcategory $\mathcal{P}rim(A) \subset \mathcal{S}ei(A)$ of primitive modules is trivial. See Proposition 6.3 and part 2. of Lemma 6.4.

The definitions of Blanchfield-Duval form invariants in Section 3 parallel the author's Seifert form invariants in [51, Ch2]. The three steps outlined above to analyze $W^\pm(\mathcal{F}lk(\mathbb{Q}))$ were applied to $W^\pm(\mathcal{S}ei(\mathbb{Q}))$ in [51]. Theorem 1.4 is proved in Section 6.2 by checking that the duality-preserving functor $(B, \Phi, -1)$ respects each of these three steps.

Let us summarize the contents of this paper. Section 2 discusses F_μ-link modules over an arbitrary ring A and uses universal localization (cf Vogel [54, 55] and Duval [14]) to describe hermitian structure in the category $\mathcal{F}lk(A)$. We define the Witt groups $W^\pm(\mathcal{F}lk(A))$ of Blanchfield-Duval forms.

In Section 3 we set $A = \mathbb{Q}$ and define invariants of $W^\pm(\mathcal{F}lk(\mathbb{Q}))$, obtain-

---

[4]His context differs slightly in that the free algebra $k\langle X \rangle$ on a set $X = \{x_1, \cdots, x_\mu\}$ takes the place of the group ring $k[F_\mu]$ in the present paper; the category denoted $\mathcal{Z}$ in [2] plays the role of $\mathcal{F}lk(k)$. Consequently, there is only one kind of "trivially primitive" module (denoted M_0 in [2]) as compared with the two kinds in [17] and Section 5.5 below. Ara also related the modules in $\mathcal{Z}$ to modules over the Leavitt algebra L. By definition, L is the universal localization of $k\langle X \rangle$ which makes invertible the map $(x_1 \cdots x_\mu) : k\langle X \rangle^\mu \to k\langle X \rangle$ (compare the Sato condition, Lemma 2.3 below). The category $\mathcal{Z}$ turns out to be equivalent to the category of finitely presented L-modules of finite length.

ing intrinsic cobordism invariants for F_μ-links. We discuss each of the steps 1-3. listed above, reformulating and proving Theorem 1.1 and Corollary 1.2.

In Section 4 we define Seifert modules and Seifert forms with coefficients in an arbitrary ring A. We treat a worked example of the invariants defined in [51] and we define a forgetful functor $U : \mathcal{F}lk_\infty(A) \to \mathcal{S}ei_\infty(A)$.

Section 5 begins to study the functor $B : \mathcal{S}ei_\infty(A) \to \mathcal{F}lk_\infty(A)$ from Seifert modules to F_μ-link modules. We prove that B is left adjoint to $U : \mathcal{F}lk_\infty(A) \to \mathcal{S}ei_\infty(A)$ and show that B factors through a category equivalence $\mathcal{S}ei_\infty(A)/\mathcal{P}rim_\infty(A) \to \mathcal{F}lk_\infty(A)$. We describe the structure of the primitive modules – those which are sent to zero by B – and outline a construction of the localization $\mathcal{S}ei_\infty(A) \to \mathcal{S}ei_\infty(A)/\mathcal{P}rim_\infty(A)$. We construct a duality-preserving functor $(B, \Phi, -1) : \mathcal{S}ei(A) \to \mathcal{F}lk(A)$ which is natural in A and factors through an equivalence $\mathcal{S}ei(A)/\mathcal{P}rim_\infty(A) \to \mathcal{F}lk(A)$ of hermitian categories. If A is a semi-simple Artinian ring we give a simplified description of the primitive modules and the universal localization of hermitian categories.

Section 6 contains a proof of Theorem 1.3 and a reformulation and proof of Theorem 1.4.

Acknowledgments: The invariants in Section 3 came into being during my thesis work at the University of Edinburgh under the guidance of Andrew Ranicki. I am also indebted to Andrew for several more recent conversations and e-mails and for encouragement to complete this paper. I am grateful to John Baez and James Dolan for helpful discussions in category theory, and to Pere Ara and to the referee for their comments and corrections.

I thank the London Mathematical Society for financial support to attend the ICMS workshop on "Noncommutative Localization in Algebra and Topology" in April 2002 and the Edinburgh Mathematical Society who financed my visit to Edinburgh in August 2002.

2   The Blanchfield-Duval form

2.1 F_μ-link Modules

Let A be an associative ring with 1. Modules will be left modules except where otherwise stated. Let $A[F_\mu]$ denote the group ring of the free group F_μ; an element of $A[F_\mu]$ is a formal sum of elements of F_μ, with coefficients in A. Note that elements of the group F_μ commute with elements of A and $A[F_\mu] \cong A \otimes_\mathbb{Z} \mathbb{Z}[F_\mu]$.

The symbols ϵ and j will be used for three slightly distinct purposes but the meaning will be clear from the context. Firstly, j denotes the inclusion of A in $A[F_\mu]$. Secondly j denotes the functor $V \mapsto A[F_\mu] \otimes_A V$ from the category of A-modules to the category of $A[F_\mu]$-modules. For brevity we write $V[F_\mu]$ in place of $A[F_\mu] \otimes_A V$. Thirdly, we use j to denote the inclusion of a module V in $V[F_\mu]$ given by

$$V \cong A \otimes_A V \xrightarrow{j \otimes 1} A[F_\mu] \otimes_A V = V[F_\mu].$$

In the opposite direction $\epsilon : A[F_\mu] \to A$ denotes the ring morphism which sends every element of F_μ to $1 \in A$ and is the identity on A. We also write ϵ for the functor $A \otimes_{A[F_\mu]} \underline{}$ from the category of $A[F_\mu]$-modules to the category of A-modules. Thirdly, $\epsilon : V[F_\mu] \to V$ denotes the morphism

$$V[F_\mu] = A[F_\mu] \otimes_A V \xrightarrow{\epsilon \otimes 1} A \otimes_A V \cong V.$$

Note that the composite ϵj of ring morphisms is the identity id_A and the composite ϵj of module morphisms is the identity on V. The composite ϵj of functors is naturally isomorphic to the identity functor on A-modules and we sometimes suppress the natural isomorphism identifying $A \otimes_{A[F_\mu]} (A[F_\mu] \otimes_A V)$ with V.

Definition 2.1. An F_μ-*link module* is an $A[F_\mu]$-module M which lies in an exact sequence:

$$0 \to V[F_\mu] \xrightarrow{\sigma} V[F_\mu] \to M \to 0 \tag{8}$$

such that V is an A-module and $\epsilon(\sigma) : V \to V$ is an isomorphism.

As we remarked in the introduction, the examples of F_μ-link modules in the literature are more often called "link modules". Note that if V is a finitely generated A-module then the F_μ-link module M is finitely generated as an $A[F_\mu]$-module but usually not as an A-module (see Lemma 2.3 below). It will be helpful to make the following observation about the definition of F_μ-link modules: The condition that $\epsilon(\sigma)$ is an isomorphism implies that σ is an injection (see Lemma 2.8 below).

Notation 2.2. Let $\mathcal{F}\mathrm{lk}_\infty(A)$ denote the category of F_μ-link modules and $A[F_\mu]$-module homomorphisms. Thus $\mathcal{F}\mathrm{lk}_\infty(A)$ is a full subcategory of the category of $A[F_\mu]$-modules.

Let $\mathcal{F}\mathrm{lk}(A) \subset \mathcal{F}\mathrm{lk}_\infty(A)$ denote the category of modules with a presentation (8) such that V is a finitely generated projective A-module and $\epsilon(\sigma)$ is an isomorphism. The morphisms in $\mathcal{F}\mathrm{lk}(A)$ are, as usual, the $A[F_\mu]$-module morphisms so $\mathcal{F}\mathrm{lk}(A)$ is a full subcategory of $\mathcal{F}\mathrm{lk}_\infty(A)$.

We show in Lemma 2.9 below that $\mathcal{F}lk_\infty(\_)$ and $\mathcal{F}lk(\_)$ are functorial in A. The following lemma gives an alternative characterization of F_μ-link modules. Let $z_1, \cdots, z_\mu$ denote generators for F_μ.

Lemma 2.3. *(Sato [47]) Suppose M is an $A[F_\mu]$-module which has a presentation*

$$0 \to V[F_\mu] \xrightarrow{\sigma} V'[F_\mu] \to M \to 0. \tag{9}$$

for some A-modules V and V'. The augmentation $\epsilon(\sigma) : V \to V'$ is an isomorphism if and only if the A-module homomorphism

$$\gamma : M^{\oplus\mu} \to M$$
$$(m_1, \cdots, m_\mu) \mapsto \sum_{i=1}^{\mu}(1 - z_i)m_i \tag{10}$$

is an isomorphism.

Proof. There is an exact sequence

$$0 \to (A[F_\mu])^{\oplus\mu} \xrightarrow{\gamma} A[F_\mu] \xrightarrow{\epsilon} A \to 0 \tag{11}$$

where $\gamma(l_1, \cdots, l_\mu) = \sum_{i=1}^{\mu}(1 - z_i)l_i$ for all $l_1, \cdots, l_\mu \in A[F_\mu]$. Now (11) is split (by j) when regarded as a sequence of right A-modules so the functors $\_ \otimes_A V$ and $\_ \otimes_A V'$ lead to a commutative diagram

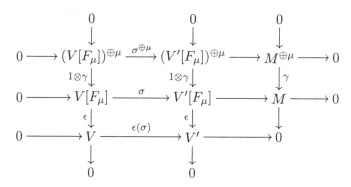

in which the first two rows and the first two columns are exact. A standard diagram chase (e.g. [36, p49]) shows that the third row is exact if and only if the third column is exact and the Lemma follows. ∎

We discuss next completions of $A[F_\mu]$-modules, which we shall need both to refine Definition 2.1 (in Lemma 2.8) and later to study the universal augmentation localization of $A[F_\mu]$ (see Lemma 2.15).

Let $I = \text{Ker}(\epsilon : A[F_\mu] \to A)$. If N is an $A[F_\mu]$-module one defines

$$\widehat{N} = \varprojlim_n \frac{N}{I^n N}.$$

An $A[F_\mu]$-module morphism $N \to N'$ maps $I^n N$ to $I^n N'$ for each n and therefore induces a homomorphism $\widehat{N} \to \widehat{N'}$.

Caveat 2.4. The natural isomorphisms $(A[F_\mu]/I^n) \otimes_{A[F_\mu]} N \to N/(I^n N)$ induce a map $\widehat{A[F_\mu]} \otimes N \to \widehat{N}$ but the latter is not in general an isomorphism.

In the examples with which we are most concerned, $N = V[F_\mu] = A[F_\mu] \otimes_A V$ for some A-module V. One can describe $\widehat{V[F_\mu]}$ as a module of power series as follows. Let $X = \{x_1, \cdots, x_\mu\}$ and let $\mathbb{Z}\langle X \rangle$ denote the free ring generated by X (in other words the ring of "polynomials" in non-commuting indeterminates $x_1, \cdots, x_\mu$). Let $A\langle X \rangle = A \otimes_{\mathbb{Z}} \mathbb{Z}\langle X \rangle$ so that the elements of A are formal sums of words in the alphabet X with coefficients in A. Let $V\langle X \rangle$ be the $A\langle X \rangle$-module $A\langle X \rangle \otimes_A V$ and denote by $X^n V\langle X \rangle \subset V\langle X \rangle$ the submodule whose elements are formal sums of words of length at least n with coefficients in V. We may now define the X-adic completion

$$V\langle\langle X \rangle\rangle = \varprojlim_n \frac{V\langle X \rangle}{X^n V\langle X \rangle}.$$

in which an element is a formal power series in non-commuting indeterminates $x_1, \cdots, x_\mu$ with coefficients in V.

Lemma 2.5. *There is a natural isomorphism $V\langle\langle X \rangle\rangle \cong \widehat{V[F_\mu]}$.*

Proof. The required isomorphism is induced by the isomorphisms

$$\frac{V[F_\mu]}{I^n V[F_\mu]} \cong \frac{V\langle X \rangle}{X^n V\langle X \rangle}$$

$$z_i \mapsto 1 + x_i$$

$$z_i^{-1} \mapsto 1 - x_i + x_i^2 \cdots + (-1)^{n-1} x_i^{n-1} \qquad \square$$

Lemma 2.6. *If V is an A-module, the canonical map $V[F_\mu] \to \widehat{V[F_\mu]}$ is an injection.*

Proof. The argument of Fox [19, Corollary 4.4] implies that $\bigcap\limits_{n=0}^{\infty} I^n V[F_\mu] = 0$ and the Lemma follows. $\qquad \square$

Lemma 2.7. *If V is an A-module and $\tau : V[F_\mu] \to V[F_\mu]$ is an $A[F_\mu]$-module morphism such that $\epsilon(\tau) = 0 : V \to V$ then $\tau(V[F_\mu]) \subset IV[F_\mu]$ and the map $1 + \tau : \widehat{V[F_\mu]} \to \widehat{V[F_\mu]}$ is invertible.*

Proof. The commutative diagram

$$
\begin{array}{ccc}
V[F_\mu] & \xrightarrow{\ \tau\ } & V[F_\mu] \\
{\scriptstyle\epsilon}\downarrow & & \downarrow{\scriptstyle\epsilon} \\
V & \xrightarrow[\epsilon(\tau)]{} & V
\end{array}
$$

implies that if $\epsilon(\tau) = 0$ then

$$\tau(V[F_\mu]) \subseteq \mathrm{Ker}(\epsilon : V[F_\mu] \to V) = \mathrm{Ker}(\epsilon \otimes 1 : A[F_\mu] \otimes_A V \to A \otimes_A V)$$

Since the surjection $\epsilon : A[F_\mu] \to A$ is split by j we obtain

$$\tau(V[F_\mu]) \subseteq I \otimes_A V = I(A[F_\mu] \otimes_A V) = IV[F_\mu].$$

By induction, $\tau^n(V[F_\mu]) \subset I^n V[F_\mu]$ for all n so $1 + \tau : \widehat{V[F_\mu]} \to \widehat{V[F_\mu]}$ has inverse

$$(1 + \tau)^{-1} = 1 - \tau + \tau^2 - \tau^3 + \cdots .$$

$\square$

Lemma 2.8. *If V and V' are A-modules and $\sigma : V[F_\mu] \to V'[F_\mu]$ is an $A[F_\mu]$-module homomorphism such that $\epsilon(\sigma) : V \to V'$ is an isomorphism then σ is an injection and the induced map $\sigma : \widehat{V[F_\mu]} \to \widehat{V'[F_\mu]}$ is an isomorphism.*

Proof. Let $\sigma' = (j\epsilon(\sigma))^{-1}\sigma : V[F_\mu] \to V[F_\mu]$. Now $\epsilon(\sigma') = 1_{V[F_\mu]}$ so we may write $\sigma' = 1 + \tau$ where $\epsilon(\tau) = 0$. Now $\sigma' : \widehat{V[F_\mu]} \to \widehat{V[F_\mu]}$ is an isomorphism by Lemma 2.7 so $\sigma : \widehat{V[F_\mu]} \to \widehat{V'[F_\mu]}$ is an isomorphism. The commutative diagram

$$
\begin{array}{ccc}
V[F_\mu] & \xrightarrow{\ \sigma'\ } & V[F_\mu] \\
\downarrow & & \downarrow \\
\widehat{V[F_\mu]} & \xrightarrow[\sigma']{} & \widehat{V[F_\mu]}
\end{array}
$$

implies that $\sigma' : V[F_\mu] \to V[F_\mu]$ and $\sigma : V[F_\mu] \to V'[F_\mu]$ are injections. $\square$

Lemma 2.9. *A homomorphism $A \to A'$ of rings induces functors*

$$A'[F_\mu] \otimes_{A[F_\mu]} \underline{\ \ } : \mathcal{F}\mathrm{lk}_\infty(A) \to \mathcal{F}\mathrm{lk}_\infty(A') \quad and$$
$$A'[F_\mu] \otimes_{A[F_\mu]} \underline{\ \ } : \mathcal{F}\mathrm{lk}(A) \to \mathcal{F}\mathrm{lk}(A')$$

Proof. If $M \in \mathcal{F}\mathrm{lk}_\infty(A)$ then M has a presentation (8) such that $\epsilon(\sigma)$ is invertible. Applying $A'[F_\mu] \otimes_{A[F_\mu]} \_$ one obtains an exact sequence

$$A'[F_\mu] \otimes_{A[F_\mu]} V[F_\mu] \xrightarrow{1 \otimes \sigma} A'[F_\mu] \otimes_{A[F_\mu]} V[F_\mu] \to A'[F_\mu] \otimes_{A[F_\mu]} M \to 0.$$

The naturality of the identifications $A' \otimes_A (A \otimes_{A[F_\mu]} V[F_\mu]) \cong A' \otimes_A V \cong A' \otimes_{A'[F_\mu]} (A'[F_\mu] \otimes_{A[F_\mu]} V[F_\mu])$ implies that $\epsilon(1 \otimes \sigma) = 1 \otimes \epsilon(\sigma) : A' \otimes_A V \to A' \otimes_A V$. But $1 \otimes \epsilon(\sigma)$ is an isomorphism so $1 \otimes \sigma$ is injective by Lemma 2.8. Thus $A'[F_\mu] \otimes_{A[F_\mu]} M \in \mathcal{F}\mathrm{lk}_\infty(A')$. The argument for $\mathcal{F}\mathrm{lk}(\_)$ is similar, for if V is a finitely generated projective A-module then $A' \otimes_A V$ is a finitely generated projective A'-module. $\qquad\square$

2.2   Hermitian Categories

Recall that an involution on a ring A is a map $A \to A; a \mapsto \overline{a}$ such that $\overline{\overline{a}} = a$, $\overline{a+b} = \overline{a} + \overline{b}$ and $\overline{(ab)} = \overline{b}\overline{a}$ for all $a, b \in A$. If A is a ring with involution then the category $\mathcal{F}\mathrm{lk}(A)$ can be endowed with a notion of duality. But let us begin with simpler examples:

Example 2.10. Suppose A is a ring with an involution. Let A-Proj denote the category of finitely generated projective (left) A-modules. There is a duality functor defined on modules by $V \mapsto V^* = \mathrm{Hom}(V, A)$ for $V \in A$-Proj and on morphisms by $f \mapsto f^* = \_ \circ f : (V')^* \to V^*$ for $f \in \mathrm{Hom}_A(V, V')$. In short,

$$\_^* = \mathrm{Hom}_A(\_, A) : A\text{-Proj} \to A\text{-Proj}.$$

Note that $\mathrm{Hom}_A(V, A)$ is a left A-module with

$$(a.\xi)(x) = \xi(x)\overline{a}$$

for all $a \in A$, $\xi \in \mathrm{Hom}_A(V, A)$ and $x \in V$.

Example 2.11. The category of finite abelian groups admits the duality functor

$$\_^\wedge = \mathrm{Ext}_{\mathbb{Z}}(\_, \mathbb{Z}). \tag{12}$$

A finite abelian group M bears similarity to an F-link module in that there exists a presentation

$$0 \to \mathbb{Z}^n \xrightarrow{\sigma} \mathbb{Z}^n \to M \to 0 \tag{13}$$

and $1 \otimes \sigma : \mathbb{Q} \otimes_{\mathbb{Z}} \mathbb{Z}^n \to \mathbb{Q} \otimes_{\mathbb{Z}} \mathbb{Z}^n$ is an isomorphism. A more explicit description of the duality functor in Example 2.11 is the following:

Lemma 2.12. *There is a natural isomorphism*

$$\operatorname{Ext}_{\mathbb{Z}}(\_\,,\mathbb{Z}) \cong \operatorname{Hom}_{\mathbb{Z}}\left(\_\,,\frac{\mathbb{Q}}{\mathbb{Z}}\right).$$

Proof. Suppose M is a finite abelian group. The short exact sequence

$$0 \to \mathbb{Z} \to \mathbb{Q} \to \mathbb{Q}/\mathbb{Z} \to 0$$

gives rise to a long exact sequence

$$0 = \operatorname{Hom}(M,\mathbb{Q}) \to \operatorname{Hom}(M,\mathbb{Q}/\mathbb{Z}) \to \operatorname{Ext}(M,\mathbb{Z}) \to \operatorname{Ext}(M,\mathbb{Q}) \to \cdots$$

which is natural in M. The presentation (13) implies that

$$\operatorname{Ext}(M,\mathbb{Q}) = \operatorname{Coker}(\operatorname{Hom}(\mathbb{Z}^n,\mathbb{Q}) \xrightarrow{\sigma^*} \operatorname{Hom}(\mathbb{Z}^n,\mathbb{Q})) = 0$$

so $\operatorname{Hom}(M,\mathbb{Q}/\mathbb{Z}) \to \operatorname{Ext}(M,\mathbb{Z})$ is an isomorphism. $\square$

The following general definition subsumes Examples 2.10 and 2.11 and the category $\mathcal{F}\mathrm{lk}(A)$ which we wish to study:

Definition 2.13. A hermitian category is a triple $(\mathcal{C},\,\_^*,i)$ where

- $\mathcal{C}$ is an additive category,

- $\_^* : \mathcal{C} \to \mathcal{C}$ is an (additive) contravariant functor and

- $(i_V)_{V \in \mathcal{C}} : id \to (\_^*)^* = \_^{**}$ is a natural isomorphism such that $i_V^* i_{V^*} = id_{V^*}$ for all $V \in \mathcal{C}$.

The functor $\_^*$ is called a duality functor. We usually abbreviate $(\mathcal{C},\,\_^*,i)$ to $\mathcal{C}$ and identify V with V^{**} via i_V. It follows from Definition 2.13 that if $\mathcal{C}$ is an abelian hermitian category then $\_^*$ is an equivalence of categories and hence respects exact sequences.

If A is a ring with involution then there is a unique involution on $A[F_\mu]$ such that $\bar{g} = g^{-1}$ for each $g \in F_\mu$ and such that the inclusion of A in $A[F_\mu]$ respects the involutions. The category of finitely generated projective $A[F_\mu]$-modules therefore admits a duality functor as in Example 2.10.

Returning to $\mathcal{F}\mathrm{lk}(A)$, duality is defined in a manner analogous to Example 2.11.

Definition 2.14. Define $\_^\wedge = \operatorname{Ext}_{A[F_\mu]}(\_\,,A[F_\mu]) : \mathcal{F}\mathrm{lk}(A) \to \mathcal{F}\mathrm{lk}(A)$.

Note that if M has presentation (8) then $M^\wedge = \mathrm{Ext}_{A[F_\mu]}(M, A[F_\mu])$ has presentation

$$0 \to (V[F_\mu])^* \xrightarrow{\sigma^*} (V[F_\mu])^* \to M^\wedge \to 0$$

where $(V[F_\mu])^* = \mathrm{Hom}_{A[F_\mu]}(V[F_\mu], A[F_\mu]) \cong \mathrm{Hom}_A(V, A)[F_\mu] = V^*[F_\mu]$; see Lemma 2.17 and Remark 2.18 below. The natural isomorphism

$$V[F_\mu] \to (V[F_\mu])^{**}$$

induces a natural isomorphism $i_M : M \to M^{\wedge\wedge}$ with $i_M^\wedge i_{M^\wedge} = \mathrm{id}_{M^\wedge}$.

There is also a more explicit description of $M^\wedge$ which is analogous to Lemma 2.12; see Lemma 2.17 below. Unlike $\mathbb{Z}$, the ring $A[F_\mu]$ is in general highly non-commutative so universal localization will be required.

2.3   Universal Localization

Let R be a ring (associative with unit) and let Σ be a set of (isomorphism classes of) triples $(P_1, P_0, \sigma : P_1 \to P_0)$ where P_0 and P_1 are finitely generated projective R-modules. In our application R will be $A[F_\mu]$ and Σ will contain the endomorphisms $\sigma : V[F_\mu] \to V[F_\mu]$ such that V is finitely generated and projective as an A-module and $\epsilon(\sigma)$ is an automorphism of V.

A homomorphism $\nu : R \to S$ is said to be Σ-*inverting* if

$$1 \otimes \sigma : S \otimes P_1 \to S \otimes P_0$$

is invertible for each morphism $\sigma \in \Sigma$. There exists a universal Σ-inverting homomorphism[5] which, for consistency with the other papers in the volume, will be denoted $i_\Sigma : R \to R_\Sigma$. The universal property is that every Σ-inverting homomorphism $\nu : R \to S$ may be written uniquely as a composite

$$R \xrightarrow{i_\Sigma} R_\Sigma \xrightarrow{\bar{\nu}} S;$$

If R is commutative and each $\sigma \in \Sigma$ is an endomorphism then the localization is the ring of fractions

$$R_\Sigma = R_S = \{p/q \mid p \in R, q \in S\}$$

whose denominators lie in the multiplicative set $S \subseteq R$ generated by the determinants of the morphisms in Σ:

$$S = \left\{ \prod_{i=1}^{r} \det(\sigma_i) \mid r \in \mathbb{Z}, r \geq 0, \sigma_i \in \Sigma \right\}$$

---

[5]The ring R_Σ was denoted $\Sigma^{-1}R$ in [51]

More general constructions of i_Σ may be found in [49, Ch4], [9, p255] or [11].

If $R = A[F_\mu]$ and Σ is defined as above, the inclusion of $A[F_\mu]$ in $\widehat{A[F_\mu]}$ is Σ-inverting; see Lemmas 2.6 and 2.8. By the universal property of $A[F_\mu]_\Sigma$ there is therefore a commutative diagram

$$A[F_\mu] \xrightarrow{\ i_\Sigma\ } A[F_\mu]_\Sigma \xrightarrow{\ \gamma\ } \widehat{A[F_\mu]} \cong A\langle\langle X \rangle\rangle \ . \tag{14}$$

where the natural isomorphism $\widehat{A[F_\mu]} \cong A\langle\langle X \rangle\rangle$ is defined as in Lemma 2.5.

The image of γ is the ring $A_{\mathrm{rat}}\langle\langle X \rangle\rangle$ of rational power series (see [50, §4]). If A is a field or a principal ideal domain then γ is known to be injective so $A[F_\mu]_\Sigma$ can be identified with $A_{\mathrm{rat}}\langle\langle X \rangle\rangle$ (Cohn and Dicks [12, p416], Dicks and Sontag [13, Thm 24], Farber and Vogel [18]). In the knot theory case $\mu = 1$, if A is commutative then the localization is a ring of fractions

$$A[F_1]_\Sigma \cong \{p/q \mid p, q \in A[z, z^{-1}], q(1) \text{ is invertible}\} \cong A_{\mathrm{rat}}[[x]].$$

and γ is injective. However, there exist non-commutative rings A such that γ is not injective [50, Prop 1.2].

Diagram (14) and Lemma 2.6 imply:

Lemma 2.15. *The localization $i_\Sigma : A[F_\mu] \to A[F_\mu]_\Sigma$ is injective.* $\qquad\square$

The following is a generalization of Lemma 2.8:

Lemma 2.16. *If $i_\Sigma : R \to R_\Sigma$ is injective then each $\sigma \in \Sigma$ is injective.*

Proof. Suppose $\sigma : P_1 \to P_0$ is in Σ. There is a commutative diagram

$$
\begin{array}{ccc}
R \otimes_R P_1 & \xrightarrow{\ i_\Sigma \otimes 1\ } & R_\Sigma \otimes_R P_1 \\
{\scriptstyle 1 \otimes \sigma}\downarrow & & \downarrow{\scriptstyle 1 \otimes \sigma} \\
R \otimes_R P_0 & \xrightarrow[\ i_\Sigma \otimes 1\ ]{} & R_\Sigma \otimes_R P_0
\end{array}
$$

and $1 \otimes \sigma : R_\Sigma \otimes_R P_1 \to R_\Sigma \otimes_R P_0$ is an isomorphism. If i_Σ is injective then $1 \otimes \sigma : R \otimes_R P_1 \to R \otimes_R P_0$ is also injective so σ is injective. $\qquad\square$

Lemma 2.17. *Suppose $i_\Sigma : R \to R_\Sigma$ is an injection and $M = \mathrm{Coker}(\sigma)$ with $\sigma \in \Sigma$.*

1. *The (right) R-module $M^\wedge = \mathrm{Ext}_R(M, R)$ is isomorphic to $\mathrm{Coker}(\sigma^*)$.*

2. *There is a natural isomorphism $\mathrm{Ext}_R(M, R) \cong \mathrm{Hom}_R(M, R_\Sigma/R)$.*

If R has involution one can regard these right modules as left modules.

Proof of Lemma 2.17. (Compare Example 2.11).
1. By Lemma 2.16, the map σ is injective so M has presentation

$$0 \to P_1 \xrightarrow{\sigma} P_0 \to M \to 0. \tag{15}$$

There is therefore an exact sequence

$$P_0^* \xrightarrow{\sigma^*} P_1^* \to \operatorname{Ext}_R(M, R) \to \operatorname{Ext}_R(P_0, R) = 0$$

2. The short exact sequence of (R, R)-bimodules

$$0 \to R \to R_\Sigma \to R_\Sigma/R \to 0$$

induces a long exact sequence of right R-modules

$$\cdots \to \operatorname{Hom}_R(M, R_\Sigma) \to \operatorname{Hom}_R(M, R_\Sigma/R)$$
$$\to \operatorname{Ext}_R(M, R) \to \operatorname{Ext}_R(M, R_\Sigma) \to \cdots$$

which is natural in M. It remains to prove that $\operatorname{Hom}_R(M, R_\Sigma) = \operatorname{Ext}_R(M, R_\Sigma) = 0$. The presentation (15) gives rise to the long exact sequence

$$0 \to \operatorname{Hom}_R(M, R_\Sigma) \to \operatorname{Hom}_R(P_0, R_\Sigma) \xrightarrow{\sigma^*} \operatorname{Hom}_R(P_1, R_\Sigma)$$
$$\to \operatorname{Ext}_R(M, R_\Sigma) \to 0.$$

There is a natural isomorphism

$$\operatorname{Hom}_{R_\Sigma}(R_\Sigma \otimes_R \_\ , R_\Sigma) \to \operatorname{Hom}_R(R \otimes_R \_\ , R_\Sigma)$$

induced by $i_\Sigma : R \to R_\Sigma$ and, in particular, a commutative diagram

$$
\begin{array}{ccc}
\operatorname{Hom}_{R_\Sigma}(R_\Sigma \otimes_R P_0, R_\Sigma) & \xrightarrow{(\mathrm{id} \otimes \sigma)^*} & \operatorname{Hom}_{R_\Sigma}(R_\Sigma \otimes_R P_1, R_\Sigma) \\
\cong \downarrow & & \downarrow \cong \\
\operatorname{Hom}_R(P_0, R_\Sigma) & \xrightarrow{\quad \sigma^* \quad} & \operatorname{Hom}_R(P_1, R_\Sigma)
\end{array}
\tag{16}
$$

The upper horizontal arrow is an isomorphism since $\mathrm{id} \otimes \sigma$ is an isomorphism so the lower horizontal arrow is also an isomorphism. Thus

$$\operatorname{Hom}_R(M, R_\Sigma) = \operatorname{Ext}_R(M, R_\Sigma) = 0$$

as required. $\square$

Remark 2.18. If $R \to R_\Sigma$ is injective and $\sigma \in \Sigma$ then the sequence $0 \to P_1 \xrightarrow{\sigma} P_0 \xrightarrow{q} M \to 0$ is exact by Lemma 2.16. Now Lemma 2.17 gives an exact sequence

$$P_0^* \xrightarrow{\sigma^*} P_1^* \xrightarrow{q'} \operatorname{Hom}_R(M, R_\Sigma/R) \to 0.$$

Let us give an explicit formula for q'. There is a short exact sequence of right R-module chain complexes

$$
\begin{array}{ccccccccc}
0 & \longrightarrow & \operatorname{Hom}_R(P_0, R) & \longrightarrow & \operatorname{Hom}_R(P_0, R_\Sigma) & \longrightarrow & \operatorname{Hom}_R(P_0, R_\Sigma/R) & \longrightarrow & 0 \\
& & \downarrow{\sigma^*} & & \downarrow{\sigma^*}\,{\cong} & & \downarrow{\sigma^*} & & \\
0 & \longrightarrow & \operatorname{Hom}_R(P_1, R) & \longrightarrow & \operatorname{Hom}_R(P_1, R_\Sigma) & \longrightarrow & \operatorname{Hom}_R(P_1, R_\Sigma/R) & \longrightarrow & 0
\end{array}
$$

and the natural isomorphism

$$\operatorname{Ext}_R(M, R) = \frac{\operatorname{Hom}_R(P_1, R)}{\operatorname{Image}(\sigma^*)} \longrightarrow \operatorname{Hom}_R(M, R_\Sigma/R)$$

is inverse to the boundary map in the induced long exact homology sequence from the kernel of the right-most σ^* to the cokernel of the left-most σ^*. Written out at length, the map $q' : P_1^* \to \operatorname{Hom}_R(M, R_\Sigma/R)$ is the composite

$$\operatorname{Hom}_R(P_1, R) \to \operatorname{Hom}_{R_\Sigma}(R_\Sigma \otimes P_1, R_\Sigma) \xleftarrow[\ (\mathrm{id} \otimes \sigma)^*\ ]{\cong} \operatorname{Hom}_{R_\Sigma}(R_\Sigma \otimes P_0, R_\Sigma)$$

$$\xrightarrow{\cong} \operatorname{Hom}_R(P_0, R_\Sigma) \to \operatorname{Hom}_R(P_0, R_\Sigma/R),$$

the image of which lies in the submodule $\operatorname{Hom}_R(M, R_\Sigma/R) \subset \operatorname{Hom}(P_0, R_\Sigma/R)$. Suppose $f \in P_1^* = \operatorname{Hom}_R(P_1, R)$ and $m \in M$. Choose $x \in P_0$ such that $q(x) = m$ and write $1 \otimes x \in R_\Sigma \otimes P_0$. Now

$$q'(f)(m) = (\mathrm{id} \otimes f)\left((\mathrm{id} \otimes \sigma)^{-1}(1 \otimes x)\right) \in R_\Sigma/R.$$

Combining Lemmas 2.17 and 2.15 we have

Proposition 2.19. *There is a natural isomorphism of contravariant functors*

$$\operatorname{Ext}_{A[F_\mu]}(\_\,, A[F_\mu]) \cong \operatorname{Hom}_{A[F_\mu]}(\_\,, A[F_\mu]_\Sigma/A[F_\mu]) : \mathcal{F}\mathrm{lk}(A) \to \mathcal{F}\mathrm{lk}(A).$$

2.4　Hermitian forms and the Witt group

As we noted in the introduction (equations (3) and (4)), the cobordism group $C_n(F_\mu)$ can be identified with a Witt group of Seifert or Blanchfield-Duval forms. Let us recall the definition of a hermitian form in a hermitian category and the appropriate definition of Witt group.

Definition 2.20. Let $\zeta = 1$ or -1. A ζ-*hermitian form* in a hermitian category $(\mathcal{C}, *, i)$ is a pair (V, ϕ) where $\phi : V \to V^*$ and $\phi^* i_V = \zeta\phi$. If ϕ is an isomorphism then ϕ is called *non-singular*.

For example, in the category $\mathcal{F}lk(A)$ a ζ-hermitian form is an $A[F_\mu]$-module isomorphism $\phi : M \to M^\wedge = \mathrm{Hom}(M, A[F_\mu]_\Sigma / A[F_\mu])$ such that $\phi^\wedge = \zeta\phi$ (we suppress the natural isomorphism i_M).

Definition 2.21. An object V in a hermitian category $(\mathcal{C}, *, i)$ is called *self-dual* if V is isomorphic to V^*. If there exists a non-singular ζ-hermitian form (V, ϕ) then V is called ζ-*self-dual*.

When one has a suitable notion of exact sequences in a hermitian category one can define the Witt group of the category. For simplicity, suppose $\mathcal{C}$ is a full subcategory of an abelian category $\mathcal{A}$, so that every morphism in $\mathcal{C}$ has kernel, image and cokernel in $\mathcal{A}$. Suppose further that $\mathcal{C}$ is admissible in $\mathcal{A}$ in the following sense: If there is an exact sequence $0 \to V \to V' \to V'' \to 0$ and the modules V and V'' lie in $\mathcal{C}$ then V' lies in $\mathcal{C}$. In Section 4 below we consider a Serre subcategory of an abelian category which is defined by $V' \in \mathcal{C}$ if and only if $V \in \mathcal{C}$ and $V'' \in \mathcal{C}$. For the present we maintain greater generality; in particular an admissible subcategory $\mathcal{C}$ is not required to be an abelian category.

Definition 2.22. Let $\zeta = 1$ or -1. A non-singular ζ-hermitian form (V, ϕ) is called metabolic if there is a submodule $L \subset V$ such that i) L and V/L are in $\mathcal{C}$ and ii) $L = L^\perp$. By definition $L^\perp = \mathrm{Ker}(j^*\phi : V \to L^*)$ where $j : L \to V$ is the inclusion.

Definition 2.23. The Witt group $W^\zeta(\mathcal{C})$ is the abelian group with one generator $[V, \phi]$ for each isomorphism class of non-singular ζ-hermitian forms $(V, \phi) \in \mathcal{C}$ subject to relations

$$\begin{cases} [V', \phi'] = [V, \phi] + [V'', \phi''], & \text{if } (V', \phi') \cong (V, \phi) \oplus (V'', \phi'') \\ [V, \phi] = 0, & \text{if } (V, \phi) \text{ is metabolic.} \end{cases}$$

Two forms represent the same Witt class $[V, \phi] = [V', \phi']$ if and only if there exist metabolic forms (H, η) and (H', η') such that

$$(V \oplus H, \phi \oplus \eta) \cong (V' \oplus H', \phi' \oplus \eta').$$

For example, a non-singular ζ-hermitian form $\phi : \mathbb{Z}^{2n} \to \mathbb{Z}^{2n}$ in the category $\mathbb{Z}$-proj (see Example 2.10) is metabolic if there exists a summand $L \cong \mathbb{Z}^n$ with $\phi(L)(L) = 0$.

In the category $\mathcal{F}\mathrm{lk}(A)$ a metabolizer L for a form $\phi : M \to M^\wedge$ need not be a summand but L and M/L must lie in $\mathcal{F}\mathrm{lk}(A)$ and one must have $\phi(L)(L) = 0$ and $\phi(x)(L) \neq 0$ if $x \notin L$. Now that the notation is defined, we repeat equation (4):

$$C_{2q-1}(F_\mu) \cong W^{(-1)^{q+1}}(\mathcal{F}\mathrm{lk}(\mathbb{Z})). \quad (q \geq 3) \tag{17}$$

2.5 Duality-preserving functors

A functor between hermitian categories which respects their structure is called duality-preserving. Our first examples will be functors induced by a morphism of rings with involution. Duality-preserving functors will also play an essential role in later sections (see Theorem 1.3 above and Theorem 3.19 below).

Definition 2.24. A duality-preserving functor from $(\mathcal{C}, \_^*, i)$ to $(\mathcal{D}, \_^*, i)$ is a triple (G, Ψ, η) where

- $G : \mathcal{C} \to \mathcal{D}$ is a functor,

- $\Psi = (\Psi_V)_{V \in \mathcal{C}} : G(\_^*) \to G(\_)^*$ is a natural isomorphism,

- $\eta = 1$ or -1

such that

$$\Psi_V^* i_{G(V)} = \eta \Psi_{V^*} G(i_V) : G(V) \to G(V^*)^* \tag{18}$$

for all $V \in \mathcal{C}$.

We sometimes abbreviate (G, Ψ, η) to G.

Definition 2.25. The composite of duality-preserving functors is defined by

$$(G, \Psi, \eta) \circ (G', \Psi', \eta') = (GG', \Psi G(\Psi'), \eta \eta'). \tag{19}$$

Example 2.26. A homomorphism $\nu : A \to A'$ of rings with involution induces a duality-preserving functor $(A' \otimes_A \_ , \Pi, 1)$ from the category A-Proj of finitely generated projective A-modules to the category A'-Proj of finitely generated projective A'-modules. Explicitly

$$\Pi_V : A' \otimes_A (V^*) \xrightarrow{\cong} (A' \otimes_A V)^* \tag{20}$$
$$a'_1 \otimes \theta \mapsto (a'_2 \otimes x \mapsto a'_2 \nu(\theta(x))\overline{a'_1}).$$

for all $a'_1, a'_2 \in A'$, $\theta \in V^*$ and $x \in V$.

We are particularly concerned with the category $\mathcal{F}\mathrm{lk}(A)$:

Lemma 2.27. *A homomorphism $\nu : A \to A'$ of rings with involution induces a canonical duality-preserving functor*

$$(A'[F_\mu] \otimes_{A[F_\mu]} \_ , \Upsilon, 1) : \mathcal{F}\mathrm{lk}(A) \to \mathcal{F}\mathrm{lk}(A').$$

The natural isomorphism $\Upsilon : A'[F_\mu] \otimes_{A[F_\mu]} \_^\wedge \to (A'[F_\mu] \otimes_{A[F_\mu]} \_)^\wedge$ will be defined in the course of the proof.

Proof. We saw in Lemma 2.9 that there is a functor $A'[F_\mu] \otimes_{A[F_\mu]} \_$ from $\mathcal{F}\mathrm{lk}(A)$ to $\mathcal{F}\mathrm{lk}(A')$. The dual $M^\wedge$ of a module $M \in \mathcal{F}\mathrm{lk}(A)$ has presentation

$$0 \to (V[F_\mu])^* \xrightarrow{\sigma^*} (V[F_\mu])^* \to M^\wedge \to 0$$

(see Remark 2.18). Applying Example 2.26 to $\nu : A[F_\mu] \to A'[F_\mu]$ one obtains a natural isomorphism

$$\Pi_{V[F_\mu]} : A'[F_\mu] \otimes_{A[F_\mu]} (V[F_\mu])^* \cong (A'[F_\mu] \otimes_{A[F_\mu]} V[F_\mu])^*$$

and hence a commutative diagram

$$\begin{array}{ccccccccc}
0 \to & A'[F_\mu] \otimes_{A[F_\mu]} (V[F_\mu])^* & \to & A'[F_\mu] \otimes_{A[F_\mu]} (V[F_\mu])^* & \longrightarrow & A'[F_\mu] \otimes_{A[F_\mu]} M^\wedge & \longrightarrow 0 \\
& \downarrow \cong & & \downarrow \cong & & \downarrow \Upsilon_M & \\
0 \to & (A'[F_\mu] \otimes_{A[F_\mu]} V[F_\mu])^* & \to & (A'[F_\mu] \otimes_{A[F_\mu]} V[F_\mu])^* & \to & (A'[F_\mu] \otimes_{A[F_\mu]} M)^\wedge & \to 0.
\end{array}$$

We must check that the induced isomorphism

$$\Upsilon_M : A'[F_\mu] \otimes_{A[F_\mu]} M^\wedge \to (A'[F_\mu] \otimes_{A[F_\mu]} M)^\wedge$$

is independent of the choice of presentation σ and that Υ is natural with respect to M. If C denotes the chain complex $V[F_\mu] \xrightarrow{\sigma} V[F_\mu]$ and C' denotes an alternative choice of resolution for M, say $C' = (V'[F_\mu] \xrightarrow{\sigma'} V'[F_\mu])$ then the identity map id : $M \to M$ lifts to a chain equivalence $C \to C'$. The naturality of Π in Example 2.26 implies that the diagram

$$
\begin{array}{ccc}
A'[F_\mu] \otimes C^* & \longrightarrow & A'[F_\mu] \otimes (C')^* \\
\downarrow & & \downarrow \\
(A'[F_\mu] \otimes C)^* & \to & (A'[F_\mu] \otimes C')^*
\end{array}
$$

commutes. The horizontal arrows induce the identity map on $A'[F_\mu] \otimes M^\wedge$ and $(A'[F_\mu] \otimes M)^\wedge$ respectively, so the vertical arrows induce the same map Υ_M. The naturality of Υ follows similarly from the naturality of the transformation Π in Example 2.26. $\qquad\square$

Remark 2.28. Identifying $M^\wedge$ with $\mathrm{Hom}(M, A[F_\mu]_\Sigma/A[F_\mu])$ by Proposition 2.19, an explicit formula for Υ_M is

$$
\Upsilon_M : A'[F_\mu] \otimes_{A[F_\mu]} \mathrm{Hom}\left(M, \frac{A[F_\mu]_\Sigma}{A[F_\mu]}\right) \to \mathrm{Hom}\left(A'[F_\mu] \otimes_{A[F_\mu]} M, \frac{A'[F_\mu]_\Sigma}{A'[F_\mu]}\right)
$$
$$
a_1' \otimes \theta \mapsto (a_2' \otimes m \mapsto a_2'\nu(\theta(m))\overline{a_1'})
$$

We conclude this section by noting the effect of duality-preserving functor on Witt groups:

Lemma 2.29. *A duality-preserving functor $(G, \Psi, \eta) : \mathcal{C} \to \mathcal{D}$ which respects exact sequences induces a homomorphism of Witt groups*

$$
G : W^\zeta(\mathcal{C}) \to W^{\zeta\eta}(\mathcal{D})
$$
$$
[V, \phi] \mapsto [G(V), \Psi_V G(\phi)]
$$

Proof. See for example [51, p41-42]. $\qquad\square$

3 Intrinsic Invariants

In [51] the author defined invariants of the cobordism group $C_{2q-1}(F_\mu)$ of F_μ-links using the identification (3) due to Ko [27] and Mio [38] of $C_{2q-1}(F_\mu)$ with a Witt group of μ-component Seifert forms, denoted $W^{(-1)^q}(\mathcal{S}\mathrm{ei}(\mathbb{Z}))$ below ($q \geq 3$). To distinguish F_μ-links one first chooses a Seifert surface for each and then computes invariants of the associated Seifert forms.

In the present section we define F_μ-link cobordism invariants via Duval's identification $C_{2q-1}(F_\mu) \cong W^{(-1)^q}(\mathcal{F}\mathrm{lk}(\mathbb{Z}))$. The definitions will parallel those in [51] and we shall prove in Section 6 that the invariants obtained are equivalent. Whereas Seifert forms are convenient for computing the invariants in explicit examples, the Blanchfield-Duval form has the advantage that it is defined without making a choice of Seifert surface.

3.1   Overview

Let $\zeta = 1$ or -1. The inclusion $\mathbb{Z} \subset \mathbb{Q}$ induces a duality-preserving functor $\mathcal{F}\mathrm{lk}(\mathbb{Z}) \to \mathcal{F}\mathrm{lk}(\mathbb{Q})$ (see Lemma 2.27 above) and hence a homomorphism of Witt groups

$$W^\zeta(\mathcal{F}\mathrm{lk}(\mathbb{Z})) \to W^\zeta(\mathcal{F}\mathrm{lk}(\mathbb{Q})). \qquad (21)$$

It follows from Theorem 1.3 that (21) is an injection; see the proof of Corollary 3.3 below. We proceed to compute $W^\zeta(\mathcal{F}\mathrm{lk}(\mathbb{Q}))$ in three steps, which were outlined in the introduction. We list them again here in more detail:

1. *Devissage.* We prove that $\mathcal{F}\mathrm{lk}(\mathbb{Q})$ is an abelian category with ascending and descending chain conditions. Recall that a module $M \in \mathcal{F}\mathrm{lk}(\mathbb{Q})$ is called simple (or irreducible) if $M \neq 0$ and there are no submodules of M in $\mathcal{F}\mathrm{lk}(\mathbb{Q})$ other than 0 and M. If M is a simple module then $\mathcal{F}\mathrm{lk}(\mathbb{Q})|_M \subset \mathcal{F}\mathrm{lk}(\mathbb{Q})$ denotes the full subcategory in which the objects are direct sums of copies of M. Recall that M is called ζ-self-dual if there is an isomorphism $b : M \to M^\wedge$ such that $b^\wedge = \zeta b$. We obtain, by "hermitian devissage", the decomposition

$$W^\zeta(\mathcal{F}\mathrm{lk}(\mathbb{Q})) \cong \bigoplus W^\zeta(\mathcal{F}\mathrm{lk}(\mathbb{Q})|_M) \qquad (22)$$

with one summand for each isomorphism class of ζ-self-dual simple F_μ-link modules M. Let p_M denote the projection of $W^\zeta(\mathcal{F}\mathrm{lk}(\mathbb{Q}))$ onto $W^\zeta(\mathcal{F}\mathrm{lk}(\mathbb{Q})|_M)$.

2. *Morita equivalence.* For each ζ-self-dual simple module M we choose a non-singular ζ-hermitian form $b : M \to M^\wedge$. We obtain by hermitian Morita equivalence an isomorphism

$$\Theta_{M,b} : W^\zeta(\mathcal{F}\mathrm{lk}(\mathbb{Q})|_M) \to W^1(E) \qquad (23)$$

where $E = \mathrm{End}_{\mathbb{Q}[F_\mu]} M$ is the endomorphism ring of M and is endowed with the involution $f \mapsto b^{-1}f^\wedge b$. By Schur's Lemma E is a division ring and, as we discuss, E turns out to be of finite dimension over $\mathbb{Q}$.

3. We recall from the literature invariants of each group $W^1(E)$. In most cases some combination of dimension modulo 2, signatures, discriminant and Hasse-Witt invariant are sufficient to distinguish the elements

of $W^1(E)$ (see the table (29) below). One class of division algebra with involution requires a secondary invariant, such as the Lewis θ, which is defined only if all the other invariants vanish.

Let us make two remarks about the modules M which appear in item 1. Firstly, every simple module $M \in \mathcal{F}lk(\mathbb{Q})$ such that $M \cong M^\wedge$ is either 1-self-dual or (-1)-self-dual (or both). Secondly, M is both 1-self-dual and (-1)-self-dual if and only if the involution $f \mapsto b^{-1}f^\wedge b$ induced on $E = \mathrm{End}_{\mathbb{Q}[F_\mu]}(M)$ is not the identity map for some (and therefore every) ς-self-dual form $b : M \to M^\wedge$. See lemmas 5.5 and 5.6 of [51] for details.

Example 3.1. In the knot theory case $\mu = 1$ we can add simplifying remarks to each of the three steps:

1. A simple self-dual module $M \in \mathcal{F}lk(\mathbb{Q})$ may be written

$$M = \mathbb{Q}[z, z^{-1}]/(p)$$

   where p is an irreducible polynomial and $(p(z^{-1})) = (p(z)) \lhd \mathbb{Q}[z, z^{-1}]$.

2. The endomorphism ring $E = \mathrm{End}_{\mathbb{Q}[z,z^{-1}]}(M)$ may also be written as a quotient $\mathbb{Q}[z, z^{-1}]/(p)$ and is an algebraic number field of finite dimension over $\mathbb{Q}$. The involution on E is given by $z \mapsto z^{-1}$ and does not depend on the choice of form b.

   Setting aside the case $M = \mathbb{Q}[z, z^{-1}]/(1+z)$, the involution on E is not the identity so every self-dual M is both 1-self-dual and (-1)-self-dual. The exceptional module $M = \mathbb{Q}[z, z^{-1}]/(1 + z)$ is only 1-self-dual but plays little role since -1 is not a root of any polynomial $p \in \mathbb{Z}[z, z^{-1}]$ such that $p(1) = \pm 1$. In other words, the projection of $W^1(\mathcal{F}lk(\mathbb{Z}))$ on this exceptional summand of $W^1(\mathcal{F}lk(\mathbb{Q}))$ is zero.

3. As discussed in 2., one need only consider the Witt groups $W^1(E)$ of number fields with non-trivial involution, or in other words, hermitian forms over number fields. The dimension modulo 2, signatures and discriminant are sufficient to distinguish the elements of $W^1(E)$.

Equation (1) can be derived as a consequence of this analysis.

Returning to the general case $\mu \geq 1$ and putting together steps 1-3. we obtain the following restatement of Theorem 1.1.

Theorem 3.2. *Let $\varsigma = 1$ or -1. An element $\alpha \in W^\varsigma(\mathcal{F}lk(\mathbb{Q}))$ is equal to zero if and only if for each finite-dimensional ς-self-dual simple F_μ-link* *module M and non-singular ζ-hermitian form $b : M \to M^\wedge$, the dimension modulo 2, the signatures, the discriminant, the Hasse-Witt invariant and the Lewis θ-invariant of*

$$\Theta_{M,b}\, p_M(\alpha) \in W^1(\mathrm{End}_{\mathbb{Q}[F_\mu]} M)$$

are trivial (if defined). □

Note that if the invariants corresponding to one form $b : M \to M^\wedge$ are trivial then $p_M(\alpha) = 0$ so the invariants are trivial for any other choice $b' : M \to M^\wedge$. We now restate Corollary 1.2:

Corollary 3.3. *Suppose (L^0, θ^0) and (L^1, θ^1) are $(2q-1)$-dimensional F_μ-links, where $q > 1$. Let $\overline{X}_i$ denote the free cover of the complement of L_i, let $N_i = H_q(\overline{X}_i)/(\mathbb{Z}-\text{torsion})$ and let $\phi^i : N_i \to \mathrm{Hom}(N_i, \mathbb{Z}[F_\mu]_\Sigma/\mathbb{Z}[F_\mu])$ denote the Blanchfield-Duval form for (L^i, θ^i).*

The F_μ-links (L^0, θ^0) and (L^1, θ^1) are cobordant if and only if for each finite-dimensional ζ-self-dual simple F_μ-link module M and each non-singular ζ-hermitian form $b : M \to M^\wedge$, the dimension modulo 2, the signatures, the discriminant, the Hasse-Witt invariant and the Lewis θ-invariant of

$$\Theta_{M,b}\, p_M[\mathbb{Q} \otimes_{\mathbb{Z}} (N^0 \oplus N^1, \phi^0 \oplus -\phi^1)] \in W^1(\mathrm{End}_{\mathbb{Q}[F_\mu]} M)$$

are trivial (if defined). □

Proof. We deduce Corollary 3.3 from Theorem 1.1 (=Theorem 3.2) and Theorem 1.3. As we remarked in the introduction, Corollary 3.3 also follows from Theorem 1.4 and Theorem B of [51].

Proposition 5.7 below says that the duality-preserving functor

$$(B, \Phi, -1) : \mathcal{S}ei(A) \to \mathcal{F}lk(A)$$

of Section 5 respects coefficient change so there is a commutative diagram

$$
\begin{array}{ccc}
W^{(-1)^q}(\mathcal{S}ei(\mathbb{Z})) & \xrightarrow{\ B\ } & W^{(-1)^{q+1}}(\mathcal{F}lk(\mathbb{Z})) \\
\downarrow & & \downarrow \\
W^{(-1)^q}(\mathcal{S}ei(\mathbb{Q})) & \xrightarrow[\ B\ ]{} & W^{(-1)^{q+1}}(\mathcal{F}lk(\mathbb{Q})).
\end{array}
\tag{24}
$$

The category $\mathcal{S}ei(A)$ is defined in Section 4.1. The lower horizontal map in (24) is an isomorphism by Theorem 1.3 and the upper horizontal map is an isomorphism by (3) and (4) above (see also Remark 5.6). It is easy to prove that the left hand vertical map is an injection (see for example Lemma 11.1 of [51]). Thus the right-hand vertical map is also an injection. Corollary 3.3 therefore follows from Theorem 3.2 □

3.2 Step 1: Devissage

Let us briefly recall some definitions. If $\mathcal{A}$ is an additive category and M is an object in $\mathcal{A}$ the symbol $\mathcal{A}|_M$ denotes the full subcategory such that $N \in \mathcal{A}$ if and only if N is a summand of some finite direct sum of copies of M. If $\mathcal{A}$ is a hermitian category and M is self-dual then $\mathcal{A}|_M$ is a hermitian subcategory.

Suppose now that $\mathcal{A}$ is an abelian category. A non-zero module M in $\mathcal{A}$ is called simple (or irreducible) if there are no submodules of M in $\mathcal{A}$ other than 0 and M. The category $\mathcal{A}$ has both ascending and descending chain conditions if and only if every module M in $\mathcal{A}$ has a finite composition series

$$0 = M_0 \subset M_1 \subset M_2 \subset M_3 \subset \cdots \subset M_s = M \qquad (25)$$

where M_i/M_{i-1} is simple for $i = 1, \cdots, s$. If M is simple then every module in $\mathcal{A}|_M$ is a direct sum of copies of M. Let $\zeta = 1$ or -1. If $(\mathcal{A}, *, i)$ is a hermitian category then a module M is called ζ-self-dual if there is an isomorphism $\phi : M \to M^*$ such that $\phi^* i_M = \zeta \phi$.

The general decomposition theorem we need is the following:

Theorem 3.4 (Devissage). *Suppose $\mathcal{A}$ is an abelian hermitian category with ascending and descending chain conditions. There is an isomorphism of Witt groups*

$$W^\zeta(\mathcal{C}) \cong \bigoplus W^\zeta(\mathcal{C}|_M)$$

with one summand for each isomorphism class of simple ζ-self-dual modules in $\mathcal{C}$.

Proof. See [51, Theorem 5.3] or [41]. $\qquad \square$

To prove equation (22) it therefore suffices to show:

Proposition 3.5. *If k is a (commutative) field then the category $\mathcal{F}\mathrm{lk}(k)$ is an abelian category with ascending and descending chain conditions.*

We take coefficients in a field for simplicity. With little extra work one can show that Proposition 3.5 holds when k is replaced by any semi-simple Artinian ring. See also Remark 3.18 below.

Caveat 3.6. When $\mu \geq 2$ a simple module in the category $\mathcal{F}\mathrm{lk}(k)$ is not a simple module in the category of $k[F_\mu]$-modules.

We prove Proposition 3.5 as follows: We first show that $\mathcal{F}\mathrm{lk}(k)$ is a subcategory of the category $\mathcal{T}_{k[F_\mu]}$ of "torsion" modules which P.M.Cohn introduced and proved to be an abelian category with ascending and descending chain conditions (see Proposition 3.13 below). After giving details of Cohn's work, we conclude the proof of Proposition 3.5 by checking that $\mathcal{F}\mathrm{lk}(k)$ is closed under direct sums and that the kernel and cokernel of every morphism in $\mathcal{F}\mathrm{lk}(k)$ again lie in $\mathcal{F}\mathrm{lk}(k)$.

3.2.1 Firs and torsion modules

To describe Cohn's results we must state some properties of the group ring $k[F_\mu]$.

Definition 3.7. A ring R has invariant basis number (IBN) if $R^n \cong R^m$ implies $n = m$. In other words R has IBN if every finitely generated free left R-module has unique rank.

The existence of the augmentation $\epsilon : k[F_\mu] \to k$ implies that $k[F_\mu]$ has IBN for if $k[F_\mu]^n \cong k[F_\mu]^m$ then

$$k^n \cong k \otimes_{k[F_\mu]} k[F_\mu]^n \cong k \otimes_{k[F_\mu]} k[F_\mu]^m \cong k^m$$

so $m = n$.

Note that if R has IBN then one can use the duality functor $\mathrm{Hom}(\_\,, R)$ to prove that finitely generated free right R-modules also have unique rank.

Definition 3.8. An associative ring R is called a *free ideal ring (fir)* if R has IBN, every left ideal in R is a free left R-module and every right ideal is a free right R-module.

Cohn showed in [8, Corollary 3] that if k is a field then the group ring $k[F_\mu]$ of the free group F_μ is a fir.

If R is a fir then every submodule of a free R-module is free (Cohn [10, p71]). Hence every R-module has a presentation $0 \to F_1 \to F_0 \to M \to 0$ where F_1 and F_0 are free.

Definition 3.9. If an R-module M has a presentation

$$0 \to R^n \xrightarrow{\sigma} R^m \xrightarrow{p} M \to 0 \tag{26}$$

the *Euler characteristic* of M is $\chi(M) = m - n$.

If a finite presentation exists the Euler characteristic is independent of the choice of presentation by Schanuel's Lemma. Note also that an exact sequence $0 \to M \to M' \to M'' \to 0$ of finitely presented modules implies the equation $\chi(M') = \chi(M) + \chi(M'')$ (compare the diagram (27) below).

If R is a fir then the category of finitely presented R-modules (and R-module maps) is an abelian category; in other words direct sums of finitely presented modules are finitely presented and the cokernel and kernel of a map between finitely presented modules are finitely presented. In fact, if R is any ring such that every finitely generated one-sided ideal is finitely related then the finitely presented R-modules form an abelian category (e.g. Cohn [10, p554-556]).

Definition 3.10. A morphism $\sigma : R^n \to R^n$ between free left R-modules is called *full* if every factorization

$$R^n \xrightarrow[\sigma_2]{} F \xrightarrow[\sigma_1]{\sigma} R^n$$

where F is a free module has $\mathrm{Rank}(F) \geq n$.

Lemma 3.11. *Suppose R is a fir and M is a finitely presented R-module with $\chi(M) = 0$. The following are equivalent:*

1. *In every finite presentation (26) of M, the map σ is full.*

2. *There exists a presentation (26) such that σ is full.*

3. *$\chi(N) \geq 0$ for all finitely generated submodules N of M.*

4. *$\chi(M/N) \leq 0$ for all finitely generated submodules N of M.*

Proof. The implication $1 \Rightarrow 2$ is immediate. To show 2 implies 3, suppose we are given a presentation (26) such that $m = n$ and σ is full. If N is a finitely generated submodule of M then $\sigma(R^n) \subset p^{-1}(N) \subset R^n$. Now $p^{-1}(N)$ is a free module because R is a fir and $p^{-1}(N)$ has rank at least n since σ is full. The exact sequence

$$0 \to R^n \to p^{-1}(N) \to N \to 0$$

implies that $\chi(N) \geq 0$. This completes the proof that 2 implies 3.

The equation $\chi(M) = \chi(N) + \chi(M/N)$ implies that 3 and 4 are equivalent so we can conclude the proof of the Lemma by showing that 3 implies 1. The equation $\chi(M) = 0$ says that every finite presentation (26) has $m = n$.

We must prove that σ is full. Suppose σ can be written as a composite $R^n \xrightarrow{\sigma_2} R^k \xrightarrow{\sigma_1} R^n$. We aim to show $k \geq n$. Now $\sigma(R^n)$ and $\sigma_1(R^k)$ are free modules since R is a fir and

$$\frac{\sigma_1(R^k)}{\sigma(R^n)} \subseteq \frac{R^n}{\sigma(R^n)} = M.$$

By statement 3, $\chi(\sigma_1(R^k)/\sigma(R^n)) \geq 0$ so $\mathrm{Rank}(\sigma_1(R^k)) \geq \mathrm{Rank}(\sigma(R^n)) = n$. Thus $k \geq n$ and hence σ is full. $\qquad\square$

A module which satisfies the equivalent conditions in Lemma 3.11 is called a *torsion* module. The symbol $\mathcal{T}_R$ denote the category of torsion modules and R-module maps.

Lemma 3.12. *(Cohn [10, p166]) If $0 \to M \to M' \to M'' \to 0$ is an exact sequence and M and M'' are torsion modules then M' is a torsion module. In particular a direct sum of torsion modules is again a torsion module.*

Proof. If M and M'' are finitely presented then M' is also finitely presented (compare (27) below). Since $\chi(M) = \chi(M'') = 0$ we have $\chi(M') = 0$. Now if $N \leq M'$ is finitely generated it suffices by Lemma 3.11 to show that $\chi(N) \geq 0$. Note first that $\chi(N) = \chi(N \cap M) + \chi(N/(N \cap M))$. Now $N \cap M \leq M$ so $\chi(N \cap M) \geq 0$ and $N/(N \cap M) \cong (N + M)/M \leq M''$ so $\chi(N/(N \cap M)) \geq 0$. Thus $\chi(N) \geq 0$. $\qquad\square$

Proposition 3.13. *(Cohn [10, p167,234]) Suppose R is a fir.*

1. *The category $\mathcal{T}_R$ of torsion modules is an abelian category.*

2. *Every module in $\mathcal{T}_R$ has a finite composition series (25) in $\mathcal{T}_R$.*

Proof. To establish statement 1, it suffices to show that if M and M' lie in $\mathcal{T}_R$ then $M \oplus M' \in \mathcal{T}_R$ and the kernel and cokernel of every morphism $f : M \to M'$ are in $\mathcal{T}_R$. Lemma 3.12 gives $M \oplus M' \in \mathcal{T}_R$. Suppose then that $f : M \to M'$ is an R-module morphism. Since the finitely presented R-modules are an abelian category the kernel, image and cokernel of f are finitely presented. Now $f(M)$ is a submodule of M' and a quotient module of M so $\chi(f(M)) \geq 0$ and $\chi(f(M)) \leq 0$ by Lemma 3.11. Thus $\chi(f(M)) = 0$ and it follows that $\chi(\mathrm{Ker}(f)) = 0$ and $\chi(\mathrm{Coker}(f)) = 0$. Every finitely generated submodule $N \leq \mathrm{Ker}(f)$ is a submodule of M so $\chi(N) \geq 0$ and hence $\mathrm{Ker}(f) \in \mathcal{T}_R$. Similarly every quotient N' of $\mathrm{Coker}(f)$ is a quotient of M' and hence has $\chi(N') \leq 0$. Thus $\mathrm{Coker}(f) \in \mathcal{T}_R$ also.

To prove part 2. of Proposition 3.13 we note first that it is sufficient to check the ascending chain condition. Indeed, if $M \in \mathcal{T}_R$ then $M^\wedge = \text{Ext}_R(M, R)$ is a torsion right R-module and a descending chain

$$M = M_0 \supseteq M_1 \supseteq M_2 \supseteq \cdots$$

in $\mathcal{T}_R$ gives rise to an ascending chain

$$\left(\frac{M}{M_0}\right)^\wedge \subseteq \left(\frac{M}{M_1}\right)^\wedge \subseteq \left(\frac{M}{M_2}\right)^\wedge \subseteq \cdots$$

In fact Cohn showed that a larger class of modules, the finitely related bound modules, have the ascending chain condition.

Definition 3.14. An R-module M is *bound* if $\text{Hom}(M, R) = 0$.

Lemma 3.15. *Every torsion module over a fir is bound.*

Proof. If M is an R-module and $\theta : M \to R$ then $\theta(M)$ is a free module so $M \cong \text{Ker}(\theta) \oplus \theta(M)$. Now $\theta(M)$ is a quotient module of M so if M is a torsion module then $\chi(\theta(M)) \leq 0$. It follows that $\theta(M) = 0$ so $\theta = 0$. Thus $\text{Hom}_R(M, R) = 0$. $\qquad\square$

Lemma 3.16. *(Cohn [10, p231]) If R is a fir and M is a finitely related R-module then every bound submodule of M is finitely presented.*

Proof. There is an exact sequence $0 \to R^n \to F \xrightarrow{p} M \to 0$ where F is a free R-module. A submodule $B \leq M$ has presentation

$$0 \to R^n \to p^{-1}(B) \to B \to 0$$

and $p^{-1}(B)$ is a free module since R is a fir. The image of R^n is contained in a finitely generated summand of $p^{-1}(B)$ so B is a direct sum of a free module and a finitely presented module. If B is bound then B does not have any non-zero free summand so B itself is finitely presented. $\qquad\square$

Thus every torsion submodule of a torsion module M is finitely generated so we have the ascending chain condition on torsion submodules. It follows that $\mathcal{T}_R$ has both ascending and descending chain conditions and the proof of Proposition 3.13 is complete. $\qquad\square$

We are now in a position to deduce Proposition 3.5.

Lemma 3.17. *The category $\mathcal{F}\text{lk}(k)$ is a full subcategory of the category $\mathcal{T}_{k[F_\mu]}$ of torsion modules.*

Proof. It suffices to show that every module in $\mathcal{F}\mathrm{lk}(k)$ is a torsion module. By definition, a module in $\mathcal{F}\mathrm{lk}(k)$ has a presentation (8) where $\epsilon(\sigma)$ is an isomorphism. Since $\epsilon(\sigma)$ is full we can deduce that σ is full since each factorization of σ induces a corresponding factorization of $\epsilon(\sigma)$. $\square$

Proof of Proposition 3.5. Since $\mathcal{T}_{k[F_\mu]}$ has ascending and descending chain conditions and $\mathcal{F}\mathrm{lk}(k) \subset \mathcal{T}_{k[F_\mu]}$ we need only show that $\mathcal{F}\mathrm{lk}(k)$ is an abelian category.

Suppose M and M' are in $\mathcal{F}\mathrm{lk}(k)$. It follows directly from the definition of $\mathcal{F}\mathrm{lk}(k)$ that $M \oplus M' \in \mathcal{F}\mathrm{lk}(k)$. We must show that the kernel and cokernel of each map $f : M \to M'$ lie in $\mathcal{F}\mathrm{lk}(k)$. By Proposition 3.13 the kernel, image and cokernel all lie in $\mathcal{T}_{k[F_\mu]}$ so

$$\chi(\mathrm{Ker}(f)) = \chi(f(M)) = \chi(\mathrm{Coker}(f)) = 0.$$

After choosing presentations σ and σ' for $\mathrm{Ker}(f)$ and $f(M)$ respectively one can fill in the dotted arrows below to obtain a commutative diagram with exact rows and exact columns:

$$
\begin{array}{ccccccccc}
& & 0 & & 0 & & 0 & & \\
& & \vdots & & \vdots & & \downarrow & & \\
0 & \longrightarrow & k[F_\mu]^n & \overset{\sigma}{\longrightarrow} & k[F_\mu]^n & \longrightarrow & \mathrm{Ker}(f) & \longrightarrow & 0 \\
& & \vdots & & \vdots & & \downarrow & & \\
0 & \dashrightarrow & k[F_\mu]^{n+n'} & \overset{\sigma''}{\dashrightarrow} & k[F_\mu]^{n+n'} & \dashrightarrow & M & \longrightarrow & 0 \\
& & \vdots & & \vdots & & \downarrow & & \\
0 & \longrightarrow & k[F_\mu]^{n'} & \overset{\sigma'}{\longrightarrow} & k[F_\mu]^{n'} & \longrightarrow & f(M) & \longrightarrow & 0 \\
& & \vdots & & \vdots & & \downarrow & & \\
& & 0 & & 0 & & 0 & &
\end{array}
\qquad (27)
$$

The map σ'' is given by $\begin{pmatrix} \sigma & \tau \\ 0 & \sigma' \end{pmatrix}$ for some τ. Since $M \in \mathcal{F}\mathrm{lk}(k)$, the augmentation $\epsilon(\sigma'')$ is an isomorphism by Lemma 2.3 above. It follows that $\epsilon(\sigma)$ and $\epsilon(\sigma'')$ are isomorphisms and hence $\mathrm{Ker}(f)$ and $f(M)$ are in $\mathcal{F}\mathrm{lk}(k)$. The same argument, applied to the exact sequence $0 \to f(M) \to M' \to \mathrm{Coker}(f) \to 0$ shows that $\mathrm{Coker}(f)$ is also in $\mathcal{F}\mathrm{lk}(k)$. $\square$

Remark 3.18. The arguments above can be adapted to generalize Propositions 3.13 and 3.5 as follows. Suppose every finitely generated one-sided ideal of a ring R is a projective module (i.e. R is semi-hereditary). Suppose S is a ring with the property that $S^n \oplus P \cong S^n$ implies $P = 0$ (i.e. S is weakly finite) and $\nu : R \to S$ is a ring homomorphism. The Grothendieck group $K_0(S)$ admits a partial order in which $x \leq y$ if and only $y - x$ lies in the positive cone

$$\{[P] \mid P \in S\text{-Proj}\} \subset K_0(S).$$

Now the following category $\mathcal{T}_\nu$ is an abelian category: An R-module M lies in $\mathcal{T}_\nu$ if M has a finite presentation by projective modules $0 \to P_1 \xrightarrow{\sigma} P_0 \to M \to 0$ such that $1 \otimes \sigma : S \otimes_R P_1 \to S \otimes P_0$ is full with respect to the partial order on $K_0(S)$. The subcategory of modules for which $1 \otimes \sigma$ is invertible is also an abelian category. Under the additional hypotheses that all the one-sided ideals of R are projective modules (i.e. R is hereditary) and that the equation $S \otimes_R P = 0$ implies $P = 0$ for projective R-modules P, one can also conclude that these abelian categories have ascending and descending chain conditions. One recovers Proposition 3.13 when R is a fir by setting $\nu = \mathrm{id} : R \to R$ and one recovers Proposition 3.5 by setting $\nu = \epsilon : k[F_\mu] \to k$.

3.3 Step 2: Morita Equivalence

Having reduced $W^\varsigma(\mathcal{F}\mathrm{lk}(\mathbb{Q}))$ to a direct sum of Witt groups $W^\varsigma(\mathcal{F}\mathrm{lk}(\mathbb{Q})|_M)$ where M is simple and ς-self-dual (see equation (22)) we pass next from $W^\varsigma(\mathcal{F}\mathrm{lk}(\mathbb{Q})|_M)$ to the Witt group $W^1(\mathrm{End}_{\mathbb{Q}[F_\mu]} M)$ of the endomorphism ring of M.

Recall that if $\mathcal{C}$ is an additive category then an object in $\mathcal{C}|_M$ is a summand of a direct sum of copies of M. The general theorem we employ in this section is the following:

Theorem 3.19 (Hermitian Morita Equivalence). *Let $\eta = +1$ or -1. Suppose that $b : M \to M^*$ is a non-singular η-hermitian form in a hermitian category $\mathcal{C}$, and assume further that every idempotent endomorphism in the hermitian subcategory $\mathcal{C}|_M$ splits. Let $E = \mathrm{End}_{\mathcal{C}} M$ be endowed with the involution $f \mapsto \overline{f} = b^{-1} f^* b$. Then there is an equivalence of hermitian categories*

$$\Theta_{M,b} = (\mathrm{Hom}(M, \_), \Omega^b, \eta) : \mathcal{C}|_M \to E\text{-Proj}$$

where for $N \in \mathcal{C}|_M$ the map $\Omega^b_N(\gamma) = (\alpha \mapsto \eta b^{-1} \alpha^* \gamma)$ is the composite of natural isomorphisms

$$\mathrm{Hom}_{\mathcal{C}}(M, N^*) \to \mathrm{Hom}_{\mathcal{C}}(N, M) \to \mathrm{Hom}_E(\mathrm{Hom}_{\mathcal{C}}(M, N), E)$$
$$\gamma \mapsto b^{-1}\gamma^*; \qquad \delta \mapsto (\alpha \mapsto \overline{\delta\alpha}). \tag{28}$$

Proof. See [51, Theorem 4.7], [26, §I.9,ch.II] or [41]. $\qquad\square$

The following is a corollary of Theorem 3.19 and Lemma 2.29.

Corollary 3.20. *If M is a simple module in $\mathcal{F}\mathrm{lk}(k)$ and $b : M \to M^\wedge$ is a non-singular ζ-hermitian form then the duality-preserving functor $\Theta_{M,b}$ of Theorem 3.19 induces an isomorphism of Witt groups*

$$W^\zeta(\mathcal{F}\mathrm{lk}(k)|_M) \to W^1(\mathrm{End}_{k[F_\mu]} M)$$
$$[N, \phi] \mapsto [\mathrm{Hom}(M, N), \Omega_N^b \phi_*]$$

where $\phi_* : \mathrm{Hom}(M, N) \to \mathrm{Hom}(M, N^*)$ and

$$(\Omega_N^b \phi_*)(\alpha)(\beta) = \zeta b^{-1} \beta^* \phi \alpha \in \mathrm{End}_{k[F_\mu]} M$$

for all $\alpha, \beta \in \mathrm{Hom}(M, N)$. Equation (22) implies that

$$W^\zeta(\mathcal{F}\mathrm{lk}(k)) \cong \bigoplus_M W^1(\mathrm{End}_{k[F_\mu]} M) \ .$$

with one summand for each isomorphism class of ζ-self-dual simple F_μ-link modules M.

Proof. Every exact sequence in $\mathcal{F}\mathrm{lk}(k)|_M$ splits, so $\Theta_{M,b}$ is exact and hence induces a morphism of Witt groups

$$W^\zeta(\mathcal{F}\mathrm{lk}(k)|_M) \to W^1(\mathrm{End}_{k[F_\mu]}(M)).$$

Since $\Theta_{M,b}$ is an equivalence of hermitian categories it follows by Lemma A.2 of Appendix A that the induced map of Witt groups is an isomorphism. $\square$

Equation (23) above is a special case of corollary 3.20 so the following proposition completes step 2:

Proposition 3.21. *The endomorphism ring of every module in $\mathcal{F}\mathrm{lk}(k)$ is of finite dimension over k.*

Proposition 3.21 follows from Theorem 5.17 and Lemma 5.34 below. Theorem 5.17 can be considered analogous to the geometric fact that one can choose a Seifert surface for an F_μ-link. Since the homology of a Seifert surface is finite-dimensional the endomorphism ring of the associated Seifert module is finite-dimensional. Using chapter 12 of [51], part 2. of Lemma 6.1 and part 2. of Lemma 6.2 we can also deduce that every division ring with involution which is finite-dimensional over $\mathbb{Q}$ arises as

$$(\mathrm{End}_{\mathbb{Q}[F_\mu]} M \ , \ f \mapsto \overline{f} = b^{-1} f^\wedge b)$$

for some pair (M, b) where M is a simple module in $\mathcal{F}\mathrm{lk}(\mathbb{Q})$.

The proofs of the results cited in the previous paragraph do not use Proposition 3.21 (i.e. the arguments presented are not circular). However, the spirit of this section is to define invariants of F_μ-links by studying the category $\mathcal{F}\mathrm{lk}(\mathbb{Q})$ directly so we desire a proof of Theorem 3.21 which avoids any choice of Seifert surface or Seifert module. One such proof is due to Lewin [34]. In a subsequent paper we shall give a proof which applies when $k[F_\mu]$ is replaced by a wider class of rings.

Before leaving the subject of Morita equivalence we pause to note the following "naturality" statement which we will need in Section 6 to prove Theorem 1.4. The reader may refer to equation (19) for the definition of composition for duality-preserving functors.

Proposition 3.22. *Suppose $(G, \Psi, \eta') : \mathcal{C} \to \mathcal{D}$ is a duality-preserving functor between hermitian categories and $b : M \to M^*$ is an η-hermitian form in $\mathcal{C}$. Let $E = \mathrm{End}_\mathcal{C}(M)$ and let $E' = \mathrm{End}_\mathcal{D} G(M)$. The following diagram of duality-preserving functors commutes up to natural isomorphism:*

$$
\begin{array}{ccc}
\mathcal{C}|_M & \xrightarrow{\ (G,\Psi,\eta')\ } & \mathcal{D}|_{G(M)} \\
{\scriptstyle (\mathrm{Hom}(M,\_),\Omega^b,\eta)}\Big\downarrow & & \Big\downarrow{\scriptstyle (\mathrm{Hom}(G(M),\_),\Omega^{\eta'\Psi G(b)},\eta\eta')} \\
E\text{-Proj} & \xrightarrow[\ (E'\otimes_E \_,\Pi,1)\ ]{} & E'\text{-Proj.}
\end{array}
$$

Proof. See Section A.2 of Appendix A. $\qquad\square$

Corollary 3.23. *If G is exact then the following square also commutes:*

$$
\begin{array}{ccc}
W^\zeta(\mathcal{C}|_M) & \xrightarrow{\ G\ } & W^{\zeta\eta'}(\mathcal{D}|_{G(M)}) \\
{\scriptstyle \Theta_{M,b}}\Big\downarrow{\scriptstyle \cong} & & {\scriptstyle \cong}\Big\downarrow{\scriptstyle \Theta_{G(M),\eta'\Psi G(b)}} \\
W^{\zeta\eta}(E) & \xrightarrow[\ G\ ]{} & W^{\zeta\eta}(E').
\end{array}
$$

3.4 Step 3: Invariants

Equation (22) leads one to consider invariants to distinguish Witt classes of forms over division algebras E of finite dimension over $\mathbb{Q}$. Such division algebras are well understood (see Albert [1, p149,p161], Scharlau [48] or our earlier summary in [51, §11.1,11.2]).

One considers five distinct classes of division algebras with involution. Firstly a division algebra E may be commutative or non-commutative. Secondly, if I is an involution on E let $\mathrm{Fix}(I) = \{a \in E \mid I(a) = a\}$. The involution is said to be "of the first kind" if $\mathrm{Fix}(I)$ contains the center $K = Z(E)$ of E. Otherwise, the involution is "of the second kind". Finally, one of these four classes is further partitioned. A non-commutative division algebra with involution of the first kind is necessarily a quaternion algebra, with presentation $K\langle i, j \mid i^2 = a, j^2 = b, ij = -ji\rangle$ for some number field K and some elements $a, b \in K$. If $\mathrm{Fix}(I) = Z(E) = K$ then the involution is called "standard". On the other hand if $\mathrm{Fix}(I)$ strictly larger than K then the involution is called "non-standard".

Table (29) below lists sufficient invariants to distinguish the Witt classes of forms over each class of division algebras with involution. The symbol m (2) denotes dimension modulo 2. The letter σ signifies all signature invariants (if any) each of which takes values in $\mathbb{Z}$. The discriminant Δ is the determinant with a possible sign adjustment and takes values in the group of "square classes"

$$\frac{\mathrm{Fix}(I) \cap K}{\{aI(a) \mid a \in K^\bullet\}}$$

where K is the center of E and $K^\bullet = K \setminus 0$.

Kind	Commutative?	Involution	Invariants
1st	Yes	(Trivial)	m (2), σ, Δ, c
1st	No	Standard	m (2), σ
1st	No	Non-standard	m (2), σ, Δ, θ
2nd	Yes	(Non-trivial)	m (2), σ, Δ
2nd	No	(Non-trivial)	m (2), σ, Δ

(29)

Two symbols in the table have not yet been mentioned. The Hasse-Witt invariant, c, which appears in the first row takes values in a direct sum of copies of $\{1, -1\}$, one copy for each prime of the number field K. Finally, if E is a quaternion algebra with non-standard involution of the first kind then the local-global principle fails and one requires a secondary invariant such as the Lewis θ which is defined if all the other invariants vanish. The value group for θ is the quotient $\{1, -1\}^S / \sim$ where S is the set of primes $\mathfrak{p}$ of K such that the completion $E_\mathfrak{p}$ is a division algebra and the relation $\sim$ identifies each element $\{\epsilon_\mathfrak{p}\}_{\mathfrak{p} \in S}$ with its antipode $\{-\epsilon_\mathfrak{p}\}_{\mathfrak{p} \in S}$.

4   Seifert forms

In this section we describe algebraic structures arising in the study of a Seifert surface of an F_μ-link. We define in Section 4.1 a category $\mathcal{S}ei_\infty(A)$

of "Seifert modules" and a full subcategory $\mathcal{S}ei(A) \subset \mathcal{S}ei_\infty(A)$ in which the objects are finitely generated and projective as A-modules. The category $\mathcal{S}ei(A)$ was denoted $(P_\mu - A)$-Proj in [51].

In Section 4.2 we prove that every F_μ-link module is also a Seifert module in a canonical way (cf Farber [16]) and obtain a "forgetful" functor

$$ U : \mathcal{F}lk_\infty(A) \to \mathcal{S}ei_\infty(A). $$

The image of $\mathcal{F}lk(A)$ is usually not contained in $\mathcal{S}ei(A)$ which explains our motivation for introducing $\mathcal{S}ei_\infty(A)$. We construct later (Section 5) a functor B from $\mathcal{S}ei(A)$ to $\mathcal{F}lk(A)$ which extends in an obvious way to a functor $B : \mathcal{S}ei_\infty(A) \to \mathcal{F}lk_\infty(A)$ and is left adjoint to U.

In Section 4.3 we put hermitian structure on $\mathcal{S}ei(A)$. We will see in Section 6 below that the functor B induces an isomorphism of Witt groups $W^\varsigma(\mathcal{S}ei(A)) \to W^{-\varsigma}(\mathcal{F}lk(A))$ when A is semi-simple and Artinian and, in particular, when $A = \mathbb{Q}$.

4.1   Seifert modules

Suppose V is a finite-dimensional vector space over a field k and $\alpha : V \to V$ is an endomorphism. A time-honoured technique in linear algebra regards the pair (V, α) as a module over a polynomial ring $k[s]$ in which the action of s on V is given by α. Equivalently, (V, α) is a representation of $\mathbb{Z}[s]$ in the category of finite-dimensional vector spaces over k. We shall use the words "module" and "representation" interchangeably.

Given a Seifert surface $U^{n+1} \subset S^{n+2}$ for a knot $S^n \subset S^{n+2}$, small translations in the directions normal to U induce homomorphisms

$$ f^+, f^- : H_i(U) \to H_i(S^{n+2} \setminus U). $$

Using Alexander duality one finds that $f^+ - f^-$ is an isomorphism for $i \neq 0, n+1$, so $H_i(U)$ is endowed with an endomorphism $(f^+ - f^-)^{-1}f^+$ and may therefore be regarded as a representation of a polynomial ring $\mathbb{Z}[s]$. The homology of a Seifert surface for a μ-component boundary link has, in addition to the endomorphism $(f^+ - f^-)^{-1}f^+$, a system of μ orthogonal idempotents which express the component structure of the Seifert surface. Following Farber [17] (see also [51]) we regard $H_i(U)$ as a representation of the ring

$$P_\mu = \mathbb{Z}\left\langle s, \pi_1, \cdots, \pi_\mu \;\middle|\; \pi_i^2 = \pi_i;\; \pi_i\pi_j = 0 \text{ for } i \neq j;\; \sum_{i=1}^{\mu} \pi_i = 1 \right\rangle.$$

$$\cong \mathbb{Z}[s] *_{\mathbb{Z}} \left(\prod_{\mu} \mathbb{Z} \right).$$

P_μ is also the path ring of a quiver which we illustrate in the case $\mu = 2$:

Notation 4.1. Let $\mathcal{S}ei_\infty(A)$ denote the category of representations of P_μ by A-modules. An object in $\mathcal{S}ei_\infty(A)$ (called a Seifert module) is a pair (V, ρ) where V is an A-module and $\rho : P_\mu \to \mathrm{End}_A(V)$ is a ring homomorphism.

Let $\mathcal{S}ei(A)$ denote the category of representations of P_μ by finitely generated projective A-modules. In other words, $\mathcal{S}ei(A)$ is the full subcategory of pairs (V, ρ) such that V is finitely generated and projective.

We sometimes omit ρ, confusing an element of P_μ with its image in $\mathrm{End}_A(V)$.

4.2   Seifert structure on F_μ-link modules

Every F_μ-link module has a canonical Seifert module structure which we describe next. In fact $\mathcal{F}lk_\infty(A)$ can be regarded as a full subcategory of $\mathcal{S}ei_\infty(A)$. Note however that a module in $\mathcal{F}lk(A)$ is in general neither finitely generated nor projective as an A-module (e.g. see Lemma 2.3 above) and therefore does not lie in $\mathcal{S}ei(A)$.

If $M \in \mathcal{F}lk_\infty(A)$ then Lemma 2.3 implies that the A-module map

$$\gamma : M^{\oplus\mu} \to M$$

$$(m_1, \cdots, m_\mu) \mapsto \sum_{i=1}^{\mu}(1 - z_i)m_i$$

is an isomorphism. Let p_i denote the projection of $M^{\oplus\mu}$ onto its ith component and let

$$\omega : M^{\oplus\mu} \to M$$

$$(m_1, \cdots, m_\mu) \mapsto \sum_{i=1}^{\mu} m_i$$

denote addition. Define $\rho : P_\mu \to \mathrm{End}_A M$ by

$$\rho(\pi_i) = \gamma p_i \gamma^{-1}$$
$$\rho(s) = \omega \gamma^{-1}. \tag{30}$$

We denote by $U(M)$ the A-module M with the Seifert module structure ρ. As remarked in the introduction, in the case $\mu = 1$ the Seifert module structure $\rho : P_\mu \to \mathrm{End}_A(M)$ can be described more simply by the equation

$$\rho(s) = (1 - z)^{-1}.$$

The following lemma says that $U : \mathcal{F}\mathrm{lk}_\infty(A) \to \mathcal{S}\mathrm{ei}_\infty(A)$ is a full and faithful functor, so $\mathcal{F}\mathrm{lk}_\infty(A)$ can be regarded as a full subcategory of $\mathcal{S}\mathrm{ei}_\infty(A)$.

Lemma 4.2. *An A-module morphism $f : M \to M'$ between F_μ-link modules M and M' is an $A[F_\mu]$-module morphism if and only if f is a morphism of Seifert modules. In other words, $f \in \mathrm{Hom}_{\mathcal{S}\mathrm{ei}_\infty(A)}(U(M), U(M'))$ if and only if $f \in \mathrm{Hom}_{\mathcal{F}\mathrm{lk}_\infty(A)}(M, M')$.*

Proof. If $f : M \to M'$ is an $A[F_\mu]$-module morphism then the diagram

$$
\begin{array}{ccc}
M^{\oplus \mu} & \xrightarrow{f^{\oplus \mu}} & M'^{\oplus \mu} \\
\gamma \downarrow & & \downarrow \gamma \\
M & \xrightarrow{f} & M'
\end{array}
\tag{31}
$$

is commutative. Conversely, if the diagram (31) commutes then the equation $f((1 - z_i)x) = (1 - z_i)f(x)$ holds for each $x \in M$ and $i = 1, \cdots, \mu$, so $f(z_i x) = z_i f(x)$ for each i and hence f is an $A[F_\mu]$-module morphism.

It remains to show that (31) commutes if and only if f is a Seifert morphism. If (31) commutes then

$$f(\gamma p_i \gamma^{-1} x) = \gamma p_i \gamma^{-1} f(x) \quad \text{and} \quad f(\omega \gamma^{-1} x) = \omega \gamma^{-1} f(x)$$

for each $x \in M$ so f is a Seifert morphism. Conversely, suppose f is a Seifert morphism. Now

$$
\gamma^{-1} = \begin{pmatrix} \omega p_1 \gamma^{-1} \\ \omega p_2 \gamma^{-1} \\ \vdots \\ \omega p_\mu \gamma^{-1} \end{pmatrix} = \begin{pmatrix} (\omega \gamma^{-1})(\gamma p_1 \gamma^{-1}) \\ (\omega \gamma^{-1})(\gamma p_2 \gamma^{-1}) \\ \vdots \\ (\omega \gamma^{-1})(\gamma p_\mu \gamma^{-1}) \end{pmatrix} : M \to M^{\oplus \mu}
$$

so the diagram (31) commutes. $\qquad\square$

4.3   Seifert forms

Let us make $\mathcal{S}ei(A)$ a hermitian category, assuming A is a ring with involution. Recall that if V is a finitely generated projective left A-module then
$V^* = \mathrm{Hom}(V, A)$ is a left A-module with $(a.\theta)(x) = \theta(x)\bar{a}$ for all $a \in A$,
$\theta \in V^*$ and $x \in V$.

If (V, ρ) is an object in $\mathcal{S}ei(A)$ define $(V, \rho)^* = (V^*, \rho^*)$ where

$$\rho^*(\pi_i) = \rho(\pi_i)^* : V^* \to V^* \quad \text{and} \quad \rho^*(s) = 1 - \rho(s)^* : V^* \to V^*.$$

It is easy to see that if $f : V \to V'$ is a morphism in $\mathcal{S}ei(A)$ then the
dual $f^* : (V')^* \to V^*$ again lies in $\mathcal{S}ei(A)$. Equivalently, if one gives the
ring P_μ the involution defined by $\bar{s} = 1 - s$ and $\bar{\pi}_i = \pi_i$ for each i then
$\rho^* : P_\mu \to \mathrm{End}(V^*)$ is given by $\rho^*(r)(\theta)(x) = \theta(\rho(\bar{r})x)$ for all $x \in V$, $\theta \in V^*$
and $r \in P_\mu$.

If $V = H_q(U^{2q})/\text{torsion}$ then the intersection form $\phi : V \to V^*$ is a
morphism in $\mathcal{S}ei(\mathbb{Z})$. In other words, the intersection form respects the projections π_i and respects the endomorphism $(f^+ - f^-)^{-1}f^+$ (see section 4.1)
in the sense that

$$\phi((f^+ - f^-)^{-1}f^+x)(y) = \phi(x)((1 - (f^+ - f^-)^{-1}f^+)y).$$

for all $x, y \in V$. Furthermore, ϕ is an isomorphism by Poincaré duality and
is $(-1)^q$-hermitian. This form ϕ will be called the Seifert form associated
to a Seifert surface. Kervaire [25, p94] or Lemma 3.31 of [51] shows that by
an elementary change of variables, this form ϕ is equivalent to the Seifert
matrix of linking numbers more commonly encountered in knot theory.

By Ko [27] and Mio [38] (see also Lemma 3.31 of [51]), the association
of this form ϕ to a Seifert surface induces the isomorphism

$$C_{2q-1}(F_\mu) \cong W^{(-1)^q}(\mathcal{S}ei(\mathbb{Z})) \qquad (q \geq 3).$$

mentioned in the introduction. Every $(-1)^q$-hermitian Seifert form is associated to some $2q$-dimensional Seifert surface. Although there are many
possible Seifert surfaces for a given F_μ-link, all are cobordant and the corresponding Seifert forms lie in the same Witt class.

In [51] the author applied to $\mathcal{S}ei(\mathbb{Q})$ the steps 1-3. described in Section 3.1
obtaining explicit invariants to distinguish F_μ-link cobordism classes. Although the Blanchfield-Duval form is more intrinsic, the advantage of the
Seifert form is that it is easier to compute the numerical invariants. For
illustration, we treat a worked example:

Example 4.3. Setting $\mu = 2$, consider the Seifert module $V = \mathbb{Z}^6$ with the endomorphism s and (-1)-hermitian form ϕ given by

$$
s = \left(\begin{array}{cccc|cc}
1 & 0 & 1 & 0 & 0 & 0 \\
0 & 1 & -1 & -1 & -1 & 0 \\
0 & 1 & 0 & 0 & 0 & -1 \\
0 & 0 & 0 & 0 & 0 & 0 \\
\hline
0 & 1 & 0 & 0 & 1 & -1 \\
0 & 0 & 1 & 0 & 1 & 0
\end{array}\right)
\quad\text{and}\quad
\phi = \left(\begin{array}{cccc|cc}
0 & 0 & 0 & 1 & 0 & 0 \\
0 & 0 & -1 & 0 & 0 & 0 \\
0 & 1 & 0 & 0 & 0 & 0 \\
-1 & 0 & 0 & 0 & 0 & 0 \\
\hline
0 & 0 & 0 & 0 & 0 & -1 \\
0 & 0 & 0 & 0 & 1 & 0
\end{array}\right)
$$

The horizontal and vertical lines indicate the component structure of the Seifert form. In other words, π_1 projects onto the span of the first four basis elements while π_2 projects onto the span of the last two. The corresponding Seifert matrix of linking numbers is

$$
\phi s = \left(\begin{array}{cccc|cc}
0 & 0 & 0 & 0 & 0 & 0 \\
0 & -1 & 0 & 0 & 0 & 1 \\
0 & 1 & -1 & -1 & -1 & 0 \\
-1 & 0 & -1 & 0 & 0 & 0 \\
\hline
0 & 0 & -1 & 0 & -1 & 0 \\
0 & 1 & 0 & 0 & 1 & -1
\end{array}\right)
$$

but we shall work directly with s and ϕ.

The first step is to pass from $\mathbb{Z}$ to $\mathbb{Q}$ so we regard the entries in the matrices as rational numbers. Devissage is next; let $e_1, \cdots, e_6$ denote the standard basis of $\mathbb{Q}^6$. Now $\mathbb{Q}e_1$ is s-invariant and $\phi(e_1)(e_1) = 0$ so our Seifert form is Witt-equivalent to the induced form on $e_1^\perp / \mathbb{Q}e_1 \cong \mathbb{Q}\{e_2, e_3, e_5, e_6\}$. We have reduced s and ϕ to

$$
s' = \left(\begin{array}{cc|cc}
1 & -1 & -1 & 0 \\
1 & 0 & 0 & -1 \\
1 & 0 & 1 & -1 \\
0 & 1 & 1 & 0
\end{array}\right)
\quad\text{and}\quad
\phi' = \left(\begin{array}{cc|cc}
0 & -1 & 0 & 0 \\
1 & 0 & 0 & 0 \\
0 & 0 & 0 & -1 \\
0 & 0 & 1 & 0
\end{array}\right)
$$

The two-dimensional representation (over $\mathbb{Q}$) of $\mathbb{Z}[s]$ given by the matrix $r = \begin{pmatrix} 1 & -1 \\ 1 & 0 \end{pmatrix}$ is simple (=irreducible) since there do not exist eigenvalues in $\mathbb{Q}$. It follows that the Seifert module $V' = \mathbb{Q}^4$ with the action s' and π_1, π_2 shown is simple. The devissage process is therefore complete.

Turning to the Morita equivalence step, the endomorphism ring of this module V' has $\mathbb{Q}$-basis consisting of the identity and $\left(\begin{array}{c|c} r & 0 \\ \hline 0 & r \end{array}\right)$. The minimum polynomial of r is $x^2 - x + 1$ so $\mathrm{End}_{\mathcal{S}ei(\mathbb{Q})}(V')$ is isomorphic to $\mathbb{Q}(\sqrt{-3})$.

We may choose $b = -\phi' : V' \to (V')^*$. It is easy to verify that the involution $f \mapsto b^{-1}f^*b$ is not the identity map so it must send $\sqrt{-3}$ to $-\sqrt{-3}$. Morita equivalence sends the form $\phi' : V' \to V'^*$ to the composite

$$\mathrm{Hom}_{\mathcal{S}\mathrm{ei}(\mathbb{Q})}(V', V') \xrightarrow{\phi'} \mathrm{Hom}_{\mathcal{S}\mathrm{ei}(\mathbb{Q})}(V', V'^*)$$

$$\xrightarrow{\Omega^b_{V'}} \mathrm{Hom}_{\mathbb{Q}(\sqrt{-3})}(\mathrm{Hom}(V', V'^*), \mathrm{Hom}(V', V'))$$

which is given by

$$\Omega^b_{V'}\phi'_*(\alpha)(\beta) = -b^{-1}\beta^*\phi'\alpha = \bar{\beta}\alpha.$$

for $\alpha, \beta \in \mathrm{Hom}(V', V')$. This form may be written $\langle 1 \rangle$ as a form over $\mathbb{Q}(\sqrt{-3})$.

Reading the fourth line of the table (29), an element of the Witt group $W(\mathbb{Q}(\sqrt{-3}))$ for non-trivial involution is determined by signatures and discriminant (and rank modulo 2 if there are no signatures). Up to complex conjugation there is precisely one embedding of $\mathbb{Q}(\sqrt{-3})$ in $\mathbb{C}$ (with the complex conjugate involution), so there is in fact one signature, which takes value $1 \in \mathbb{Z}$ with our choice of b. The discriminant is

$$1 \in \frac{\mathbb{Q} \setminus 0}{\mathbb{Q}(\sqrt{-3})\overline{\mathbb{Q}(\sqrt{-3})}} = \frac{\mathbb{Q} \setminus 0}{\{a^2 + 3b^2 \mid a, b \in \mathbb{Q}\}}.$$

5   The Covering construction

In this section we introduce a functor $B : \mathcal{S}\mathrm{ei}_\infty(A) \to \mathcal{F}\mathrm{lk}_\infty(A)$ which is the algebraic analogue of the geometric construction of the free cover of an F_μ-link complement from a Seifert surface. (illustrated in Figure 1 on page 150). The restriction of B to $\mathcal{S}\mathrm{ei}(A)$ takes values in $\mathcal{F}\mathrm{lk}(A)$ and extends to a duality-preserving functor

$$(B, \Phi, -1) : \mathcal{S}\mathrm{ei}(A) \to \mathcal{F}\mathrm{lk}(A)$$

which is natural in A (see Propositions 5.4 and 5.7).

We show that $B : \mathcal{S}\mathrm{ei}_\infty(A) \to \mathcal{F}\mathrm{lk}_\infty(A)$ is left adjoint to the full and faithful functor $U : \mathcal{F}\mathrm{lk}_\infty(A) \to \mathcal{S}\mathrm{ei}_\infty(A)$; in other words, among functors $\mathcal{S}\mathrm{ei}_\infty(A) \to \mathcal{F}\mathrm{lk}_\infty(A)$, the geometrically motivated functor B satisfies a universal property with respect to U (see Definition 5.8). In particular there is a natural transformation $\theta_V : V \to UB(V)$ for $V \in \mathcal{S}\mathrm{ei}(A)$ and a natural isomorphism $\psi_M : BU(M) \to M$ for $M \in \mathcal{F}\mathrm{lk}_\infty(A)$. The reader is referred to chapter 3 of Borceux [4] or chapter IV of Mac Lane [35] for a detailed treatment of adjoint functors.

We use the adjunction in Section 5.3 to show that the covering construction $B : \mathcal{S}ei_\infty(A) \to \mathcal{F}lk_\infty(A)$ is equivalent to a universal localization $\mathcal{S}ei_\infty(A) \to \mathcal{S}ei_\infty(A)/\mathcal{P}rim_\infty(A)$ of categories. We describe the structure of the "primitive" modules $V \in \mathcal{P}rim_\infty(A)$ in Section 5.5 and outline a construction of the quotient category in Section 5.6.

We show in Sections 5.3 and 5.4 that $B : \mathcal{S}ei(A) \to \mathcal{F}lk(A)$ is equivalent to a localization $\mathcal{S}ei(A) \to \mathcal{S}ei(A)/\mathcal{P}rim_\infty(A)$ of hermitian categories. In the case where A is a semi-simple Artinian ring we simplify the descriptions of the quotient and primitive modules in Section 5.7.

5.1 Definition

To simplify notation in this section and Section 5.2, we suppress the symbol ρ which appears in the definition of a Seifert module (V, ρ), identifying an element $r \in P_\mu$ with $\rho(r) \in \mathrm{End}_A(V)$. We shall extend Seifert structure from an A-module V to the induced module $A[F_\mu] \otimes_A V$ by $s(\alpha \otimes v) = \alpha \otimes s(v)$ and $\pi_i(\alpha \otimes v) = \alpha \otimes \pi_i(v)$ for $\alpha \in A[F_\mu]$.

Recall that $z_1, \cdots, z_\mu$ are distinguished generators of F_μ; let us now write $z = \sum z_i \pi_i$.

Definition 5.1. If V is a module in $\mathcal{S}ei_\infty(A)$ let

$$B(V) = \mathrm{Coker}\left((1 - s(1 - z)) : V[F_\mu] \to V[F_\mu]\right).$$

Since $\epsilon(1 - s(1 - z)) = 1$, it is clear that $B(V)$ lies in $\mathcal{F}lk_\infty(A)$. To achieve more symmetric notation we write $z_i = y_i^2$ and deduce $z = y^2$ where $y = \sum y_i \pi_i$. We write $F_\mu = F_\mu(y^2)$ when we wish to indicate that elements of F_μ are to be written as words in the symbols $y_i^{\pm 2}$. The $A[F_\mu]$-module $\bigoplus_{i=1}^{\mu} \pi_i V[F_\mu(y^2)y_i]$ is isomorphic to $V[F_\mu]$ and will be written $V[F_\mu(y^2)y]$ for brevity. Now we have

$$\begin{aligned}
B(V) &\cong \mathrm{Coker}\left((1 - s(1 - y^2))y^{-1} : V[F_\mu(y^2)y] \to V[F_\mu(y^2)]\right) \\
&= \mathrm{Coker}\left((1 - s)y^{-1} + sy : V[F_\mu(y^2)y] \to V[F_\mu(y^2)]\right).
\end{aligned} \tag{32}$$

In detail,

$$(1 - s)y^{-1} + sy = \sum_{i=1}^{\mu}(1 - s)\pi_i y_i^{-1} + s\pi_i y_i : vwy_i \mapsto (1 - s)(v)w + s(v)wy_i^2$$

for $v \in \pi_i(V)$, $w \in F_\mu(y^2)$ and $i \in \{1, \cdots, \mu\}$. A morphism $f : V \to V'$ induces a commutative diagram

$$
\begin{array}{ccccccccc}
0 & \longrightarrow & V[F_\mu] & \xrightarrow{\ \sigma\ } & V[F_\mu] & \xrightarrow{\ q\ } & B(V) & \longrightarrow & 0 \\
& & \downarrow{\scriptstyle f} & & \downarrow{\scriptstyle f} & & \downarrow{\scriptstyle B(f)} & & \\
0 & \longrightarrow & V'[F_\mu] & \xrightarrow{\ \sigma'\ } & V'[F_\mu] & \xrightarrow{\ q'\ } & B(V') & \longrightarrow & 0
\end{array}
\tag{33}
$$

where $\sigma = 1 - s(1 - z)$ and hence induces an $A[F_\mu]$-module map $B(f)$ as shown.

Lemma 5.2. *The functor B is exact. In other words, if $V \to V' \to V''$ is an exact sequence in $\mathcal{S}ei_\infty(A)$ then the sequence $B(V) \to B(V') \to B(V'')$ induced is also exact.*

Proof. It suffices to show that B preserves short exact sequences. Suppose $0 \to V \to V' \to V'' \to 0$ is exact. There is a commutative diagram

$$
\begin{array}{ccccccccc}
& & 0 & & 0 & & 0 & & \\
& & \downarrow & & \downarrow & & \downarrow & & \\
0 & \to & V[F_\mu] & \longrightarrow & V[F_\mu] & \longrightarrow & B(V) & \to & 0 \\
& & \downarrow & & \downarrow & & \downarrow & & \\
0 & \to & V'[F_\mu] & \to & V'[F_\mu] & \to & B(V') & \to & 0 \\
& & \downarrow & & \downarrow & & \downarrow & & \\
0 & \to & V''[F_\mu] & \to & V''[F_\mu] & \to & B(V'') & \to & 0 \\
& & \downarrow & & \downarrow & & \downarrow & & \\
& & 0 & & 0 & & 0 & &
\end{array}
$$

in which the rows and the two left-most columns are exact. It follows that the right-hand column is also exact. $\qquad\square$

The category $\mathcal{S}ei_\infty(A)$ has limits and colimits. For example, the coproduct of a system of modules is the direct sum. Since B is exact and respects arbitrary direct sums B respects all colimits:

Lemma 5.3. *The functor B is cocontinuous.* $\qquad\square$

In particular if V is a direct limit $V = \varinjlim V_i$ then $B(V) = \varinjlim B(V_i)$. On the other hand B does not respect infinite limits. For example one finds $B(\prod V_i) \not\cong \prod(B(V_i))$ because $(\prod V_i)[F_\mu] \not\cong \prod(V_i[F_\mu])$. However, B does respect finite limits.

The idea behind the proof of the following proposition is due to A.Ranicki.

Proposition 5.4. *The functor B extends to a duality-preserving functor $(B, \Phi, -1) : \mathcal{S}ei(A) \to \mathcal{F}lk(A)$.*

Proof. For each finitely generated projective A-module V there is a natural isomorphism $\Pi_V : V^*[F_\mu] \to (V[F_\mu])^*$ by Example 20. Naturality asserts that for each morphism $\alpha : V \to W$ the diagram

$$\begin{array}{ccc} W^*[F_\mu] & \xrightarrow{\alpha^*} & V^*[F_\mu] \\ \Pi_W \downarrow & & \downarrow \Pi_V \\ (W[F_\mu])^* & \xrightarrow{\alpha^*} & (V[F_\mu])^* \end{array}$$

is commutative. Moreover one can check commutativity of

$$\begin{array}{ccc} V^*[F_\mu] & \xrightarrow{z_i^{-1}} & V^*[F_\mu] \\ \Pi_V \downarrow & & \downarrow \Pi_V \\ (V[F_\mu])^* & \xrightarrow{z_i^*} & (V[F_\mu])^* \end{array}$$

where, as usual, $z_i : V[F_\mu] \to V[F_\mu]$ and $z_i : V^*[F_\mu] \to V^*[F_\mu]$ denote multiplication on the right by z_i.

Now if $V \in \mathcal{S}\mathrm{ei}(A)$ then there is a commutative diagram:

$$\begin{array}{ccccccccc} 0 & \longrightarrow & V^*[F_\mu] & \xrightarrow{\sigma(V^*)} & V^*[F_\mu] & \longrightarrow & B(V^*) & \longrightarrow & 0 \\ & & {\scriptstyle -\Pi_V(1-z)}\downarrow & & \downarrow{\scriptstyle \Pi_V(1-z^{-1})} & & \downarrow{\scriptstyle \Phi_V} & & \\ 0 & \longrightarrow & (V[F_\mu])^* & \xrightarrow{\sigma(V)^*} & (V[F_\mu])^* & \longrightarrow & B(V)^\wedge & \longrightarrow & 0 \end{array} \quad (34)$$

where $\sigma(V^*) = 1 - (1 - s^*)(1 - z)$ and $\sigma(V)^* = (1 - s(1 - z))^*$. By definition, $\Phi_V : B(V^*) \to B(V)^\wedge$ is the induced morphism. Plainly Φ_V is a natural transformation.

The duality-preserving functor Π has the property

$$\Pi_V^* i_{V[F_\mu]} = \Pi_{V^*} i_V : V[F_\mu] \to (V^*[F_\mu])^*$$

(indeed, this equation features in Definition 2.24). The equations

$$(\Pi_V(1 - z^{-1}))^* i_{V[F_\mu]} = \Pi_{V^*}(1 - z)i_V$$
$$(\Pi_V(1 - z))^* i_{V[F_\mu]} = \Pi_{V^*}(1 - z^{-1})i_V$$

imply that

$$\Phi_V^\wedge i_{B(V)} = -\Phi_{V^*} B(i_V).$$

To show that $(B, \Phi, -1)$ is a duality-preserving functor it remains to check that Φ is an isomorphism. There is a commutative diagram

$$\begin{array}{ccccccccc} 0 & \longrightarrow & (V[F_\mu])^* & \xrightarrow{\sigma(V)^*} & (V[F_\mu])^* & \longrightarrow & B(V)^\wedge & \longrightarrow & 0 \\ & & {\scriptstyle -(1-s^*)\Pi_V^{-1}}\downarrow & & \downarrow{\scriptstyle -(1-s^*)z\Pi_V^{-1}} & & \downarrow & & \\ 0 & \longrightarrow & V^*[F_\mu] & \xrightarrow{\sigma(V^*)} & V^*[F_\mu] & \longrightarrow & B(V^*) & \longrightarrow & 0 \end{array}$$

and the composite morphisms of chain complexes

$$
\begin{array}{ccc}
V^*[F_\mu] & \xrightarrow{\sigma(V^*)} & V^*[F_\mu] \\
{\scriptstyle (1-s^*)(1-z)}\Big\downarrow & {\scriptstyle 1} & \Big\downarrow{\scriptstyle (1-s^*)(1-z)} \\
V^*[F_\mu] & \xrightarrow[\sigma(V^*)]{} & V^*[F_\mu]
\end{array}
$$

and

$$
\begin{array}{ccc}
(V[F_\mu])^* & \xrightarrow{\sigma(V)^*} & (V[F_\mu])^* \\
{\scriptstyle (1-z^{-*})(1-s^*)}\Big\downarrow & {\scriptstyle z^{-*}} & \Big\downarrow{\scriptstyle -(1-z^{-*})(1-s^*)z^{-*}} \\
(V[F_\mu])^* & \xrightarrow[\sigma(V)_*]{} & (V[F_\mu])^*
\end{array}
$$

are chain homotopic to the identity by the indicated chain homotopies. [Chain complexes are drawn horizontally and morphisms of chain complexes are given by vertical arrows. The symbol z^{-*} is shorthand for $(z^{-1})^*$]. These composite chain maps therefore induce the identity on $B(V^*)$ and $B(V)^\wedge$ respectively so Φ_V is an isomorphism and $(B, \Phi, -1)$ is a duality-preserving functor.   □

Since B is an exact functor we have:

Corollary 5.5. *The functor* $(B, \Phi, -1)$ *induces a homomorphism of Witt groups*

$$
B : W^\zeta(\mathcal{S}ei(A)) \to W^{-\zeta}(\mathcal{F}lk(A)).
$$

If (V, ϕ) is a ζ-hermitian form in $\mathcal{S}ei(A)$ then the covering construction gives $B(V, \phi) = (B(V), \Phi_V B(\phi))$ (Lemma 2.29) which can be described explicitly as follows. The morphism $\phi : V \to V^*$ induces $\phi : V[F_\mu] \to V^*[F_\mu]$. Let $\tilde{\phi} : V[F_\mu] \to \mathrm{Hom}_{A[F_\mu]_\Sigma}(V[F_\mu]_\Sigma, A[F_\mu]_\Sigma)$ denote the composition of ϕ with

$$
\Pi_V : V^*[F_\mu] \to (V[F_\mu])^* = \mathrm{Hom}_{A[F_\mu]}(V[F_\mu], A[F_\mu])
$$

(see Example 2.26) and the localization

$$
\mathrm{Hom}_{A[F_\mu]}(V[F_\mu], A[F_\mu]) \to \mathrm{Hom}_{A[F_\mu]_\Sigma}(V[F_\mu]_\Sigma, A[F_\mu]_\Sigma).
$$

If $m, m' \in B(V)$ we may write $m = q(x)$, $m' = q(x')$ for some $x, x' \in V[F_\mu]$.

It follows from Remark 2.18 that in $\dfrac{A[F_\mu]_\Sigma}{A[F_\mu]}$ we have

$$\begin{aligned}
\Phi_V B(\phi)(m)(m') &= \Phi_V B(\phi)(q(x))(q(x')) \\
&= q'(\Pi_V(1-z^{-1})\phi(x))(q(x')) \quad \text{(using (34))} \\
&= (\mathrm{id}\otimes\Pi_V(1-z^{-1})\phi(x))(\mathrm{id}\otimes\sigma)^{-1}(1\otimes x') \\
&= \widetilde{\phi}(x)((1-z)(\mathrm{id}\otimes\sigma)^{-1}(1\otimes x')).
\end{aligned} \tag{35}$$

Remark 5.6. If $V \in \mathcal{S}\mathrm{ei}(\mathbb{Z})$ and $\phi : V \to V^*$ is the Seifert form corresponding to a Seifert surface for an F_μ-link then by (35) $\Phi_V B(\phi) : B(V) \to B(V)^\wedge$ is the corresponding Blanchfield-Duval form for the F_μ-link; compare Kearton [23], Levine [33, Prop 14.3], Cochran and Orr [7, Thm4.2] and Ranicki [43, Defn32.7]. For example, setting $r = \phi(x)$, $s = \phi(x')$, $\Gamma = z$, $\theta = \phi s$, and $\epsilon = \zeta$ one obtains from (35) the equations appearing immediately prior to Theorem 4.2 in [7].

For the proof of Theorem 1.4 in Section 6 we will need the observation that $(B, \Phi, -1)$ respects a change of coefficients from $\mathbb{Z}$ to $\mathbb{Q}$. Let us make a more general statement. Recall from equation (19) the definition of composition for duality-preserving functors.

Proposition 5.7. *A ring homomorphism $A \to A'$ induces a diagram of duality-preserving functors*

$$\begin{array}{ccc}
\mathcal{S}\mathrm{ei}(A) & \xrightarrow{(A'\otimes_A -,\,\Pi,1)} & \mathcal{S}\mathrm{ei}(A') \\
{\scriptstyle (B,\Phi,-1)}\Big\downarrow & & \Big\downarrow{\scriptstyle (B,\Phi,-1)} \\
\mathcal{F}\mathrm{lk}(A) & \xrightarrow[(A'[F_\mu]\otimes_{A[F_\mu]} -,\,\Upsilon,1)]{} & \mathcal{F}\mathrm{lk}(A').
\end{array}$$

which commutes up to natural isomorphism. Consequently, there is a commutative diagram of Witt groups

$$\begin{array}{ccc}
W^\zeta(\mathcal{S}\mathrm{ei}(A)) & \longrightarrow & W^\zeta(\mathcal{S}\mathrm{ei}(A')) \\
{\scriptstyle B}\downarrow & & \downarrow{\scriptstyle B} \\
W^{-\zeta}(\mathcal{F}\mathrm{lk}(A)) & \to & W^{-\zeta}(\mathcal{F}\mathrm{lk}(A')).
\end{array} \tag{36}$$

Proof. See Appendix A. $\qquad\qquad\qquad\qquad\qquad\qquad\qquad\qquad\square$

5.2   Adjunction

We leave duality structures behind for the present and prove that the functor $B : \mathcal{S}ei_\infty(A) \to \mathcal{F}lk_\infty(A)$ is left adjoint to $U : \mathcal{F}lk_\infty(A) \to \mathcal{S}ei_\infty(A)$.

Definition 5.8. Suppose $F : \mathcal{C} \to \mathcal{D}$ is a functor. A functor $G : \mathcal{D} \to \mathcal{C}$ is called *left adjoint* to F if there exists a natural transformation $\theta : \mathrm{id}_\mathcal{D} \to FG$ such that for every object $D \in \mathcal{D}$ the morphism $\theta_D : D \to FG(D)$ has the following universal property: For every morphism d in $\mathcal{D}$ of the form $d : D \to F(C)$ there is a unique morphism $c : G(D) \to C$ in $\mathcal{C}$ such that $d = F(c)\theta_D$.

$$
\begin{array}{ccc}
D & \xrightarrow{\quad d \quad} & F(C) \\
& \theta_D \searrow \quad \nearrow F(c) & \\
& FG(D) &
\end{array}
\tag{37}
$$

Let us recall a few examples: 1) The inclusion of the category of abelian groups in the category of groups has left adjoint known as "abelianization" which sends a group G to $G/[G,G]$. 2) The inclusion of the category of compact Hausdorff topological spaces in the category of (all) topological spaces has a left adjoint known as "Stone-Čech compactification". 3) Colimit constructions (e.g. direct limit or coproduct) can be expressed via a left adjoint as follows. Suppose $\mathcal{C}$ is a category, J is a small category and $\mathcal{C}^J$ denotes the category of functors $J \to \mathcal{C}$. If there is a left adjoint to the constant functor $\mathcal{C} \to \mathcal{C}^J$ then that left adjoint sends each functor $J \to \mathcal{C}$ to its colimit in $\mathcal{C}$ (and the colimit exists).

Proposition 5.9. *The functor* $B : \mathcal{S}ei_\infty(A) \to \mathcal{F}lk_\infty(A)$ *is left adjoint to* $U : \mathcal{F}lk_\infty(A) \to \mathcal{S}ei_\infty(A)$.

The required map $\theta_V : V \to UB(V)$ is the restriction of the map $q : V[F_\mu] \to B(V)$ in the diagram (33) above. In symbols $\theta_V = q| : V \to UB(V)$. During the proof of Proposition 5.9 below we show that θ_V is a morphism of Seifert modules. It follows from the diagram (33) that $\theta : \mathrm{id} \to UB$ is a natural transformation.

Before proving Proposition 5.9, we note some consequences:

Corollary 5.10. *Let* $V \in \mathcal{S}ei_\infty(A)$ *and* $M \in \mathcal{F}lk_\infty(A)$. *There is a natural isomorphism* $\psi_M : BU(M) \to M$ *and the composites*

$$
U(M) \xrightarrow{\theta_{U(M)}} UBU(M) \xrightarrow{U(\psi_M)} U(M)
$$

$$
B(V) \xrightarrow{B(\theta_V)} BUB(V) \xrightarrow{\psi_{B(V)}} B(V)
$$

are identity morphisms. In particular $\theta_{U(M)}$ *and* $B(\theta_V)$ *are isomorphisms.*

The existence of a natural transformation $\psi_M : BU(M) \to M$ follows from Proposition 5.9 alone. To prove that ψ is an isomorphism one requires the additional information that U is full and faithful. We are not claiming that $\theta_V : V \to UB(V)$ is an isomorphism. Indeed, U and B are not equivalences of categories.

Proof of Corollary 5.10. Let $\psi_M : BU(M) \to M$ be the unique morphism such that $\mathrm{id}_{U(M)} = U(\psi_M)\theta_{U(M)}$. One can check that ψ_M is a natural transformation and that $\psi_{B(V)}B(\theta_V) = \mathrm{id}_{B(V)}$ (see for example Theorem 3.1.5 of [4]). The functor $U : \mathcal{F}\mathrm{lk}_\infty(A) \to \mathcal{S}\mathrm{ei}_\infty(A)$ is full and faithful by Lemma 4.2 so ψ_M is an isomorphism (see Theorem 3.4.1 of [4]). It follows that $\theta_{U(V)}$ and $B(\theta_V)$ are isomorphisms. □

Proof of Proposition 5.9. By Definition 5.8 there are two statements to prove:

1. The map $\theta_V : V \to UB(V)$ is a morphism of Seifert modules.

2. If $M \in \mathcal{F}\mathrm{lk}_\infty(A)$ and $f : V \to U(M)$ is a morphism in $\mathcal{S}\mathrm{ei}_\infty(A)$ then there is a unique morphism $g : B(V) \to M$ such that $f = U(g)\theta_V$.

As we remarked above, it follows from the diagram (33) that $\theta : \mathrm{id} \to UB$ is a natural transformation. We shall need the following lemma which is proved shortly below:

Lemma 5.11. *Suppose $V \in \mathcal{S}\mathrm{ei}_\infty(A)$, $M \in \mathcal{F}\mathrm{lk}_\infty(A)$ and $f : V \to M$ is an A-module morphism. Let $\widetilde{f} : V[F_\mu] \to M$ denote the induced $A[F_\mu]$-module morphism. The map $f : V \to U(M)$ is a morphism in $\mathcal{S}\mathrm{ei}_\infty(A)$ if and only if $f(x) = \widetilde{f}(s(1-z)x)$ for all $x \in V$.*

Let us deduce statement 1. above. By the definition of $B(V)$ there is an exact sequence

$$0 \to V[F_\mu] \xrightarrow{1-s(1-z)} V[F_\mu] \xrightarrow{q} B(V) \to 0$$

so $q(x) = q(s(1-z)x)$ for all $x \in V$. By Lemma 5.11, $\theta_V = q|_V$ is a morphism of Seifert modules.

We turn now to statement 2. Since V generates $B(V)$ as an $A[F_\mu]$-module, an $A[F_\mu]$-module morphism $g : B(V) \to M$ satisfies $f = U(g)\theta_V$ if and only if g fits into the diagram

$$0 \longrightarrow V[F_\mu] \xrightarrow{1-s(1-z)} V[F_\mu] \xrightarrow{q} B(V) \longrightarrow 0$$
$$\searrow_{\widetilde{f}} \quad \downarrow_{g}$$
$$M$$

Since f is a morphism of Seifert modules we have $f(x) = \widetilde{f}(s(1-z)x)$ for all $x \in V$ by Lemma 5.11. So $\widetilde{f}\sigma = 0$, and therefore there exists unique $g : B(V) \to M$ such that $gq = \widetilde{f}$. It follows that there exists unique g such that $f = U(g)\theta_V$. Thus we have established both 1. and 2. assuming Lemma 5.11.

Proof of Lemma 5.11. The Seifert module structure on $U(M)$ is given by (30) above so f is a Seifert morphism if and only if

a) $\omega\gamma^{-1}f(x) = f(sx)$ and

b) $\gamma p_i \gamma^{-1}f(x) = f(\pi_i x)$ for each $x \in V$.

To prove the 'if' part of Lemma 5.11, suppose $f(x) = \widetilde{f}(s(1-z)x)$.

a) The equations $\omega\gamma^{-1}\left(\sum_{i=1}^{\mu}(1-z_i)x_i\right) = \sum_{i=1}^{\mu} x_i$ and

$$f(x) = \widetilde{f}(s(1-z)x) = \sum_{i=1}^{\mu} \widetilde{f}(s(1-z_i)\pi_i x) = \sum_{i=1}^{\mu}(1-z_i)f(s\pi_i x).$$

imply that $\omega\gamma^{-1}f(x) = \omega\gamma^{-1}\sum_{i=1}^{\mu}(1-z_i)f(s\pi_i x) = \sum_{i=1}^{\mu} f(s\pi_i x) = f(sx)$.

b) Observe that

$$f(\pi_i x) = \widetilde{f}(s(1-z)\pi_i x) = \sum_{j=1}^{\mu} \widetilde{f}(s(1-z_j)\pi_j \pi_i x) = (1-z_i)f(s\pi_i x).$$

while

$$\gamma p_i \gamma^{-1} f(x) = \gamma p_i \gamma^{-1}\widetilde{f}(s(1-z)x) = \gamma p_i \gamma^{-1}\sum_{j=1}^{\mu}(1-z_j)f(s\pi_j x)$$
$$= (1-z_i)f(s\pi_i x).$$

Thus $f(\pi_i x) = \gamma p_i \gamma^{-1}f(x)$.

To prove the "only if" part of Lemma 5.11, suppose we have a) and b) above. Now

$$\widetilde{f}(s(1-z)x) = \sum_{i=1}^{\mu}(1-z_i)f(s\pi_i x)$$

$$= \sum_{i=1}^{\mu}(1-z_i)(\omega\gamma^{-1})(\gamma p_i \gamma^{-1})f(x)$$

$$= \sum_{i=1}^{\mu}(1-z_i)\omega p_i \gamma^{-1}f(x)$$

$$= f(x). \qquad \qquad \square$$

This completes the proof of Proposition 5.9. $\square$

5.3   Localization

When one passes from Seifert modules to F_μ-link modules, certain Seifert modules disappear altogether; following Farber we shall call such modules primitive.

Definition 5.12. Let $\mathcal{P}\mathrm{rim}_\infty(A)$ denote the full subcategory of $\mathcal{S}\mathrm{ei}_\infty(A)$ containing precisely the modules V such that $B(V) = 0$. Modules in $\mathcal{P}\mathrm{rim}_\infty(A)$ will be called primitive.

For example, if V is a Seifert module with $\rho : P_\mu \to \mathrm{End}_A V$ such that $\rho(s) = 0$ or $\rho(s) = 1$ then

$$(1 - \rho(s))y^{-1} + \rho(s)y : V[F_\mu(y^2)y] \to V[F_\mu(y^2)]$$

is an isomorphism and therefore has zero cokernel. A module in $\mathcal{S}\mathrm{ei}_\infty(A)$ with $\rho(s) = 0$ or 1 will be called *trivially primitive*. We show in Section 5.5 that all the primitive Seifert modules in $\mathcal{S}\mathrm{ei}_\infty(A)$ can be "built" from trivially primitive modules. If A is semi-simple Artinian then a similar result applies when one restricts attention to the category $\mathcal{S}\mathrm{ei}(A)$ of representations of P_μ by finitely generated projective A-modules: Every primitive in $\mathcal{S}\mathrm{ei}(A)$ can be "built" from a finite number of trivially primitive modules in $\mathcal{S}\mathrm{ei}(A)$ (see Proposition 5.33). This statement is not true for all rings A; one must consider primitives which exhibit a kind of nilpotence. Such primitives were described by Bass, Heller and Swan when $\mu = 1$ (see also Ranicki [44]). The general case $\mu \geq 1$ will be analyzed in a subsequent paper (joint work with A.Ranicki) [46].

In the present section we construct an equivalence between $\mathcal{F}\mathrm{lk}_\infty(A)$ and a quotient category $\mathcal{S}\mathrm{ei}_\infty(A)/\mathcal{P}\mathrm{rim}_\infty(A)$. This quotient is an example of universal localization for categories; the objects in $\mathcal{S}\mathrm{ei}_\infty(A)/\mathcal{P}\mathrm{rim}_\infty(A)$ are the same as the objects in $\mathcal{S}\mathrm{ei}_\infty(A)$ but the morphisms are different. The universal property is that a morphism in $\mathcal{S}\mathrm{ei}_\infty(A)$ whose kernel and cokernel are in $\mathcal{P}\mathrm{rim}_\infty(A)$ has an inverse in $\mathcal{S}\mathrm{ei}_\infty(A)/\mathcal{P}\mathrm{rim}_\infty(A)$. A more detailed construction of the quotient appears in Section 5.6. We proceed to derive an equivalence between $\mathcal{F}\mathrm{lk}(A)$ and a corresponding quotient of $\mathcal{S}\mathrm{ei}(A)$.

Definition 5.13. The functor $F : \mathcal{S}\mathrm{ei}_\infty(A) \to \mathcal{S}\mathrm{ei}_\infty(A)/\mathcal{P}\mathrm{rim}_\infty(A)$ is the universal functor which makes invertible all morphisms whose kernel and cokernel are primitive. In other words, any functor which makes these morphisms invertible factors uniquely through F.

We outline in Section 5.6 one construction of F which will be convenient for our purposes; see Gabriel [20] or Srinivas [53, Appendix B.3] for further details. A more general construction can be found in Gabriel and Zisman [21] or Borceux [4, Ch5]. It follows directly from the definition that the localization F is unique (up to unique isomorphism).

Applying Definition 5.13 to the functor $B : \mathcal{S}\mathrm{ei}_\infty(A) \to \mathcal{F}\mathrm{lk}_\infty(A)$, there is a unique functor $\overline{B}$ such that $B = \overline{B}F$:

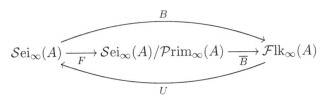

Proposition 5.9 stated that $B : \mathcal{S}\mathrm{ei}_\infty(A) \to \mathcal{F}\mathrm{lk}_\infty(A)$ is left adjoint to the forgetful functor U. We deduce in the next proposition that B satisfies the same universal property as F, but only "up to natural isomorphism". If f is a morphism in $\mathcal{S}\mathrm{ei}_\infty(A)$ let us write $f \in \Xi_\infty$ if the kernel and cokernel of f both lie in $\mathcal{P}\mathrm{rim}_\infty(A)$.

Proposition 5.14. *If $G : \mathcal{S}\mathrm{ei}_\infty(A) \to \mathcal{B}$ is a functor which sends every morphism in Ξ_∞ to an invertible morphism then there is a functor*

$$\widetilde{G} : \mathcal{F}\mathrm{lk}_\infty(A) \to \mathcal{B}$$

such that $\widetilde{G}B$ is naturally isomorphic to G. The functor $\widetilde{G}$ is unique up to natural isomorphism.

Proof. We prove uniqueness first. If there is a natural isomorphism $G \simeq \widetilde{G}B$ then $GU \simeq \widetilde{G}BU \simeq \widetilde{G}$ by Corollary 5.10.

To prove existence we must show that if $\widetilde{G} = GU$ then $\widetilde{G}B \simeq G$. Indeed, by Corollary 5.10 $B(\theta_V) : B(V) \to BUB(V)$ is an isomorphism for each $V \in \mathcal{S}ei_\infty(A)$. Since B respects exact sequences we have $\theta_V \in \Xi_\infty$. It follows that $G(\theta) : G \to GUB = \widetilde{G}B$ is a natural isomorphism. $\qquad\square$

The following is an immediate consequence of the fact that F and B have the same universal property (up to natural isomorphism):

Corollary 5.15. *The functor* $\overline{B} : \mathcal{S}ei_\infty(A)/\mathcal{P}rim_\infty(A) \to \mathcal{F}lk_\infty(A)$ *is an equivalence.* $\qquad\square$

We turn now to the categories $\mathcal{S}ei(A)$ and $\mathcal{F}lk(A)$.

Definition 5.16. Let $\mathcal{S}ei(A)/\mathcal{P}rim_\infty(A) \subset \mathcal{S}ei_\infty(A)/\mathcal{P}rim_\infty(A)$ denote the full subcategory whose objects are precisely the modules in $\mathcal{S}ei(A)$ (i.e. the modules which are finitely generated and projective as A-modules).

There is a commutative diagram of functors

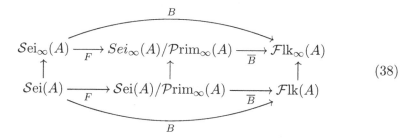

$$(38)$$

in which all the vertical arrows are inclusions of full subcategories.

Theorem 5.17. *The functor* $\overline{B} : \mathcal{S}ei(A)/\mathcal{P}rim_\infty(A) \to \mathcal{F}lk(A)$ *is an equivalence of categories.*

We will use the following general lemma in the proof of Theorem 5.17. Recall that a functor $G : \mathcal{C} \to \mathcal{D}$ is called full and faithful if it induces an isomorphism $\mathrm{Hom}_\mathcal{C}(V, V') \to \mathrm{Hom}_\mathcal{D}(G(V), G(V'))$ for every pair of objects $V, V' \in \mathcal{C}$.

Lemma 5.18. *A functor* $G : \mathcal{C} \to \mathcal{D}$ *is an equivalence of categories if and only if G is full and faithful and every object in $\mathcal{D}$ is isomorphic to $G(V)$ for some $V \in \mathcal{C}$.*

Proof. See for example Borceux [4, Prop 3.4.3]. $\qquad\square$

It follows from Corollary 5.15 and Lemma 5.18 that

$$\overline{B} : \mathcal{S}ei(A)/\mathcal{P}rim_\infty(A) \to \mathcal{F}lk(A)$$

is full and faithful. Theorem 5.17 is therefore a consequence of the following proposition:

Proposition 5.19. *Every module in $\mathcal{F}lk(A)$ is isomorphic to $B(V)$ for some $V \in \mathcal{S}ei(A)$.*

Proof. By definition, every module $M \in \mathcal{F}lk(A)$ has presentation

$$0 \to V[F_\mu] \xrightarrow{\sigma} V[F_\mu] \to M \to 0$$

where V is a finitely generated projective A-module and $\epsilon(\sigma) : V \to V$ is an isomorphism. Given any A-module W there is a canonical isomorphism

$$\mathrm{Hom}_{A[F_\mu]}(V[F_\mu], W[F_\mu]) \cong \mathrm{Hom}_A(V, W)[F_\mu]$$

and in particular σ can be expressed uniquely as a sum $\sum_{w \in F_\mu} \sigma_w w$ with each $\sigma_w \in \mathrm{Hom}_A(V, V)$.

Lemma 5.20. *Every $M \in \mathcal{F}lk(A)$ is isomorphic to the cokernel of an endomorphism $\sigma : V[F_\mu] \to V[F_\mu]$ of the form*

$$\sigma = 1 + \sigma_1(1 - z_1) + \cdots + \sigma_\mu(1 - z_\mu)$$

where V is finitely generated and projective and $\sigma_1, \cdots, \sigma_\mu \in \mathrm{Hom}_A(V, V)$.

Proof of Lemma. By the definition of $\mathcal{F}lk(A)$, the module M is isomorphic to the cokernel of some map $\sigma : V[F_\mu] \to V[F_\mu]$ where V is finitely generated and projective. The idea of this proof is to reduce the support of σ to $\{1, z_1, \cdots, z_\mu\} \subset F_\mu$ at the expense of replacing V by a larger finitely generated projective module. Note first that

$$\mathrm{Coker}(\sigma) \cong \mathrm{Coker} \begin{pmatrix} \sigma & 0 \\ 0 & 1 \end{pmatrix} : (V \oplus V')[F_\mu] \to (V \oplus V')[F_\mu]$$

where V' is any A-module and 1 denotes the identity morphism. The equation

$$\begin{pmatrix} 1 & -b \\ 0 & 1 \end{pmatrix} \begin{pmatrix} a + bc & 0 \\ 0 & 1 \end{pmatrix} \begin{pmatrix} 1 & 0 \\ c & 1 \end{pmatrix} = \begin{pmatrix} a & -b \\ c & 1 \end{pmatrix}, \tag{39}$$

therefore implies that $\mathrm{Coker} \begin{pmatrix} a & -b \\ c & 1 \end{pmatrix}$ is isomorphic to $\mathrm{Coker}(a + bc)$. Re-
peated application of equation (39) implies that M is isomorphic to the cok-
ernel of an endomorphism $\sigma = \sigma_0 + \sum_{i=1}^{\mu} \sigma_i^+ z_i + \sum_{i=1}^{\mu} \sigma_i^- z_i^{-1}$ with σ_0, σ_i^+
and σ_i^- in $\mathrm{Hom}_A(V, V)$ for some V. For each of the indices $i = 1, \cdots, \mu$
in turn, one can apply the identity $\mathrm{Coker}(\sigma) = \mathrm{Coker}(\sigma z_i)$ followed by
further equations (39). One obtains an identity $M \cong \mathrm{Coker}(\beta)$ where
$\beta = \beta_0 + \beta_1 z_1 + \cdots + \beta_\mu z_\mu$ and $\beta_i \in \mathrm{Hom}_A(V, V)$ for some finitely generated projective module V over A. By Lemma 2.3, $\epsilon(\beta)$ is an isomorphism.
Let $\sigma = \epsilon(\beta)^{-1}\beta$. Now $\epsilon(\sigma) = 1$ and so

$$\sigma = 1 + \sigma_1(z_1 - 1) + \cdots \sigma_\mu(z_\mu - 1)$$

for some $\sigma_1, \cdots, \sigma_\mu \in \mathrm{Hom}_A(V, V)$. This completes the proof of Lemma 5.20.
□

We may now finish the proof of Proposition 5.19. If

$$\sigma = 1 + \sum_i \sigma_i(1 - z_i)$$

then the equation

$$\begin{pmatrix} 1 & 0 & \cdots & 0 \\ 1 & 1 & \cdots & 0 \\ \vdots & \vdots & \ddots & \vdots \\ 1 & 0 & \cdots & 1 \end{pmatrix} \begin{pmatrix} 1 & \sigma_2(1-z_2) & \cdots & \sigma_\mu(1-z_\mu) \\ 0 & 1 & \cdots & 0 \\ \vdots & \vdots & \ddots & \vdots \\ 0 & 0 & \cdots & 1 \end{pmatrix} \begin{pmatrix} \sigma & 0 & \cdots & 0 \\ 0 & 1 & \cdots & 0 \\ \vdots & \vdots & \ddots & \vdots \\ 0 & 0 & \cdots & 1 \end{pmatrix} \begin{pmatrix} 1 & 0 & \cdots & 0 \\ -1 & 1 & \cdots & 0 \\ \vdots & \vdots & \ddots & \vdots \\ -1 & 0 & \cdots & 1 \end{pmatrix}$$
$$= \begin{pmatrix} 1+\sigma_1(1-z_1) & \sigma_2(1-z_2) & \cdots & \sigma_\mu(1-z_\mu) \\ \sigma_1(1-z_1) & 1+\sigma_2(1-z_2) & \cdots & \sigma_\mu(1-z_\mu) \\ \vdots & \vdots & \ddots & \vdots \\ \sigma_1(1-z_1) & \sigma_2(1-z_2) & \cdots & 1+\sigma_\mu(1-z_\mu) \end{pmatrix}$$

implies that

$$\mathrm{Coker}(\sigma) \cong \mathrm{Coker}\left(1 - s(1 - z) : V^{\oplus\mu}[F_\mu] \to V^{\oplus\mu}[F_\mu]\right)$$

where π_i acts as projection on the ith component of $V^{\oplus\mu}$ and s acts as

$$\begin{pmatrix} \sigma_1 & \sigma_2 & \cdots & \sigma_\mu \\ \sigma_1 & \sigma_2 & \cdots & \sigma_\mu \\ \vdots & \vdots & \ddots & \vdots \\ \sigma_1 & \sigma_2 & \cdots & \sigma_\mu \end{pmatrix}.$$

Thus $M \cong B(V^{\oplus\mu})$. □

This completes the proof of Theorem 5.17. □

5.4 Duality in the quotient

Having established that $\overline{B} : \mathcal{S}ei(A)/\mathcal{P}rim_\infty(A) \to \mathcal{F}lk(A)$ is an equivalence, we may use $\overline{B}$ to give duality structure to $\mathcal{S}ei(A)/\mathcal{P}rim_\infty(A)$ and make the lower part of (38) a commutative diagram of duality-preserving functors. Since the objects in $\mathcal{S}ei(A)/\mathcal{P}rim_\infty(A)$ coincide with those in $\mathcal{S}ei(A)$ we define $F(V)^* = F(V^*)$ and $i_{F(V)} = F(i_V) : F(V) \to F(V)^{**}$ where

$$F : \mathcal{S}ei(A) \to \mathcal{S}ei(A)/\mathcal{P}rim_\infty(A)$$

is the canonical functor. If $f : V \to V'$ is a morphism in $\mathcal{S}ei(A)/\mathcal{P}rim_\infty(A)$ let

$$f^* = \overline{B}^{-1}(\Phi_V^{-1}\overline{B}(f)^\wedge\Phi_{V'}) : V'^* \to V^*. \tag{40}$$

It is easy to see that $\_^*$ is a contravariant functor and that $i_V^* i_{V^*} = id_{V^*}$ for all V so $\mathcal{S}ei(A)/\mathcal{P}rim_\infty(A)$ is a hermitian category.

Recall that the composite of duality-preserving functors is defined by

$$(G, \Psi, \eta) \circ (G', \Psi', \eta') = (GG', \Psi G(\Psi'), \eta\eta').$$

Proposition 5.21. *The duality-preserving functor*

$$(B, \Phi, -1) : \mathcal{S}ei(A) \to \mathcal{F}lk(A)$$

coincides with the composite $(\overline{B}, \Phi, -1) \circ (F, id, 1)$.

Proof. It follows from equation (40) and Proposition 5.4 that $(\overline{B}, \Phi, -1)$ is a duality-preserving functor.

By definition $F(V)^* = F(V^*)$ and $i_{F(V)} = F(i_V)$; to show that

$$(F, id, 1) : \mathcal{S}ei(A) \to \mathcal{S}ei(A)/\mathcal{P}rim_\infty(A)$$

is a duality-preserving functor we must check that $F(f)^* = F(f^*)$ for each morphism $f : V \to V'$ in $\mathcal{S}ei(A)$. Indeed,

$$\begin{aligned}
F(f)^* &= \overline{B}^{-1}(\Phi_V^{-1}\overline{B}F(f)^\wedge\Phi_{V'}) \\
&= \overline{B}^{-1}(\Phi_V^{-1}B(f)^\wedge\Phi_{V'}) \\
&= \overline{B}^{-1}(B(f^*)) \qquad \text{(since } \Phi \text{ is natural)} \\
&= F(f^*).
\end{aligned}$$

It is easy to verify that $(B, \Phi, -1) = (\overline{B}, \Phi, -1) \circ (F, id, 1)$. □

Proposition 5.22. *Let $\zeta = 1$ or -1. The duality-preserving functor*

$$(\overline{B}, \Phi, -1) : \mathcal{S}\text{ei}(A)/\mathcal{P}\text{rim}_\infty(A) \to \mathcal{F}\text{lk}(A)$$

is an equivalence of hermitian categories and induces an isomorphism of Witt groups

$$\overline{B} : W^\zeta\left(\mathcal{S}\text{ei}(A)/\mathcal{P}\text{rim}_\infty(A)\right) \to W^{-\zeta}(\mathcal{F}\text{lk}(A)). \tag{41}$$

Proof. Since $\overline{B}$ is an equivalence of categories (by Theorem 5.17 above) it follows that $(\overline{B}, \Phi, -1)$ is an equivalence of hermitian categories (see Proposition II.7 of [51]). It also follows that $\overline{B}$ preserves limits and colimits so $\overline{B}$ preserves exact sequences and hence induces a homomorphism (41) of Witt groups. By Lemma A.2 of Appendix A this homomorphism is an isomorphism (41). □

5.5 Structure of Primitives

Recall that a module (V, ρ) in $\mathcal{S}\text{ei}_\infty(A)$ is called trivially primitive if $\rho(s) = 0$ or $\rho(s) = 1$. In this section we prove that every primitive module in $\mathcal{S}\text{ei}_\infty(A)$ is composed of trivially primitive modules.

Lemma 5.23. *If $(V, \rho) \in \mathcal{S}\text{ei}_\infty(A)$ and there exists a non-zero element $x \in V$ such that $\rho(s\pi_i)x = 0$ for all i then V has a non-zero submodule (V', ρ') such that $\rho'(s) = 0$.*

Proof. Note that $x = \sum \pi_i x$ and at least one of the terms $\pi_i x$ must be non-zero. Choose non-zero $V' = A\pi_i x$ and define ρ' by

$$\rho'(\pi_j) = \begin{cases} 1 \text{ if } j = i \\ 0 \text{ if } j \neq i \end{cases} , \quad \rho'(s) = 0.$$

Now (V', ρ') is the required non-zero submodule of (V, ρ). □

Lemma 5.24. *If (V, ρ) is primitive and non-zero then there exists a non-zero trivially primitive submodule (V', ρ').*

Proof. (compare Lemma 7.10c in Farber [17]) Since (V, ρ) is primitive,

$$\rho(1 - s)y^{-1} + \rho(s)y : V[F_\mu(y^2)y] \to V[F_\mu(y^2)]$$

is an isomorphism with inverse α say. Now α can be written as a finite sum $\sum_{w \in S} \alpha_w w$ where S is a finite subset of $\bigcup_{i=1}^\mu F_\mu(y^2)y_i$ and $\alpha_w : V \to V$ has non-zero image in $\pi_i V$ for each $w \in S$. Choose an element $w \in S$ whose expression in reduced form as a product of letters $y_i^{\pm}$ is of maximal length. We consider two cases:

Case 1: $w = w'y_i$ for some $w' \in F_\mu(y^2)$ and some i. The equation

$$((1 - \rho(s))y^{-1} + \rho(s)y)\alpha = 1 \tag{42}$$

implies that $\rho(s\pi_j)\alpha_w = 0$ for each j. Any element x in the image of α_w satisfies the conditions of Lemma 5.23 so (V, ρ) has a non-zero submodule (V', ρ') with $\rho'(s) = 0$.

Case 2: $w = w'y_i^{-1}$ for some $w' \in F_\mu(y^2)$. The equation (42) implies that $\rho((1-s)\pi_i)\alpha_w = 0$ for each i. By Lemma 5.23 there is a non-zero submodule (V', ρ') with $\rho'(1 - s) = 0$ or in other words $\rho'(s) = 1$. $\quad\square$

Recall that a module V is called simple if there are no submodules other than 0 and V. The following remark is a consequence of Lemma 5.24.

Remark 5.25. Every simple primitive module is trivially primitive. $\quad\square$

Definition 5.26. If $\mathcal{A}$ is an abelian category then a non-empty full subcategory $\mathcal{E} \subset \mathcal{A}$ is called a *Serre subcategory* if for every exact sequence $0 \to V \to V' \to V'' \to 0$ in $\mathcal{A}$ one has

$$V' \in \mathcal{E} \iff V \in \mathcal{E} \text{ and } V'' \in \mathcal{E} .$$

Note that every Serre subcategory of an abelian category is again an abelian category. Since B preserves exact sequences and arbitrary direct sums $\mathcal{P}\text{rim}_\infty(A)$ is a Serre subcategory of $\mathcal{S}\text{ei}_\infty(A)$ and is closed under direct sums.

Lemma 5.27. *Suppose $\mathcal{E} \subset \mathcal{A}$ is a Serre subcategory of an abelian category and $\mathcal{E}$ is closed under arbitrary direct sums. Every module $V \in \mathcal{A}$ contains a unique maximal submodule $U \leq V$ such that $U \in \mathcal{E}$. If $U' \leq V$ and $U' \in \mathcal{E}$ then $U' \leq U$.*

Proof. Let U be the sum in V of all the submodules $U_i \leq V$ with $U_i \in \mathcal{E}$. Since U is a factor module of $\bigoplus U_i$, one finds $U \in \mathcal{E}$. $\quad\square$

Proposition 5.28. *The category $\mathcal{P}\text{rim}_\infty(A)$ is the smallest Serre subcategory of $\mathcal{S}\text{ei}_\infty(A)$ which a) contains the trivially primitive modules and b) is closed under arbitrary direct sums.*

Proof. Let $\mathcal{P}_\infty(A)$ denote the smallest full subcategory of $\mathcal{S}ei_\infty(A)$ satisfying the conditions of the Proposition. Now $\mathcal{P}rim_\infty(A)$ satisfies these conditions so $\mathcal{P}_\infty(A) \subset \mathcal{P}rim_\infty(A)$.

Conversely, we must show that $\mathcal{P}rim_\infty(A) \subset \mathcal{P}_\infty(A)$. Suppose $B(V) = 0$. Let $W \leq V$ be the maximal submodule such that $W \in \mathcal{P}_\infty(A)$ (the module W exists by Lemma 5.27). Now $B(V/W) = 0$ since B respects exact sequences so Lemma 5.24 implies that either $V/W = 0$ or there is a non-zero submodule V' of V/W which lies in $\mathcal{P}_\infty(A)$. In the latter case, let $p : V \to V/W$ denote the projection and note the exact sequence

$$0 \to W \to p^{-1}(V') \xrightarrow{p|} V' \to 0.$$

Since $W \in \mathcal{P}_\infty(A)$ and $V' \in \mathcal{P}_\infty(A)$ we have $p^{-1}(V) \in \mathcal{P}_\infty(A)$ which contradicts the maximality of W. Thus $V/W = 0$ and hence $V = W$, so V lies in $\mathcal{P}_\infty(A)$. □

5.6 Construction of the quotient

We outline next a construction of $\mathcal{S}ei_\infty(A)/\mathcal{P}rim_\infty(A)$. We will use this construction in Section 6 to show that $B : W^\varsigma(\mathcal{S}ei(A)) \to W^{-\varsigma}(\mathcal{F}lk(A))$ is an isomorphism when A is a semi-simple Artinian ring. The notion of Serre subcategory was defined in the preceding section. Let us note some basic properties:

Lemma 5.29. *Suppose $\mathcal{A}$ is an abelian category and $\mathcal{E}$ is a Serre subcategory. Suppose $V \in \mathcal{A}$, $W \leq V$ and $W' \leq V$.*

1. If $W \in \mathcal{E}$ and $W' \in \mathcal{E}$ then $W + W' \in \mathcal{E}$.

2. If $V/W \in \mathcal{E}$ and $V/W' \in \mathcal{E}$ then $V/(W \cap W') \in \mathcal{E}$.

Proof. 1. There is an exact sequence

$$0 \to W \to W + W' \to (W + W')/W \to 0$$

and $(W + W')/W$ is isomorphic to $W'/(W \cap W') \in \mathcal{E}$. Hence $W + W' \in \mathcal{E}$.
2. There is an exact sequence

$$0 \to W/(W \cap W') \to V/(W \cap W') \to V/W \to 0.$$

Now $W/(W \cap W')$ is isomorphic to $(W + W')/W'$ which is contained in V/W' and so $W/(W \cap W') \in \mathcal{E}$ and hence $V/W \cap W' \in \mathcal{E}$. □

We may now recall a construction for the quotient of an abelian category by a Serre subcategory. See Gabriel [20] or Srinivas [53, Appendix B.3] for further details.

Suppose $\mathcal{A}$ is an abelian category and $\mathcal{E}$ is a Serre subcategory. The symbol $\mathcal{A}/\mathcal{E}$ will denote a category with the same objects as $\mathcal{A}$ but different groups of morphisms. To define $\mathrm{Hom}_{\mathcal{A}/\mathcal{E}}(V, V')$, consider the pairs (W, U') where $W \leq V$, $U' \leq V'$, $V/W \in \mathcal{E}$ and $U' \in \mathcal{E}$. One says that $(W_1, U_1') \leq (W_2, U_2')$ if $W_2 \leq W_1$ and $U_1' \leq U_2'$ [note the directions of inclusion]. Lemma 5.29 above implies that these pairs are a directed set. Indeed, given pairs (W_1, U_1') and (W_2, U_2') one finds $(W_1, U_1') \leq (W_1 \cap W_2, U_1' + U_2')$ and $(W_2, U_2') \leq (W_1 \cap W_2, U_1' + U_2')$. The following definition can now be made:

$$\mathrm{Hom}_{\mathcal{A}/\mathcal{E}}(V, V') = \varinjlim_{(W,U')} \mathrm{Hom}_{\mathcal{A}}(W, V'/U'). \tag{43}$$

We leave to the reader the definition of composition of morphisms and the canonical functor $F : \mathcal{A} \to \mathcal{A}/\mathcal{E}$. Proofs of the following statements can be found in the references cited above:

(a) The quotient category $\mathcal{A}/\mathcal{E}$ is an abelian category and F is an exact additive functor.

(b) If f is a morphism in $\mathcal{A}$ then $F(f)$ is an isomorphism if and only if $\mathrm{Coker}(f) \in \mathcal{E}$ and $\mathrm{Ker}(f) \in \mathcal{E}$.

In particular if V is an object in $\mathcal{A}$ then $F(V) \cong 0$ if and only if $V \in \mathcal{E}$.

As we indicated in earlier sections, the functor $F : \mathcal{A} \to \mathcal{A}/\mathcal{E}$ is universal with respect to property (b). In detail, if $G : \mathcal{A} \to \mathcal{B}$ makes invertible every morphism whose kernel and cokernel lie in $\mathcal{E}$ then there is a unique functor $\widetilde{G} : \mathcal{A}/\mathcal{E} \to \mathcal{B}$ such that $\widetilde{G}F = G$. In particular the functor

$$F : \mathcal{S}ei_\infty(A) \to \mathcal{S}ei_\infty(A)/\mathcal{P}rim_\infty(A)$$

satisfies Definition 5.13. Let us be explicit about $\widetilde{G}$:

If V is an object in $\mathcal{A}/\mathcal{E}$ then one writes $\widetilde{G}(V) = G(V)$. Every morphism $f \in \mathrm{Hom}_{\mathcal{A}/\mathcal{E}}(V, V')$ is represented by some $\overline{f} \in \mathrm{Hom}_{\mathcal{A}}(W, V'/U')$ with $U' \in \mathcal{E}$ and $V/W \in \mathcal{E}$ If $i : W \to V$ and $p : V' \to V'/U'$ denote the canonical monomorphism and epimorphism respectively one must define

$$\widetilde{G}(f) = G(p)^{-1}G(\overline{f})G(i)^{-1} : G(V) \to G(V').$$

In our particular example Lemma 5.27 provides one simplification in our description of the quotient category $\mathcal{S}ei_\infty(A)/\mathcal{P}rim_\infty(A)$. If $V \in \mathcal{S}ei_\infty(A)$ let us call a submodule $W \leq V$ *coprimitive* if $V/W \in \mathcal{P}rim_\infty(A)$.

Lemma 5.30. *If $V, V' \in \mathcal{S}ei_\infty(A)$ and U' denotes the maximal primitive submodule of V' then*

$$\mathrm{Hom}_{\mathcal{S}ei_\infty(A)/\mathcal{P}rim_\infty(A)}(V, V') = \varinjlim_W \mathrm{Hom}_{\mathcal{S}ei_\infty(A)}(W, V'/U')$$

where the direct limit is over coprimitive submodules W of V. $\qquad\square$

Note that there is not in general a minimal coprimitive in V; the functor B does not respect infinite limits and an infinite intersection of coprimitives is not in general coprimitive (but see Lemma 5.31 below).

5.7 Global dimension zero

In this section the ring A will be assumed semi-simple and Artinian or, in other words, a finite product of matrix rings over division rings. The basic theory of semi-simple Artinian rings can be found in many algebra textbooks (e.g. Lam [29, §1-4] or Lang [30, Ch.XVII]). In particular, all A-modules are projective and $\mathcal{S}ei(A)$ is an abelian category with ascending and descending chain conditions; these facts lead to simplifications of results in Sections 5.5 and 5.6 above. We show that the primitive modules in $\mathcal{S}ei(A)$ are composed of a finite number of simple trivially primitive modules (Proposition 5.33) and give a simplified description of the hermitian category $\mathcal{S}ei(A)/\mathcal{P}rim_\infty(A)$. We shall consider semi-simple Artinian rings again in Section 6 but it is not essential to read the present section before Section 6.

The key lemma we will need is the following:

Lemma 5.31. *Suppose $\mathcal{A}$ is an abelian category with ascending and descending chain conditions and $\mathcal{E}$ is a Serre subcategory.*

1. *Every module $V \in \mathcal{A}$ contains a unique maximal submodule in $\mathcal{E}$ which contains all others in $\mathcal{E}$.*

2. *Every module $V \in \mathcal{A}$ contains a unique submodule $W \leq V$ which is minimal with respect to the property $V/W \in \mathcal{E}$. If $V/W' \in \mathcal{E}$ then $W \leq W'$.*

Proof. 1. Since $\mathcal{A}$ has the ascending chain condition there is a submodule $U \leq V$ which is maximal with respect to the property $U \in \mathcal{E}$. In other words, if $U \leq U' \leq V$ and $U' \in \mathcal{E}$ then $U' = U$. If U' is any other submodule in $\mathcal{E}$ then $U + U' \in \mathcal{E}$ by Lemma 5.29 so $U + U' = U$ and hence $U' \leq U$.
2. Since $\mathcal{A}$ has the descending chain condition there is a submodule $W \leq V$ which is minimal with respect to the property $V/W \in \mathcal{E}$ (i.e. if $W' \leq W \leq V$

and $V/W' \in \mathcal{E}$ then $W' = W$). If $V/W' \in \mathcal{E}$ then $V/(W \cap W') \in \mathcal{E}$ by Lemma 5.29 so $W \cap W' = W$ and hence $W \leq W'$. $\qquad\qquad\square$

Recall that if $V \in \mathcal{S}\mathrm{ei}(A)$, a submodule $W \leq V$ is called coprimitive if V/W is primitive. Since A is Artinian and Noetherian, Lemma 5.31 implies that there is a maximal primitive submodule $U \leq V$ and a minimal coprimitive submodule $W \leq V$ for each $V \in \mathcal{S}\mathrm{ei}(A)$.

5.7.1   Structure of Primitives

Definition 5.32. Let $\mathcal{P}\mathrm{rim}(A)$ denote the intersection of $\mathcal{P}\mathrm{rim}_\infty(A)$ and $\mathcal{S}\mathrm{ei}(A)$. In other words, $\mathcal{P}\mathrm{rim}(A) \subset \mathcal{S}\mathrm{ei}(A)$ is the full subcategory containing those modules V such that $B(V) = 0$.

Note that $\mathcal{P}\mathrm{rim}(A)$ is both a Serre subcategory and a hermitian subcategory of $\mathcal{S}\mathrm{ei}(A)$. Moreover, $\mathcal{S}\mathrm{ei}(A)/\mathcal{P}\mathrm{rim}(A) = \mathcal{S}\mathrm{ei}(A)/\mathcal{P}\mathrm{rim}_\infty(A)$.

Proposition 5.33. *The category $\mathcal{P}\mathrm{rim}(A)$ is the smallest Serre subcategory of $\mathcal{S}\mathrm{ei}(A)$ which contains the trivially primitive modules in $\mathcal{S}\mathrm{ei}(A)$.*

Proof. We proceed as in the proof of Proposition 5.28, using Lemma 5.31 in place of Lemma 5.27. Let $\mathcal{P}(A)$ denote the smallest Serre subcategory of $\mathcal{S}\mathrm{ei}(A)$ which contains all the trivially primitive modules. To show that $\mathcal{P}(A) \subset \mathcal{P}\mathrm{rim}(A)$ it suffices to observe that $\mathcal{P}\mathrm{rim}(A)$ is a Serre subcategory which contains these modules.

Conversely, to show $\mathcal{P}\mathrm{rim}(A) \subset \mathcal{P}(A)$ suppose $V \in \mathcal{P}\mathrm{rim}(A)$. There exists, by Lemma 5.31, a maximal submodule $U \leq V$ such that $U \in \mathcal{P}(A)$. Now $B(V/U) = 0$ so either $V = U$ or by Lemma 5.24 V/U has a non-zero trivially primitive submodule U'. If $p : V \twoheadrightarrow V/U$ is the canonical map then the exact sequence $0 \to U \to p^{-1}(U') \to U' \to 0$ implies that $p^{-1}(U') \in \mathcal{P}(A)$ contradicting the maximality of U. Thus $V = U$ and $V \in \mathcal{P}(A)$. $\qquad\qquad\square$

5.7.2   Construction of the quotient

With the benefit of Lemma 5.31 we can give a simpler description of the quotient category than Lemma 5.30.

Lemma 5.34. *The morphisms in $\mathcal{S}\mathrm{ei}(A)/\mathcal{P}\mathrm{rim}(A)$ are*

$$\mathrm{Hom}_{\mathcal{S}\mathrm{ei}(A)/\mathcal{P}\mathrm{rim}(A)}(V, V') = \mathrm{Hom}_{\mathcal{S}\mathrm{ei}(A)}(W, V'/U'). \qquad (44)$$

where $W \leq V$ is the minimal coprimitive and $U' \leq V'$ is the maximal primitive. $\qquad\qquad\square$

We simplify next the hermitian structure on $\mathcal{S}\mathrm{ei}(A)/\mathcal{P}\mathrm{rim}(A)$. We have seen that the duality-preserving functor $(B, \Phi, -1) : \mathcal{S}\mathrm{ei}(A) \to \mathcal{F}\mathrm{lk}(A)$ factors through an equivalence of hermitian categories $(\overline{B}, \Phi, -1)$:

$$
\mathcal{S}\mathrm{ei}(A) \xrightarrow[(F,\mathrm{id},1)]{} \mathcal{S}\mathrm{ei}(A)/\mathcal{P}\mathrm{rim}(A) \xrightarrow[(\overline{B},\Phi,-1)]{} \mathcal{F}\mathrm{lk}(A)
$$

with the curved arrow labelled $(B,\Phi,-1)$.

(Theorem 5.17 and Proposition 5.21 above). The duality functor on the quotient $\mathcal{S}\mathrm{ei}(A)/\mathcal{P}\mathrm{rim}(A)$ was defined in Section 5.4 above by $F(V)^* = F(V^*)$ and by equation (40). Using the assumption that A is Artinian we can re-interpret equation (40). Suppose that $f \in \mathrm{Hom}_{\mathcal{S}\mathrm{ei}(A)/\mathcal{P}\mathrm{rim}(A)}(V, V')$. As usual, let W denote the minimal coprimitive submodule of V and let U' denote the maximal primitive submodule of V'. The morphism f is identified with some $\overline{f} \in \mathrm{Hom}_{\mathcal{S}\mathrm{ei}(A)}(W, V'/U')$. Since $\_^*$ preserves exact sequences the following are exact

$$0 \to (V/W)^* \to V^* \to W^* \to 0 \tag{45}$$
$$0 \to (V'/U')^* \to (V')^* \to (U')^* \to 0. \tag{46}$$

Now $(B, \Phi, 1)$ is a duality-preserving functor, so for each $V \in \mathcal{S}\mathrm{ei}(A)$ one has $B(V) = 0$ if and only if $B(V^*) = 0$. Thus $\mathcal{P}\mathrm{rim}(A)$ is a hermitian subcategory of $\mathcal{S}\mathrm{ei}(A)$ and in particular $(U')^*$ and $(V/W)^*$ are primitive. It follows that $(V/W)^*$ is the maximal primitive in V^* and $(V'/U')^*$ is the minimal coprimitive in $(V')^*$. Since $(F, \mathrm{id}, 1)$ is a duality-preserving functor, f^* is represented by

$$\overline{f}^* \in \mathrm{Hom}((V'/U')^*, W^*). \tag{47}$$

6 Equivalence of Invariants

Cobordism invariants of F_μ-links have been defined in two different ways in Sections 3 and [51]. In this section we use the duality-preserving functor $(B, \Phi, -1)$ which was studied in Section 5 to relate the two approaches, proving Theorems 1.3 and 1.4. To prove Theorem 1.4 we show that the functor B respects each of the three steps laid out in Sections 3.1 and 4.3. A more detailed version of Theorem 1.4 is set out in Theorem 6.5 below.

6.1 Proof of Theorem 1.3

Suppose $\mathcal{A}$ is an abelian category with ascending and descending chain conditions and $\mathcal{E}$ is a Serre subcategory. Let $F : \mathcal{A} \to \mathcal{A}/\mathcal{E}$ denote the quotient functor. Recall that a module V in $\mathcal{A}$ is called simple if V is not isomorphic to 0 and V does not have submodules other than 0 and V.

Lemma 6.1.   *1. If $V \in \mathcal{A}$ is simple then either $V \in \mathcal{E}$ or $F(V)$ is simple.*

2. Every simple module in $\mathcal{A}/\mathcal{E}$ is isomorphic to $F(V)$ for some simple module $V \in \mathcal{A}$ which does not lie in $\mathcal{E}$.

Proof. 1. Suppose $V \in \mathcal{A}$ is simple, $V \notin \mathcal{E}$ and $i : V' \to F(V)$ is the inclusion of a submodule in $\mathcal{A}/\mathcal{E}$. Now i is represented by some morphism $\bar{i} : W' \to V$ where $W' \leq V'$ and $V'/W' \in \mathcal{E}$. Either $\bar{i} = 0$ in which case $V' \cong 0$ in $\mathcal{A}/\mathcal{E}$ or $\bar{i}$ is an epimorphism which implies that $V' = F(V)$ (recall that F is exact).
2. Every module in $\mathcal{A}/\mathcal{E}$ is $F(V)$ for some module $V \in \mathcal{A}$. Suppose $F(V)$ is simple. Now V has a finite filtration $0 = V_0 \leq V_1 \leq \cdots \leq V_n = V$ where each quotient V_i/V_{i-1} is a simple module. Since F respects exact sequences $F(V_i/V_{i-1}) = 0$ for all $i \in \{1, \cdots, n\}$ except one, for which there is an isomorphism $F(V_i/V_{i-1}) \cong F(V)$. This module V_i/V_{i-1} does not lie in $\mathcal{E}$. □

Suppose now that $\mathcal{A}$ and $\mathcal{A}/\mathcal{E}$ are hermitian categories and the quotient functor extends to a duality-preserving functor

$$(F, \mathrm{id}, 1) : \mathcal{A} \to \mathcal{A}/\mathcal{E}.$$

Lemma 6.2.   *1. The Serre subcategory $\mathcal{E}$ is a hermitian subcategory.*

2. Let $\zeta = 1$ or -1. If $V \in \mathcal{A}$ is simple and $V \notin \mathcal{E}$ then V is ζ-self-dual if and only if $F(V)$ is ζ-self-dual.

Proof. 1. If $V \in \mathcal{E}$ then $F(V^*) = F(V)^* \cong 0 \in \mathcal{A}/\mathcal{E}$ so $V^* \in \mathcal{E}$.
2. To prove the "only if" part it suffices to recall that for $\phi : V \to V^*$ one has $F(\phi^*) = F(\phi)^*$. For the "if" part, note also that

$$F : \mathrm{Hom}_{\mathcal{A}}(V, V^*) \to \mathrm{Hom}_{\mathcal{A}/\mathcal{E}}(V, V^*)$$

is an isomorphism.   □

Proposition 6.3. *Suppose $\mathcal{A}$ and $\mathcal{A}/\mathcal{E}$ are hermitian categories and*

$$(F, \mathrm{id}, 1) : \mathcal{A} \to \mathcal{A}/\mathcal{E}$$

is a duality-preserving functor. For each ζ-self-dual simple module $V \in \mathcal{A}$ such that $V \notin \mathcal{E}$ there is a canonical isomorphism

$$W^\zeta(\mathcal{A}|_V) \cong W^\zeta((\mathcal{A}/\mathcal{E})|_V).$$

If A has ascending and descending chain conditions then there is a canonical isomorphism

$$W^\varsigma(A) \cong W^\varsigma(\mathcal{E}) \oplus W^\varsigma(A/\mathcal{E}).$$

Proof. If $V \in A$ is a simple module and $V \notin \mathcal{E}$ then $F(V)$ is simple by part 1. of Lemma 6.1. Every module in $A|_V$ is a direct sum of copies of V so by equation (43) the restriction $F : A|_V \to (A/\mathcal{E})|_{F(V)}$ is a full and faithful functor and hence an equivalence of categories.

By part 2. of Lemma 6.2, V is ς-self-dual if and only if $F(V)$ is ς-self-dual, in which case $(F, \mathrm{id}, 1) : F : A|_V \to (A/\mathcal{E})|_{F(V)}$ is an equivalence of hermitian categories and induces an isomorphism

$$W^\varsigma(A|_V) \to W^\varsigma\left((A/\mathcal{E})|_{F(V)}\right). \tag{48}$$

To prove the last sentence of the Lemma, note first that by part 1. of Lemma 6.2, $\mathcal{E}$ is a hermitian subcategory of A. Theorem 3.4 provides canonical decompositions

$$W^\varsigma(A) \cong \bigoplus W^\varsigma(A|_V)$$
$$W^\varsigma(\mathcal{E}) \cong \bigoplus W^\varsigma(\mathcal{E}|_V)$$
$$W^\varsigma(A/\mathcal{E}) \cong \bigoplus W^\varsigma\left((A/\mathcal{E})|_V\right)$$

where the right hand side of each identity has one summand for each isomorphism class of ς-self-dual simple modules V.

By part 2. of Lemma 6.1 and part 2. of Lemma 6.2 every summand of $W^\varsigma(A/\mathcal{E})$ is the isomorphic image of $W^\varsigma(A|_V)$ for some simple ς-self-dual module V in A.

On the other hand, if $V \in \mathcal{E}$ is simple and ς-self-dual then $(F, \mathrm{id}, 1)$ sends $W^\varsigma(A|_V)$ to zero. The last sentence of the Lemma follows. $\square$

In our application, we set $A = \mathcal{S}\mathrm{ei}(A)$ and $\mathcal{E} = \mathcal{P}\mathrm{rim}(A)$ where A is semi-simple Artinian. Recall that $\mathcal{P}\mathrm{rim}(A) = \mathcal{P}\mathrm{rim}_\infty(A) \cap \mathcal{S}\mathrm{ei}(A)$ is an abelian category with ascending and descending chain conditions and $\mathcal{S}\mathrm{ei}(A)/\mathcal{P}\mathrm{rim}(A) = \mathcal{S}\mathrm{ei}(A)/\mathcal{P}\mathrm{rim}_\infty(A)$.

Lemma 6.4. *1. None of the simple primitive modules in $\mathcal{P}\mathrm{rim}(A)$ are self-dual.*

2. $W^\varsigma(\mathcal{P}\mathrm{rim}(A)) = 0$.

Proof. By Remark 5.25 above, every simple primitive module is trivially primitive. If $(V, \rho) \in \mathcal{P}\mathrm{rim}(A)$ then $\rho(s) = 0$ if and only if $\rho^*(s) = 1$ so none of the simple trivially primitive modules are self-dual. Thus part 1. is proved, and part 2. follows immediately from Theorem 3.4. □

Proof of Theorem 1.3. By Proposition 5.21 the duality-preserving functor $(B, \Phi, -1)$ is the composite $(\overline{B}, \Phi, -1) \circ (F, \mathrm{id}, 1)$. Setting $\mathcal{A} = \mathcal{S}\mathrm{ei}(A)$ and $\mathcal{E} = \mathcal{P}\mathrm{rim}(A)$ in Proposition 6.3, and invoking also Lemma 6.4, we learn that $(F, \mathrm{id}, 1)$ induces an isomorphism

$$W^\zeta(\mathcal{S}\mathrm{ei}(A)) \to W^\zeta\left(\mathcal{S}\mathrm{ei}(A)/\mathcal{P}\mathrm{rim}(A)\right).$$

By Theorem 5.17, $(\overline{B}, \Phi, 1)$ is an equivalence and hence induces an isomorphism

$$W^\zeta\left(\mathcal{S}\mathrm{ei}(A)/\mathcal{P}\mathrm{rim}(A)\right) \to W^{-\zeta}(\mathcal{F}\mathrm{lk}(A)).$$

(see Lemma A.2 in Appendix A). Thus $(B, \Phi, -1)$ induces an isomorphism

$$W^\zeta(\mathcal{S}\mathrm{ei}(A)) \to W^{-\zeta}(\mathcal{F}\mathrm{lk}(A)).$$

This completes the proof of Theorem 1.3. □

6.2   Proof of Theorem 1.4

We prove in this section that the functor $(B, \Phi, -1)$ identifies the invariants defined in [51] with those of Section 3. More precisely, we prove the following theorem:

Theorem 6.5. (Equivalence of invariants)

1. *If $V \in \mathcal{S}\mathrm{ei}(\mathbb{Q})$ is simple and $(-1)^q$-self-dual then $B(V) \in \mathcal{F}\mathrm{lk}(\mathbb{Q})$ is simple and $(-1)^{q+1}$-self-dual.*

2. *Every simple $(-1)^{q+1}$-self-dual module $M \in \mathcal{F}\mathrm{lk}(\mathbb{Q})$ is isomorphic to $B(V)$ for some simple $(-1)^q$-self-dual module $V \in \mathcal{S}\mathrm{ei}(\mathbb{Q})$.*

3. *If $V \in \mathcal{S}\mathrm{ei}(\mathbb{Q})$ and $B(V) \in \mathcal{F}\mathrm{lk}(\mathbb{Q})$ are simple then the functor B induces an isomorphism of rings $B : \mathrm{End}_{\mathcal{S}\mathrm{ei}(\mathbb{Q})}(V) \xrightarrow{\cong} \mathrm{End}_{\mathcal{F}\mathrm{lk}(\mathbb{Q})}(B(V))$.*

4. *Suppose $V \in \mathcal{S}\mathrm{ei}(\mathbb{Q})$ is simple and $b : V \to V^*$ is a ζ-hermitian form. The ring isomorphism in part 3. respects involutions. Explicitly, if $f \in \mathrm{End}_{\mathcal{S}\mathrm{ei}(\mathbb{Q})} V$ then $B(b^{-1}f^*b) = (\Phi_V B(b))^{-1} B(f)^\wedge \Phi_V B(b)$.*

5. *Suppose $W \in \mathcal{S}\mathrm{ei}(\mathbb{Z})$ and $\phi : W \to W^*$ is a $(-1)^q$-hermitian form. The dimension modulo 2, signatures, discriminant, Hasse-Witt invariant and Lewis θ-invariant of*

$$\Theta_{V,b}p_V \left[\mathbb{Q} \otimes_{\mathbb{Z}} (W, \phi) \right] \in W^1(\mathrm{End}_{\mathcal{S}\mathrm{ei}(\mathbb{Q})}(V))$$

coincide (if defined) with the corresponding invariants of

$$\Theta_{(B(V),-\Phi_V B(b))}p_{B(V)} \left[\mathbb{Q} \otimes_{\mathbb{Z}} (B(W), \Phi_W B(\phi)) \right] \in W^1(\mathrm{End}_{\mathcal{F}\mathrm{lk}(\mathbb{Q})}(B(V))).$$

Recall that $B = \overline{B} \circ F$ and $\overline{B} : \mathcal{S}\mathrm{ei}(\mathbb{Q})/\mathcal{P}\mathrm{rim}(\mathbb{Q}) \to \mathcal{F}\mathrm{lk}(\mathbb{Q})$ is an equivalence of categories. In parts 1. through 3. of Theorem 6.5 it therefore suffices to prove corresponding statements with the functor F in place of B and $(-1)^q$ in place of $(-1)^{q+1}$:

1. The statement follows from part 1. of Lemma 6.1, part 2. of Lemma 6.2 and part 1. of Lemma 6.4.

2. The statement follows from part 2. of Lemma 6.1 and part 2. of Lemma 6.2.

3. This is a consequence of equation (44).

4. Since $B(b^{-1}f^*b) = B(b^{-1})B(f^*)B(b)$ it suffices to prove that

$$B(f^*) = \Phi_V^{-1} B(f)^{\wedge} \Phi_V.$$

This equation is a consequence of the fact that Φ is a natural isomorphism.

The proof of part 5. of Theorem 6.5 is slightly more involved. Recall from proposition 5.7 that B respects changes of coefficients and, in particular, that the inclusion of $\mathbb{Z}$ in $\mathbb{Q}$ induces the commutative diagram (24). One must check that $(B, \Phi, -1)$ respects each of the three steps in the definitions of the F_μ-link invariants (see Section 3.1).

Devissage: Let $V \in \mathcal{S}\mathrm{ei}(\mathbb{Q})$ be a ζ-self-dual simple module. If W is isomorphic to a direct sum of copies of V then $B(W)$ is isomorphic to a direct sum of copies of $B(V)$. Hence the image of $W^\zeta(\mathcal{S}\mathrm{ei}(\mathbb{Q})|_V)$ under B lies in $W^{-\zeta}(\mathcal{F}\mathrm{lk}(\mathbb{Q})|_{B(V)})$ and there is a commutative diagram of isomorphisms

$$
\begin{array}{ccc}
W^\zeta(\mathcal{S}\mathrm{ei}(\mathbb{Q})) & \xrightarrow{\quad B \quad} & W^{-\zeta}(\mathcal{F}\mathrm{lk}(\mathbb{Q})) \\
\updownarrow & & \updownarrow \\
\displaystyle\bigoplus_V W^\zeta(\mathcal{S}\mathrm{ei}(\mathbb{Q})|_V) & \xrightarrow{\quad \oplus B| \quad} & \displaystyle\bigoplus_V W^{-\zeta}(\mathcal{F}\mathrm{lk}(\mathbb{Q})|_{B(V)}).
\end{array}
\tag{49}
$$

where the direct sums are indexed by the isomorphism classes of simple ζ-self-dual modules in $\mathcal{S}\mathrm{ei}(\mathbb{Q})$.

Morita Equivalence: Suppose $V \in \mathcal{S}\mathrm{ei}(\mathbb{Q})$ is a simple module and

$$b : V \to V^*$$

is a non-singular ζ-hermitian form. Let us denote the endomorphism rings $E = \text{End}_{\mathcal{S}\text{ei}(\mathbb{Q})} V$ and $E' = \text{End}_{\mathcal{F}\text{lk}(\mathbb{Q})} B(V)$. By Corollary 3.23 above the duality-preserving functor $(B, \Phi, -1)$ induces a commutative diagram

$$
\begin{array}{ccc}
W^{\zeta}(\mathcal{S}\text{ei}(\mathbb{Q})|_V) & \xrightarrow{\ \ B\ \ } & W^{-\zeta}(\mathcal{F}\text{lk}(\mathbb{Q})|_{B(V)}) \\
\Theta_{V,b}\Big\downarrow & & \Big\downarrow \Theta_{B(V),-\Phi_V B(b)} \\
W^1(E) & \xrightarrow[\ \ B\ \ ]{} & W^1(E').
\end{array}
\tag{50}
$$

Invariants: The isomorphism $E \to E'$ in part 4. induces isomorphisms between the target groups for the invariants in part 5. For example, if E and E' are commutative with trivial involution then the discriminant Δ of $\Theta_{V,b}p_V [\mathbb{Q} \otimes_{\mathbb{Z}} (W, \phi)]$ lies in E/E^2 and the functor B induces an isomorphism $E/E^2 \to E'/(E')^2$. The word "coincide" in part 5. is understood to mean that the image of Δ in $E'/(E')^2$ is equal to the discriminant of $\Theta_{(B(V),-\Phi_V B(b))}p_{B(V)} [\mathbb{Q} \otimes_{\mathbb{Z}} (B(W), \Phi_W B(\phi))]$.

The isomorphism $B : E \to E'$ of rings with involution induces an isomorphism $W^1(E) \to W^1(E')$. We leave it to the reader to check that if $\alpha \in W^1(E)$ then all the listed invariants of α coincide (in this sense) with the corresponding invariants of $B(\alpha) \in W^1(E')$. Further details of the invariants can be found in chapter 11 of [51].

This completes the proof of part 5. and hence of theorems 1.4 and 6.5. $\qquad\square$

A   Naturality of constructions

In this appendix we prove naturality theorems for the covering construction $(B, \Phi, -1)$ and for hermitian Morita equivalence, proving Propositions 5.7 and 3.22 above.

To compare duality-preserving functors one requires the following definition.

Definition A.1. Suppose $(G, \Psi, \eta) : \mathcal{C} \to \mathcal{D}$ and $(G', \Psi', \eta) : \mathcal{C} \to \mathcal{D}$ are duality-preserving functors between hermitian categories $\mathcal{C}$ and $\mathcal{D}$. A natural transformation $\alpha : (G, \Psi, \eta) \to (G', \Psi', \eta)$ is a natural transformation between the underlying functors, $\alpha : G \to G'$, such that

$$
\Psi_V = \alpha_V^* \Psi'_V \alpha_{V^*}
\tag{51}
$$

for each object $V \in \mathcal{C}$.

If $\alpha : G \to G'$ is a natural isomorphism between the underlying functors and α satisfies (51) then $\alpha^{-1} : G' \to G$ also satisfies (51) so α is in fact a natural isomorphism of duality-preserving functors.

We noted in Lemma 2.29 that an exact duality-preserving functor induces a homomorphism of Witt groups. The following lemma says that naturally isomorphic duality-preserving functors induce the same homomorphism on Witt groups.

Lemma A.2. *Suppose* $(G, \Psi, \eta), (G', \Psi', \eta) : \mathcal{C} \to \mathcal{D}$ *are duality-preserving functors which respect exact sequences and* $\alpha : (G, \Psi, \eta) \to (G', \Psi', \eta)$ *is a natural isomorphism. If* $(V , \phi : V \to V^*)$ *is a hermitian form in* $\mathcal{C}$ *then there is a natural isomorphism between the induced hermitian forms* $(G(V), \Psi G(\phi)) \cong (G'(V), \Psi'G'(\phi))$. *Let* $\zeta = 1$ *or* -1. *The duality-preserving functors* (G, Ψ, η) *and* (G', Ψ', η) *induce the same homomorphism of Witt groups* $W^\zeta(\mathcal{C}) \to W^{\zeta\eta}(\mathcal{D})$.

Proof. In the diagram

$$
\begin{array}{ccccc}
G(V) & \xrightarrow{G(\phi)} & G(V^*) & \xrightarrow{\Psi_V} & G(V)^* \\
\alpha_V \downarrow & & \alpha_{V^*} \downarrow & & \uparrow \alpha_V^* \\
G'(V) & \xrightarrow[G'(\phi)]{} & G'(V^*) & \xrightarrow[\Psi'_V]{} & G'(V)^*
\end{array}
$$

the left-hand square commutes by the naturality of α while the right-hand square commutes because α satisfies equation (51). The Lemma follows easily. $\qquad\square$

It is a consequence of Lemma A.2 that an equivalence of hermitian categories induces an isomorphism of Witt groups.

A.1 The covering construction

In this section we prove that the covering construction B respects changes to coefficients (Proposition 5.7). We need one more observation which is straightforward to verify:

Lemma A.3. *If* $A \to A' \to A''$ *are ring homomorphisms then the diagram*

$$
\begin{array}{ccc}
A'' \otimes_A V^* \longrightarrow A'' \otimes_{A'} (A' \otimes_A V^*) & & \\
\downarrow & \searrow & A'' \otimes_{A'} (A' \otimes_A V)^* \\
(A'' \otimes_A V)^* \leftarrow (A'' \otimes_{A'} (A' \otimes_A V))^* & \nwarrow &
\end{array}
$$

of natural isomorphisms is commutative. $\qquad\square$

Proof of Proposition 5.7. Suppose $A \to A'$ is a ring homomorphism and V is a module in $\mathcal{S}ei(A)$. The natural isomorphism

$$(A' \otimes_A V)[F_\mu] \to A'[F_\mu] \otimes_{A[F_\mu]} (V[F_\mu])$$

induces a natural isomorphism (see Lemma 2.9)

$$\{\alpha_V\}_{V \in \mathcal{S}ei(A)} : B(A' \otimes V) \to A'[F_\mu] \otimes_{A[F_\mu]} B(V).$$

We aim to show that α is a natural isomorphism between duality-preserving functors

$$(B, \Phi, -1) \circ (A' \otimes_A \_, \Pi, 1) \to (A'[F_\mu] \otimes_{A[F_\mu]} \_, \Upsilon, 1) \circ (B, \Phi, -1).$$

Applying Lemma A.3 to both composites in the commutative square

$$
\begin{array}{ccc}
A & \longrightarrow & A[F_\mu] \\
\downarrow & & \downarrow \\
A' & \longrightarrow & A'[F_\mu]
\end{array}
$$

of ring homomorphisms one obtains commutative diagrams

$$
\begin{array}{ccccc}
(A' \otimes_A V^*)[F_\mu] & \longrightarrow & (A' \otimes_A V)^*[F_\mu] & \overset{\gamma}{\longrightarrow} & (A' \otimes_A V[F_\mu])^* \\
\downarrow & & & & \uparrow \\
A'[F_\mu] \otimes_{A[F_\mu]} (V^*[F_\mu]) & \underset{\delta}{\longrightarrow} & A'[F_\mu] \otimes_{A[F_\mu]} (V[F_\mu])^* & \longrightarrow & (A'[F_\mu] \otimes_{A[F_\mu]} V[F_\mu]
\end{array}
$$

where $\gamma = \pm \Pi_{A' \otimes V}(1 - z^\pm)$ and $\delta = 1 \otimes \pm \Pi_V (1 - z^\pm)$ and hence the commutative diagram

$$
\begin{array}{ccccc}
B(A' \otimes_A V^*) & \overset{B(\Pi_V)}{\longrightarrow} & B((A' \otimes_A V)^*) & \overset{\Phi_{A' \otimes V}}{\longrightarrow} & B(A' \otimes V)^\wedge \\
\alpha_{V^*} \downarrow & & & & \uparrow (\alpha_V)^\wedge \\
A'[F_\mu] \otimes_{A[F_\mu]} B(V^*) & \underset{1 \otimes \Phi_V}{\longrightarrow} & A'[F_\mu] \otimes_{A[F_\mu]} B(V)^\wedge & \underset{\Upsilon_{B(V)}}{\longrightarrow} & (A'[F_\mu] \otimes_{A[F_\mu]} B(V))^\wedge
\end{array}
$$

Thus α is a natural transformation between duality-preserving functors as claimed. It follows by Lemma A.2 that the diagram (36) of Witt group homomorphisms commutes. The proof of Proposition 5.7 is complete. $\qquad\square$

A.2 Hermitian Morita Equivalence

In this section we prove that hermitian Morita equivalence respects duality-preserving functors (Proposition 3.22). Let $(G, \Psi, \eta') : \mathcal{C} \to \mathcal{D}$ denote a duality-preserving functor and let $M \in \mathcal{C}$, $E = \text{End}_{\mathcal{C}} M$ and $E' = \text{End}_{\mathcal{D}} G(M)$.

We shall define a natural isomorphism between the composite functors

$$\alpha : (E' \otimes_E \underline{}, \Pi, 1) \circ (\text{Hom}(M, \underline{}), \Omega^b, \eta)$$
$$\xrightarrow{\cong} (\text{Hom}(G(M), \underline{}), \Omega^{\eta' \Psi_M G(b)}, \eta \eta') \circ (G, \Psi, \eta')$$

If $N \in \mathcal{C}|_M$ then $\text{Hom}(M, N)$ is a left E-module for the action

$$f.\theta = \theta \overline{f} = \theta b^{-1} f^* b$$

where $f \in E$ and $\theta \in \text{Hom}(M, N)$. The group $\text{Hom}(G(M), G(N))$ is regarded as a left E'-module in the same way. Define

$$\alpha_N : E' \otimes_E \text{Hom}_{\mathcal{C}}(M, N) \to \text{Hom}_{\mathcal{D}}(G(M), G(N))$$
$$f \otimes \gamma \mapsto f.G(\gamma) = G(\gamma)\overline{f} = G(\gamma)(\eta' \Psi_M G(b))^{-1} f^*(\eta' \Psi_M G(b)). \tag{52}$$

Since α_N is an isomorphism in the case $N = M$ it follows that α_N is an isomorphism for all $N \in \mathcal{C}|_M$. It is easy to see that $\{\alpha\}_{N \in \mathcal{C}|_M}$ is a natural transformation

$$(E' \otimes_E \underline{}) \circ \text{Hom}(M, \underline{}) \to \text{Hom}(G(M), \underline{}) \circ G.$$

One must check that α is a natural transformation of duality-preserving functors. By equations (19) and (51) one must show that

$$\alpha_N^* \Omega_{G(N)}^{\eta' \Psi_M G(b)} \text{Hom}(G(M), \Psi_N) \alpha_{N^*} = \Pi_{\text{Hom}(M,N)}(1 \otimes \Omega_N^b)$$

This equation can be checked by direct calculation, substituting the formulae (52), (28) and (20) for α, Ω and Π respectively and applying the naturality of Φ and the equation (18). This completes the proof of Proposition 3.22.

References

[1] A. A. Albert. *Structure of Algebras*, volume 24 of *American Mathematical Society Colloquium Publications*. American Mathematical Society, New York, 1939.

[2] P. Ara. Finitely presented modules over Leavitt algebras. *Journal of Pure and Applied Algebra*, 191:1–21, 2004.

[3] R. C. Blanchfield. Intersection theory of manifolds with operators with applications to knot theory. *Annals of Mathematics (2)*, 65:340–356, 1957.

[4] F. Borceux. *Handbook of Categorical Algebra 1. Basic Category Theory*, volume 50 of *Encyclopedia of Mathematics*. Cambridge University Press, 1994.

[5] S. E. Cappell and J. L. Shaneson. The codimension two placement problem, and homology equivalent manifolds. *Annals of Mathematics (2)*, 99:277–348, March 1974.

[6] ——— and ——— Link cobordism. *Commentarii Mathematici Helvetici*, 55:20–49, 1980.

[7] T. D. Cochran and K. E. Orr. Homology boundary links and Blanchfield forms: Concordance classification and new tangle-theoretic constructions. *Topology*, 33(3):397–427, 1994.

[8] P. M. Cohn. Free ideal rings. *Journal of Algebra*, 1:47–69, 1964.

[9] ——— *Free Rings and their Relations*. London Mathematical Society Monographs, 2. Academic Press, London, 1971.

[10] ——— *Free Rings and their Relations*. London Mathematical Society Monographs, 19. Academic Press, London, 2nd edition, 1985.

[11] ——— Localization in general rings, a historical survey. Pages 5–23 in this volume.

[12] ——— and W. Dicks. Localization in semifirs. II. *J.London Math.Soc. (2)*, 13(3):411–418, 1976.

[13] W. Dicks and E. Sontag. Sylvester domains. *J. Pure Appl. Algebra*, 13(3):243–275, 1978.

[14] J. Duval. Forme de Blanchfield et cobordisme d'entrelacs bords. *Commentarii Mathematici Helvetici*, 61(4):617–635, 1986.

[15] M. Farber. The classification of simple knots. *Uspekhi Mat. Nauk*, 38(5):59–106, 1983. Russian Math. Surveys 38:5 (1983) 63-117.

[16] ——— Hermitian forms on link modules. *Commentarii Mathematici Helvetici*, 66(2):189–236, 1991.

[17] ——— Noncommutative rational functions and boundary links. *Mathematische Annalen*, 293(3):543–568, 1992.

[18] ——— and P. Vogel. The Cohn localization of the free group ring. *Mathematical Proceedings of the Cambridge Philosophical Society*, 111(3):433–443, 1992.

[19] R. H. Fox. Free Differential Calculus. I: Derivation in the Free Group Ring. *Annals of Mathematics (2)*, 57(3):547–560, 1953.

[20] P. Gabriel. Des catégories abéliennes. *Bulletin de la Société Mathématique de France*, 90:323–448, 1962.

[21] ——— and M. Zisman. *Calculus of Fractions and Homotopy Theory*, volume 35 of *Ergebnisse der Mathematik und ihrer Grenzgebiete*. Springer, New-York, 1967.

[22] C. Kearton. Blanchfield duality and simple knots. *Transactions of the American Mathematical Society*, 202:141–160, 1975.

[23] ——— Cobordism of knots and Blanchfield duality. *Journal of the London Mathematical Society (2)*, 10(4):406–408, 1975.

[24] M. A. Kervaire. Les noeuds de dimensions supérieures. *Bulletin de la Société Mathématique de France*, 93:225–271, 1965.

[25] ——— Knot cobordism in codimension two. In *Manifolds–Amsterdam 1970*, Lecture Notes in Math., 197, Springer, Berlin, 1971, pp. 83–105.

[26] M.-A. Knus. *Quadratic and Hermitian Forms over Rings*. Grundlehren der Mathematischen Wissenschaften, 294. Springer, Berlin, 1991.

[27] K. H. Ko. Seifert matrices and boundary link cobordisms. *Transactions of the American Mathematical Society*, 299(2):657–681, 1987.

[28] ——— A Seifert-matrix interpretation of Cappell and Shaneson's approach to link cobordisms. *Mathematical Proceedings of the Cambridge Philosophical Society*, 106:531–545, 1989.

[29] T. Y. Lam. *A First Course in Noncommutative Rings*. Springer, New York, 1991.

[30] S. Lang. *Algebra*. Addison-Wesley, 3rd edition, 1993.

[31] J. Levine. Invariants of knot cobordism. *Inventiones Mathematicae*, 8:98–110, 1969. Addendum, 8:355.

[32] _____ Knot cobordism groups in codimension two. *Commentarii Mathematici Helvetici*, 44:229–244, 1969.

[33] _____ Knot modules I. *Transactions of the American Mathematical Society*, 229:1–50, 1977.

[34] J. Lewin. Free modules over free algebras and free group algebras: The Schreier technique. *Transactions of the American Mathematical Society*, 145:455–465, November 1969.

[35] S. Mac Lane. *Categories for the Working Mathematician*. Number 5 in Graduate Texts in Mathematics. Springer, 1971.

[36] _____ *Homology*. Springer, 1995. Reprint of the 1975 edition.

[37] J. W. Milnor. On isometries of inner product spaces. *Inventiones Mathematicae*, 8:83–97, 1969.

[38] W. Mio. On boundary-link cobordism. *Mathematical Proceedings of the Cambridge Philosophical Society*, 101:259–266, 1987.

[39] W. Pardon. Local surgery and applications to the theory of quadratic forms. *Bulletin of the American Mathematical Society*, 82(1):131–133, 1976.

[40] _____ Local surgery and the exact sequence of a localization for Wall groups. *Memoirs of the American Mathematical Society*, 12(196):iv+171, 1977.

[41] H.-G. Quebbemann, W. Scharlau, and M. Schulte. Quadratic and Hermitian forms in additive and abelian categories. *Journal of Algebra*, 59(2):264–289, 1979.

[42] A. A. Ranicki. *Exact Sequences in the Algebraic Theory of Surgery*. Mathematical Notes 26. Princeton University Press, New Jersey; University of Tokyo Press, Tokyo, 1981.

[43] _____ *High-dimensional Knot Theory*. Springer, Berlin, 1998.

[44] _____ Blanchfield and Seifert algebra in high dimensional knot theory. Moscow Math. J. 3:1333-1367, 2003. arXiv:math.GT/0212187.

[45] _____ Noncommutative localization in topology. Pages 81–102 in this volume. arXiv:math.AT/0303046.

[46] _____ and D. Sheiham. Blanchfield and Seifert algebra in high dimensional boundary link theory I. Algebraic K-theory. arXiv:math.AT/0508405.

[47] N. Sato. Free coverings and modules of boundary links. *Transactions of the American Mathematical Society*, 264(2):499–505, April 1981.

[48] W. Scharlau. *Quadratic and Hermitian forms*. Grundlehren der Mathematischen Wissenschaften, 270. Springer, Berlin, 1985.

[49] A. H. Schofield. *Representations of rings over skew fields*, volume 92 of *London Mathematical Society Lecture Note Series*. Cambridge University Press, 1985.

[50] D. Sheiham. Non-commutative characteristic polynomials and Cohn localization. *Journal of the London Mathematical Society (2)*, 64(1):13–28, 2001. arXiv:math.RA/0104158.

[51] _____ *Invariants of Boundary Link Cobordism*, volume 165 of *Memoirs of the American Mathematical Society*. American Mathematical Society, 2003. arXiv:math.AT/0110249.

[52] J. R. Smith. Complements of codimension-two submanifolds - III - cobordism theory. *Pacific Journal of Mathematics*, 94(2):423–484, 1981.

[53] V. Srinivas. *Algebraic K-theory*, volume 90 of *Progress in Mathematics*. Birkhäuser, Boston, 1996.

[54] P. Vogel. Localisation in algebraic L-theory. In *Proc. 1979 Siegen Topology Conf.*, Lecture Notes in Mathematics, 788, pages 482–495. Springer, 1980.

[55] _____ On the obstruction group in homology surgery. *Publ. Math. I.H.E.S.*, 55:165–206, 1982.

International University Bremen
Campus Ring 1
Bremen 28759
Germany

Noncommutative localization
in noncommutative geometry

Zoran Škoda

Abstract

The aim of these notes is to collect and motivate the basic localization toolbox for the geometric study of "spaces" locally described by noncommutative rings and their categories of modules.

We present the basics of the ORE localization of rings and modules in great detail. Common practical techniques are studied as well. We also describe a counterexample to a folklore test principle for Ore sets. Localization in negatively filtered rings arising in deformation theory is presented. A new notion of the differential ORE condition is introduced in the study of the localization of differential calculi.

To aid the geometrical viewpoint, localization is studied with emphasis on descent formalism, flatness, the abelian categories of quasi-coherent sheaves and generalizations, and natural pairs of adjoint functors for sheaf and module categories. The key motivational theorems from the seminal works of GABRIEL on localization, abelian categories and schemes are quoted without proof, as well as the related statements of POPESCU, EILENBERG-WATTS, DELIGNE and ROSENBERG.

The COHN universal localization does not have good flatness properties, but it is determined by the localization map already at the ring level, like the perfect localizations are. Cohn localization is here related to the quasideterminants of GELFAND and RETAKH; and this may help the understanding of both subjects.

Contents

1   Introduction

1.1 Objectives and scope. This is an introduction to Ore localizations and generalizations, with *geometric* applications in mind.

The existing bibliography in localization theory is relatively vast, including several monographs. Localizations proliferate particularly in category theory, with flavours adapted to various situations, like bicategories, toposes, QUILLEN's model categories, triangulated categories etc. A noncommutative algebraic geometer replaces a space with a ring or more general 'algebra', or with some category whose objects mimic modules over the 'algebra', or mimic sheaves over a space, or he/she studies a more general category which is glued together from such local ingredients. This setup suggests that we may eventually need a similar toolbox to the one used by the category and

AMS classification: 16A08, 16U20, 18E35, 14A22

homotopy theorists; however the simplest cases are in the area of ring and module theory. Even so, we shift the emphasis from purely ring-theoretic questions to more geometrical ones.

A **localized ring** is typically structurally simpler than the original, but retaining some of its features. A controlled but *substantial simplification* of a ring is a useful tool for a ring theorist, often as extreme as passing to a local ring or a quotient skewfield. On the contrary, our main geometrical goals are *those localizations which may play the role of noncommutative analogues of (rings of regular functions on principal Zariski) open sets* in an (affine, say) variety. Rings of functions on these open sets may be slightly simpler than the rings of *global* functions, but not as radically as when, say, passing to a local ring. We start with the very basics of localization procedures. The geometric notion of a *cover* by localizations is studied in the noncommutative context as well. Only recent geometrically minded works include some elements of such a study, as some key features of covers, the *globalization lemma* in particular, were recognized only in the mid-eighties.

We use an *elementary method* to prove the existence and simple properties of the Ore localized rings, in line with the original 1931 paper of O. ORE [98] (who however assumed no zero divisors). Modern treatments often neglect details when sketching the elementary method. Another modern method (to prove existence etc. ([91])), following ASANO ([6]), is cheap, but does not give an equivalent experience with the Ore method. Calculations similar to the ones in our proofs appear in concrete examples when using or checking Ore conditions *in practice*. We also use this method to examine when there is an induced localization of a first order differential calculus over a noncommutative ring, and come to a condition, not previously observed, which we call the "differential Ore condition". The elementary method has the advantage of being parallel to the calculus of (left) fractions in general categories, which has an aura of being a difficult subject, but is more transparentafter learning Ore localization the elementary way.

Our next expositional goal is to obtain some practical criteria for finding and dealing with Ore localizations in practice. Folklore strategies to prove that some set is Ore are examined in detail. In a section on Ore localization in 'negatively' filtered rings, we explore similar methods.

While Ore localization is treated in a comprehensive introductory style, more general localizations are sketched in a survey style. For advanced topics there are often in place good references, assuming that the reader knows the motivation, and at least the Ore case. Both requisites may be fulfilled by reading the present notes. We emphasize facts important in geometry, often omitted, or which are only folklore. In order to clear up some sources of confusion, we sketch and compare several levels of generality; mention competing terminologies; and examine the difficulties of geometrical interpretation and usage.

We focus on localizations of the category $R - \text{Mod}$ of all *left* modules over a fixed ring R. Localizations in other specific categories, e.g. central bimodules (symmetric localization, cf. [62]), bimodules, and the standard approach via *injective hulls* are omitted. One of the reasons is that often there are too few central bimodules over a noncommutative ring and 2-sided ideals in particular. Bimodules in general are interpreted as generalizing the maps of noncommutative rings, as explained in the text. Generalities on localization in arbitrary categories, and abelian in particular, are outlined for convenience.

As COHN localization can be found in other works in this volume, we include only a short introduction with two goals: putting it in our context and, more importantly, relating it to the recent subject of quasideterminants. Anybody aware of both subjects is aware of some connection, and we try to spell it out as precisely as we can.

1.2 Prerequisites on algebraic structures. Basic notions on rings and modules are freely used: unital ring, left ideal, center, left module, bimodule, domain (ring with no zero divisors), skewfield (division ring), graded rings and modules, and operations with those.

1.3 Prerequisites on categories. The language of categories and functors is assumed, including a (universal) initial and terminal object, (projective=inverse) limit, colimit (= inductive/direct limit), (co)products, adjoint functors, Yoneda lemma, and the categorical duality (inverting arrows, dual category, dual statements about categories).

Appendix A in [145] suffices for our purposes; for more see [23, 21, 84].

1.4 A morphism $f : B \to C$ is epi (mono) if for any pair e, e' of morphisms from C to any D (resp. from any A to B), equality $ef = e'f$ (resp. $fe = fe'$) implies $e = e'$.

A **subobject** is an equivalence class of monomorphisms. A pair (F, in) consisting of a functor F, and a natural transformation of functors in : $F \hookrightarrow G$, is a **subfunctor** of functor G, if all $\text{in}_M : F(M) \hookrightarrow G(M)$ are monomorphisms. Explicitly, the naturality of in reads $\forall f : M \to N$, $\text{in}_N \circ F(f) = G(f) \circ \text{in}_M : F(M) \to G(N)$. Clearly, if F is a subfunctor of an additive (**k**-linear) functor G, between additive (**k**-linear) categories, then F is additive (**k**-linear) as well.

1.5 A (small) diagram d in the category $\mathcal{C}$ will be viewed as a functor from some (small) category D into C. For fixed D, such functors and natural transformations make a category C^D. Every object c in C gives rise to a constant diagram c^D sending each X in D into c. This rule extends to a functor $()^D : C \to C^D$. A **cone over diagram** $d : D \to C$ is a natural transformation $I_c : c^D \Rightarrow d$ for some $c \in C$. A morphism $I_c \to I_{c'}$ of cones over d is a morphism $\phi : c' \to c$ such that $I_{c'} = I_c \circ \phi^D$. A terminal among the cones over D will be called a **limiting cone** over D. A **colimiting cone** in C is a limiting cone in opposite category C^{op}. Consider a 'parallel' family of morphisms $\{f_\gamma : A \to B\}_{\gamma \in \Gamma}$ as a diagram D with 2 objects and $|\Gamma|$ arrows in an obvious way. In this case, a cone over D is given by a single map $I : c \to A$. We call the diagram $c \xrightarrow{I} A \underset{f_2}{\overset{f_1}{\rightrightarrows}} B$ a **fork diagram**. It is called an **equalizer (diagram)** if $I : c \to A$ is in addition a limiting cone; by abuse of language one often calls I, or even c an equalizer. Equalizers in C^{op} are referred to as **coequalizers**. A morphism $I \to A$ of the cone of an equalizer diagram with $\Gamma = \{1, 2\}$ is also called a **kernel** of parallel pair $f_1, f_2 : A \to B$. Cokernels are defined dually.

A **zero** object 0 is an object which is simultaneously initial and terminal. In that case, composition $X \to 0 \to Y$ is also denoted by $0 : X \to Y$. A (co)kernel of a *single* morphism $f : A \to B$ in a category with 0 is by definition a (co)kernel of pair $f, 0 : A \to B$.

1.6 A functor $F : C \to C'$ induces a (pullback) functor for diagrams $F \circ : C^D \to (C')^D$. It is defined by $d \mapsto F \circ d$ for every diagram $d : D \to C$, and $\alpha \mapsto F(\alpha)$ where $(F(\alpha))_M := F(\alpha_M)$ for $\alpha : d \Rightarrow d'$.

F **preserves** limits of some class $\mathcal{P}$ of diagrams in $\mathcal{A}$ if it sends any limiting cone $p_0 \to p$ over any $p \in \mathcal{P}$ in $\mathcal{A}$ into a limiting cone in $\mathcal{A}'$. F **reflects** limits if any cone $p_0 \to p$ over any $p \in \mathcal{P}$ in $\mathcal{A}$ must be a limiting cone if F sends it to a limiting cone in $\mathcal{A}'$. The same holds whenever the word 'limit' is replaced by 'colimit', and cone $p_0 \to p$ by a cocone $p \to p_0$.

1.7 An **Ab**-category (or **preadditive category**) is a category $\mathcal{A}$ with an abelian group operation $+$ on each set $\mathcal{A}(X, Y)$, such that each composition map $\circ : \mathcal{A}(X, Y) \times \mathcal{A}(Y, Z) \to \mathcal{A}(Y, Z)$ is bilinear. An **Ab**-category is **additive** if it contains a zero object and pairwise, hence all finite, products. Automatically then, finite products agree with finite coproducts. Recall that an additive category $\mathcal{A}$ is **abelian** if each morphism f in $\mathcal{A}$ has a kernel and a cokernel morphism, and the kernel object of a cokernel equals the cokernel object of the kernel. We assume that the reader is comfortable with elementary notions on abelian groups like exact sequences and left (right) exact functors in the greater generality of abelian categories.

1.8 Gabriel-Mitchell-Popescu embedding theorem. *Every small* *abelian category is equivalent as an abelian category to a subcategory of the category of left modules over a certain ring R. Proof:* [103].

1.9 Prerequisites on spaces of modern geometry. We expect familiarity with the notions of a *presheaf, separated presheaf* and *sheaf,* and with some examples describing a geometry via a topological space with a structure sheaf on it; as well as the idea of gluing from some sort of local models, behind the concepts of (super)manifolds, analytic spaces and schemes. Earlier exposure to commutative algebraic (or analytic) varieties and schemes is assumed in style of some sections, but no specific facts are required; an abstract sketch of the main features of scheme-like theories is supplied in the text below.

1.10 Conventions. The word *map* means a set-theoretic map unless it is accompanied with a specification of the ambient category in question when it means a (homo)morphism, e.g. a *map of rings* means a ring (homo)morphism. The word *noncommutative* means "not necessarily commutative". Though for many constructions this is not necessary, we mostly deal with unital rings and modules, unless said otherwise. *Ideal* without a modifier always means '1-sided (usually left) ideal'.

The symbol for inclusion $\subset$ may include the case of equality. The unadorned tensor symbol is over $\mathbb{Z}$, except for elements in given tensor products, like $a \otimes b := a \otimes_R b \in A \otimes_R B$. For algebras and modules over a commutative ring, this ring is usually denoted **k**. These conventions may be locally overridden by contextual remarks.

2   Noncommutative geometry

DESCARTES introduced the method of coordinates, which amounted to the identification of real vector spaces with the spaces described by the axioms of EUCLID. LAGRANGE considered more general curvilinear coordinates in analytic mechanics to obtain exhaustive treatments of space. Topological spaces do not have distinguished coordinate functions, but the whole algebra of functions suffices. The GELFAND-NEIMARK theorem (e.g. [74]) states that the category of compact Hausdorff topological spaces is equivalent to the opposite of category of commutative unital C^*-algebras. This is accomplished by assigning to a compact X the Banach $\star$-algebra $C(X)$ of all continuous $\mathbb{C}$-valued functions (with the supremum-norm and involution $f^*(x) = \overline{f(x)}$). In the other direction one (re)constructs X as a Gelfand spectrum of the algebra A, which is a space whose points are continuous characters $\chi : A \to \mathbb{C}^*$ endowed with spectral topology. These characters appear as the evaluation functionals χ_x on A at points $x \in X$, where $\chi_x(f) = f(x)$. Each annihilator $\operatorname{Ann} \chi = \{a \in A \mid \chi(a) = 0\}$ is a maximal ideal of A and all maximal ideals are of that form.

2.1 Towards noncommutative algebraic geometry. For any commutative ring R, GROTHENDIECK replaced maximal ideals from the theory of affine varieties and from the Gelfand-Neimark picture, by arbitrary prime ideals, which behave better functorially, and he endowed the resulting spectrum $\operatorname{Spec} R$ with a non-Hausdorff Zariski topology and a structure sheaf, defined with the help of commutative localization. This amounts to a fully faithful contravariant functor Spec from *CommRings* to the category l$\mathfrak{S}\mathfrak{p}$ of locally ringed spaces. In other words, the essential image of this functor, the category of geometric affine schemes Aff = Spec(*CommRings*) is equivalent to the formal category of affine schemes which is by the definition *CommRings*^{op}. Notions of points, open subspaces and sheaves are used to define l$\mathfrak{S}\mathfrak{p}$ and the functor Spec. The functor Spec takes values in a category described in local geometrical terms, translating algebraic concepts into geometric ones. The functor enables the transfer of intuition and methods between algebra and geometry. This interplay is to a large extent the basic *raison d'être* for the subject of algebraic geometry. The spaces in l$\mathfrak{S}\mathfrak{p}$ may be glued via topologies and sheaves, and certain limit constructions may be performed there which give a great flexibility in usage of a range of other subcategories in l$\mathfrak{S}\mathfrak{p}$ e.g. schemes, algebraic spaces, formal schemes, almost schemes etc. Useful constructions like blow-ups, quotients by actions of groups, flat families, infinitesimal neighborhoods etc. take us often out of the realm of affine schemes.

The dictionary between the geometric properties and abstract algebra may be partially extended to include noncommutative algebra. Noncommutative geometry means exploring the idea of faithfully extending the Spec functor (or, analogously, the Gelfand functor above) to noncommutative algebras as a domain and some geometrical universe nl$\mathfrak{S}\mathfrak{p}$ generalizing the category l$\mathfrak{S}\mathfrak{p}$ as a target, and thinking geometrically of the consequences of such construction(s). The category nl$\mathfrak{S}\mathfrak{p}$ should ideally contain more general noncommutative schemes, extending the fact that spaces in l$\mathfrak{S}\mathfrak{p}$ feature topology, enabling us to glue affine schemes over such subsets. Useful examples of noncommutative spaces are often studied in contexts which are much more restricted then what we require for nl$\mathfrak{S}\mathfrak{p}$; for instance one abandons topology or points, or one works only with Noetherian algebras close to commutative (say of finite GK-dimension, or finite dimensional over the center, when the latter is a field), or the category is not big enough to include the whole of ₗ₆p. For example, VAN OYSTAEYEN and his school ([139, 140, 142]) consider a certain class of graded rings for which they can use localizations to define their version of a noncommutative **Proj**-functor. A more restricted class of graded rings providing examples very close in behaviour to commutative projective varieties is studied by ARTIN, ZHANG, STAFFORD and others (see [5, 127] and refs. therein). Y. SOIBELMAN ([126]) advocates examples of natural compactifications of moduli spaces of commutative spaces with noncommutative spaces as points on the boundary.

Thus we often restrict ourselves either to smaller geometric realms Ns ⊂ nₗ₆p than nₗ₆p containing for example only projective "noncommutative varieties" of some sort, or to give up points, topological spaces in ordinary sense and work with a more intrinsic embedding of the category of affine schemes (now $Rings^{op}$ into some category of (pre)sheaves over $Rings^{op}$ using Yoneda lemma, Grothendieck topologies and related concepts ([69, 99, 110]). In the commutative case, while both the spectral and functorial approaches are interchangeably used (EGA prefers spectral, while SGA and DEMAZURE-GABRIEL [31] choose functorial; the latter motivated by niceties in the treatment of group schemes), the more difficult foundational constructions of theoretical nature are done using the Yoneda embedding approach (algebraic geometry over model categories ([133, 134]); THOMASON's work ([135]) on K-theory and derived categories (cf. also [7]); $\mathbb{A}^1$-homotopy theory of schemes ([92])).

2.2 One often stops consciously half way toward the construction of the functor $Rings^{op} \to$ nₗ₆p; e.g. start with rings and do nothing except for introducing a small class of open sets, e.g. commutative localizations, ignoring other natural candidates, because *it is difficult to work with them.* Unlike in the above discussed case of restricting the class of spaces Ns ⊂ nₗ₆p, we are conservative in the details of the spectral description, thus landing in some intermediate "semilocal" category **slSp** by means of a fully faithful embedding $Rings^{op} \to$ **slSp**.

2.2a An example of **slSp** is as follows. Consider the center $Z(R)$ of a ring R and construct the commutative ring Spec $Z(R)$. Then for each principal open set U in $Z(R)$, one localizes R at U (a commutative localization) and this essentially gives the structure sheaf $U \to R_U = \mathcal{O}(U)$. The problem arises if the center is small, hence Spec $Z(R)$ is small as well, hence all the information on R is kept in a few, maybe one, rings R_U, and we did not get far. In some cases the base space Spec Z' is big enough and we may glue such spectra to interesting more general "schemes" ([106]). Taking the center is not functorial, so we have to modify the categories a bit, to allow for pairs (R, C), $C \subset Z(R)$ from the start, and construct from them some "space" $\mathrm{Spec}_2(R, C)$ (see more in Ch. 10). It is argued in [106] that, when R is small relative to Z', this construction is a satisfactory geometrization and the standard tools from cohomology theory may be used. They call such a situation *semiquantum geometry*.

A fruitful method is to add a *limited class* $\{Q_i\}_i$ of other localizations on $R - \mathrm{Mod}$, and *think* of $Q_i R$ as the structure ring R_{U_i} over open subset U_i. However, now U_i is not really a subset in $\mathrm{Spec}\, Z(R)$, but rather a "geometric" label for Q_i viewed as certain open set on hypothetical noncommutative $\mathrm{Spec}\, R$. Of course the latter point of view is central ([69, 62, 111, 138, 139, 142]) for our subject far beyond the idea of a small enrichment of the Zariski topology on the spectrum of the center.

In summary, restricting sharply to a small class of localizations and/or working with small spectra, projects a coarser local description $\mathbf{slSp}(R)$ than often desired.

2.3 Alternatively, one may lose some information, for instance considering the points of spectra but not the sheaves, or types of spectra with insufficiently many points for the reconstruction of rings. We may think of such correspondences as nonfaithful functors from $\mathbf{slSp}$ into some partial geometric realms $\mathrm{Feature}_\alpha(\mathfrak{nlGp})$.

2.4 MANIN has suggested ([89]) a functor from graded rings into abelian categories: to a Noetherian ring R assign the quotient of the category of finitely generated graded R-modules by the subcategory of the finitely generated graded R-modules of finite length. In the commutative case, by a theorem of SERRE, this category is equivalent to the category of *coherent sheaves* over $\mathbf{Proj}\, R$. This is one of the candidates for *projective noncommutative geometry* and we view it as an example of functor of type $\mathbf{slSp}$. Manin here actualizes GROTHENDIECK's advice that *to do geometry one does not need the space itself but only the category of sheaves on that would-be space.* In this spirit, Grothendieck defined a **topos** ([63] and [21], vol.3) as an abstract category satisfying a list of axioms, whose consequence is that it is equivalent to the category of sheaves $\mathrm{Fas}\,\mathcal{C}$ over some site $\mathcal{C}$ (a category with a Grothendieck topology). Two different sites may give rise to the same topos, but their cohomological behaviour will be the same. Thus they are thought of as the same generalized space. Likewise, in algebraic geometry, we have examples for the same heuristics, where abelian categories of quasi-coherent sheaves of $\mathcal{O}$-modules take the place of the topos of *all* sheaves of sets. The suitable notion of a morphism between the topoi is recognized to be 'geometrical morphism' what is also an adjoint pair of functors with certain additional properties. In topos theory and applications, Grothendieck actually utilizes an interplay ('yoga') of 6 *standard* functors attached to the same 'morphism'. We shall discuss the basic pairs of adjoint functors for the categories of modules and sheaves. They appear in the disguise of maps of (noncommutative) rings (affine maps and their abstract version), as bi-modules for two rings, as direct and inverse maps for $\mathcal{O}_X$-modules, and as localization functors.

2.5 Grothendieck categories (G.c.) [62, 125]. A Grothendieck cat-egory is a cocomplete (having all small limits) abelian category, having enough injectives and a small generator. The category of left R-modules, and the category of all sheaves of left R-modules over a fixed topological space, are G.c.'s. Given a coalgebra C, the category of C-comodules is G.c. Given a bialgebra B and a B-comodule algebra E, the category of relative (E, B)-Hopf modules is a G. c. [146].

Theorem. (P. GABRIEL, [42]) *The category* $\mathfrak{Qcoh}_X$ *of quasicoherent sheaves of* $\mathcal{O}_X$*-modules over a quasicompact quasiseparated scheme X is a Grothendieck category.*

It is *not* known if $\mathfrak{Qcoh}_X$ for a general scheme X is cocomplete, or if it has enough injectives. This fact is behind our decision not to strictly require our abelian categories of noncommutative geometry to be a G.c. (which is fashionable). ROSENBERG [109] requires the weaker *property (sup)* (= *categories with exact limits*) introduced by GABRIEL ([42]): for any object M and any ascending chain of subobjects there is a supremum subobject, and taking such suprema commutes with taking the join (minimum) with a fixed subobject $N \subset M$. This holds for R-mod, Fas$\mathcal{C}$ (for a small site $\mathcal{C}$) and $\mathfrak{Qcoh}_X$ (for any scheme X).

2.6 Theorem. (P. GABRIEL for noetherian schemes ([42], Ch. VI); A.L. ROSENBERG in quasicompact case ([108]); and in general case ([115]))
Every scheme X can be reconstructed from the abelian category $\mathfrak{Qcoh}_X$ *uniquely up to an isomorphism of schemes.*

This motivates the promotion to a "space" of any member of a class of abelian categories, usually required to obey some additional axioms, allow-ing (some variant) of R-mod (R possibly noncommutative) and $\mathfrak{Qcoh}_X$ as prime examples. A distinguished object $\mathcal{O}$ in $\mathcal{A}$, corresponding to the struc-ture sheaf is often useful part of a data, even at the abstract level, hence the spaces could be actually pairs $(\mathcal{A}, \mathcal{O})$. The study of functors for the cate-gories of modules and categories of sheaves shows that there is a special role for functors having various exactness properties ([109]), e.g. having a right adjoint, hence such properties are often required in general. Gluing categories over localizations, and variants thereof, should be interpreted in good cases as gluings of spaces from local models. In noncommutative geometry, a local model is usually the full category of modules over a noncommutative ring.

2.7 The so-called *derived algebraic geometry*, treating in more natural terms and globalizing the infinitesimal picture of moduli spaces governed by the deformation theory, appeared recently ([12, 133]). Its cousin, *homotopical algebraic geometry* appeared promising in the study of homotopy theories for algebraic varieties, and also in using the reasoning of algebraic geometry for ring spectra of homotopy theory and for their globalization. In such generalizations of algebraic geometry the basic gadgets are *higher categories* (e.g. simplicially enriched, DG, Segal, A_∞, cf. [34, 65, 105, 133, 134]). The lack of smoothness in some examples of moduli spaces is now explained as an artifact of the truncation process replacing the natural and smooth 'derived moduli spaces' by ordinary moduli spaces (the '*hidden smoothness principle*' due to BONDAL, DELIGNE, DRINFELD, KAPRANOV, HINICH, KONTSEVICH...).

Part of the relevant structure here may be already expressed by replacing rings by differential graded algebras (dga-s) ([12]), or, more generally, by introducing sequences of higher (e.g. 'MASSEY') products, as in the theory of A_∞ (strongly homotopy associative) and L_∞ (strongly homotopy Lie) algebras. Such generalizations and special requirements needed to do localization in such enriched settings, are beyond the scope of the present article. A noncommutative algebraic geometry framework designed by O.A. LAUDAL ([75]), with emphasis on the problem of noncommutative deformation of moduli, implicitly includes the higher Massey products as well. In the viewpoint put forward by KONTSEVICH and FUKAYA, some of the 'dualities' of modern mathematical physics, e.g. the *homological mirror symmetry*, involve A_∞-categories defined in terms of geometric data ([70, 88, 126]). The so-called **quantization** ([30, 67, 74, 148]) in its many versions is generally of deformational and noncommutative nature. Thus it is not surprising that the formalisms combining the noncommutative and homological (or even homotopical) structures benefit from the geometrically sound models of quantum physics. MANIN suggested that a more systematic content of a similar nature exists, programmatically named *quantized algebraic geometry*, which may shed light on hidden aspects of the geometry of (commutative) varieties, including the deep subject of motives.

An interesting interplay of derived categories of coherent sheaves on varieties and their close analogues among other triangulated categories, motivates some 'noncommutative' geometry of the latter ([20, 19]). Triangulated categories are also only a "truncation" of some other higher categories.

One should also mention that some important classes of rings in *quantum algebra*, for example quantum groups, may be constructed using categories of (perverse) sheaves over certain commutative configuration spaces ([83]). Thus the structure of various sheaf categories is an ever repeating theme which relates the commutative and noncommutative world. See the essay [26] for further motivation.

3   Abstract localization

We discuss localization of 1. algebraic structures; 2. categories. These two types are related: typically a localization of a ring R induces a localization of the category $R -$ Mod of left modules over R.

A recipe $\mathcal{G}$ for a localization takes as input a structure R (monoid, lattice, ring), or a category $\mathcal{A}$, and distinguished data Σ in R (or $\mathcal{A}$). The localizing data Σ are selected from some class $\mathcal{U}(R)$ of structural data, for example elements, endomorphisms or ideals of R; similarly $\mathcal{U}(\mathcal{A})$ could be a class of subcategories or a collection of morphisms in $\mathcal{A}$. Usually not all obvious subclasses of $\mathcal{U}(R)$ may serve as distinguished data for $\mathcal{G}$, and some 'localizability' conditions apply.

A localization procedure $\mathcal{G}(R, \Sigma)$ should replace R by another object Y and a map $i : R \to Y$, which induces, for given $\mathcal{G}$, some canonical correspondence $\mathcal{G}(i) : \Sigma \rightsquigarrow \Sigma_*$ between the localization data Σ and some other data Σ_* chosen from $\mathcal{U}(Y)$. The subclass Σ_* should satisfy some natural requirement, for example that it consist of invertible elements. Pair (i, Y) should be in some sense smallest, or universal among all candidates satisfying the given requirements. For given requirements only certain collections Σ built from elements in $\mathcal{U}(R)$ give rise to a universal (i, Y). Such Σ are generically called *localizable* and the resulting Y is denoted $\Sigma^{-1}R$.

In the case of a category $\mathcal{C}$, a map i is replaced by a localization functor $Q^* : \mathcal{C} \to \Sigma^{-1}\mathcal{C}$. In this article, a localization of a category will be equivalent to an abstract localization with respect to a class of morphisms Σ in $\mathcal{C}$, often using some other equivalent data (e.g. 'localizing subcategory'). Following [43], we sketch the general case of a localization at a class of morphisms Σ, cf. also [21].

3.1 An **abstract 1-diagram** $\mathcal{E}$ is a structure weaker then a category: it consists of a class $\mathrm{Ob}\,\mathcal{E}$ of objects and a class $\mathrm{Mor}\,\mathcal{E}$ of morphisms equipped with a source and a target maps S, T : $\mathrm{Mor}\,\mathcal{E} \to \mathrm{Ob}\,\mathcal{E}$. No composition, or identity morphisms are supplied. As usual, for two objects A, B by $\mathcal{E}(A, B)$ we denote class of morphisms f with $\mathrm{S}(f) = A$ and $\mathrm{T}(f) = B$. If each $\mathcal{E}(A, B)$ is a set, one may use the word (multiple-edge) graph instead. If $\mathcal{E}, \mathcal{C}$ are diagrams, an $\mathcal{E}$-**diagram** in $\mathcal{C}$, or a morphism $d : \mathcal{E} \to \mathcal{C}$, is any pair of maps $\mathrm{Ob}\,\mathcal{E} \to \mathrm{Ob}\,\mathcal{C}$ and $\mathrm{Mor}\,\mathcal{E} \to \mathrm{Mor}\,\mathcal{C}$ which commute with the source and target maps. Small abstract 1-diagrams and their morphisms form a category $\mathfrak{Diagr}_1$. To each category one assigns its underlying abstract diagram. This correspondence induces a forgetful functor from the category $\mathfrak{Cat}$ of small categories to $\mathfrak{Diagr}_1$. The construction of a category of paths below provides the left adjoint to this functor.

If $n \geq 0$ is an integer, a **path** of length n from A to B in an abstract diagram $\mathcal{E}$ is a tuple $(A, f_1, f_2, \ldots, f_n, B)$, where A is an object and f_i are morphisms in $\mathcal{E}$, such that $\mathrm{S}(f_{i+1}) = \mathrm{T}(f_i)$ for $i = 1, \ldots, n-1$, and $\mathrm{S}(f_1) = A$, $\mathrm{T}(f_n) = B$ if $n > 0$, and $A = B$ if $n = 0$. For any abstract 1-diagram $\mathcal{E}$ define a **category** $\mathrm{Pa}\,\mathcal{E}$ **of paths** in $\mathcal{E}$ as follows. The class of objects is

$$\mathrm{Ob}\,\mathrm{Pa}\,\mathcal{E} := \mathrm{Ob}\,\mathcal{E}$$

and the morphism class $(\mathrm{Pa}\,\mathcal{E})(A, B)$ consists of all paths from A to B. One declares $\mathrm{Id}_A := (A, A)$, $\mathrm{S}'(A, f_1, \ldots, f_n, B) = A$ and $\mathrm{T}'(A, f_1, \ldots, f_n, B) = B$ to be the identity morphisms, and the source and target maps for $\mathrm{Pa}\,\mathcal{E}(A, B)$. The composition rule is

$$(A, f_1, \ldots, f_n, B) \circ (B, g_1, \ldots, g_m, C) = (A, f_1, \ldots, f_n, g_1, \ldots, g_m, C).$$

If each $(\mathrm{Pa}\,\mathcal{E})(A, B)$ is small we indeed obtain a category.

Consider the canonical $\mathcal{E}$-diagram $i_\mathcal{E} : \mathcal{E} \to \mathrm{Pa}\,\mathcal{E}$ which is tautological on objects as well as on paths of length 1. $\mathrm{Pa}\,\mathcal{E}$ has the following universal property: an $\mathcal{E}$-**diagram** d in any category $\mathcal{C}$ gives rise to a unique functor $d' : \mathrm{Pa}\,\mathcal{E} \to \mathcal{C}$ such that $d = d' \circ i_\mathcal{E}$.

3.2 Let Σ be a family of morphisms in the category $\mathcal{C}$. If $J : \mathcal{C} \to \mathcal{D}$ is any functor let $\Sigma_* := J(\Sigma)$ be the class of all morphisms $J(f)$ where $f \in \Sigma$. Given $\mathcal{C}$ and Σ, consider the diagram scheme $\mathcal{E} = \mathcal{E}(\mathcal{C}, \Sigma)$ with $\mathrm{Ob}\,\mathcal{E} := \mathrm{Ob}\,\mathcal{C}$ and $\mathrm{Mor}\,\mathcal{E} := \mathrm{Mor}\,\mathcal{C} \coprod \Sigma$, $\mathrm{S}_\mathcal{E} = \mathrm{S} \coprod \mathrm{S}|_\Sigma$, $\mathrm{T}_\mathcal{E} = \mathrm{T} \coprod \mathrm{T}|_\Sigma$. One has canonical inclusions $\mathrm{IN}_1 : \mathrm{Mor}\,\mathcal{C} \to \mathrm{Mor}\,\mathcal{E}$, $\mathrm{IN}_2 : \Sigma \hookrightarrow \mathrm{Mor}\,\mathcal{E}$. Let $\sim$ be the smallest equivalence relation on $\mathrm{Pa}\,\mathcal{E}$ such that

$$(\mathrm{IN}_1 v) \circ (\mathrm{IN}_1 u) \sim \mathrm{IN}_1(v \circ u) \text{ if } v \circ u \text{ is defined in } \mathcal{C},$$
$$\mathrm{IN}_1(\mathrm{id}_A) \sim (A, A), \quad A \in \mathcal{C},$$
$$\left. \begin{array}{l} (\mathrm{IN}_2 f) \circ (\mathrm{IN}_1 f) \sim (\mathrm{S}(f), \mathrm{S}(f)) \\ (\mathrm{IN}_1 f) \circ (\mathrm{IN}_2 f) \sim (\mathrm{T}(f), \mathrm{T}(f)) \end{array} \right\} f \in \Sigma.$$

It is direct to show that operation ∘ induces a composition on classes of morphisms with respect to this particular equivalence relation. In this way we obtain a quotient $\Sigma^{-1}\mathcal{C}$ of the category $\mathrm{Pa}\,\mathcal{E}$ together with the canonical functor $Q_\Sigma^* : \mathcal{C} \to \mathrm{Pa}\,\mathcal{E}$ which is tautological on objects and equals $i_{\mathcal{E}} \circ \mathrm{IN}_1$ followed by the projection to the classes of equivalence on morphisms.

3.3 Proposition. *If $f \in \Sigma$ then the functor $Q_\Sigma^* : \mathcal{C} \to \Sigma^{-1}\mathcal{C}$ sends f to an invertible map $Q_\Sigma^*(f)$. If $T : \mathcal{C} \to \mathcal{D}$ is any functor such that $T(s)$ is invertible for any $s \in \Sigma$ then there is a unique functor $H : \Sigma^{-1}\mathcal{A} \to \mathcal{B}$ such that $T = H \circ Q_\Sigma^*$.*

$\Sigma^{-1}\mathcal{C}$ is **category of fractions** of $\mathcal{C}$ at Σ. This construction has a defect, in that there is no general recipe to determine when two morphisms in $\mathrm{Pa}\,\mathcal{E}$ represent the same morphism in $\Sigma^{-1}\mathcal{C}$. If Σ satisfies the Ore conditions, below, there is one.

3.4 Proposition. [43] *Let $T^* \dashv T_*$ be an adjoint pair of functors (this notation means that T^* is left adjoint to T_*), where $T^* : \mathcal{A} \to \mathcal{B}$. with adjunction counit $\epsilon : T^*T_* \Rightarrow 1_\mathcal{B}$. Let $\Sigma = \Sigma(T_*)$ be the class of all morphisms f in $\mathcal{A}$ such that $T^*(f)$ is invertible, and $Q_\Sigma^* : \mathcal{A} \to \Sigma^{-1}\mathcal{A}$ the natural functor. Then the following are equivalent:*

(i) T_* is fully faithful.

(ii) $\epsilon : T^*T_* \Rightarrow 1_\mathcal{B}$ is an isomorphism of functors.

(iii) The unique functor $H : \Sigma^{-1}\mathcal{A} \to \mathcal{B}$ such that $T^* = H \circ Q_\Sigma^*$ is an equivalence; in particular Q_Σ^* has a right adjoint $Q_{\Sigma*}$.

(iv) (If $\mathcal{A}$ is small) For each category $\mathcal{X}$, functor $\mathfrak{Cat}(-, \mathcal{X}) : \mathfrak{Cat}(\mathcal{B}, \mathcal{X}) \to \mathfrak{Cat}(\mathcal{A}, \mathcal{X})$ is fully faithful.

Throughout the paper, any functor T^* agreeing with a functor $Q_\Sigma^* : \mathcal{C} \to \Sigma^{-1}\mathcal{C}$ as above up to category equivalences will be referred to as a **localization functor**. A functor T^* satisfying (i)-(iii) will be referred to as **a continuous localization functor**.

3.5 ([79]) *Suppose $Q^* : \mathcal{A} \to \mathcal{B}$ is a localization functor (cf. **3.4**), and $F : \mathcal{A} \to \mathcal{A}$ an endofunctor. If there is a functor $G : \mathcal{B} \to \mathcal{B}$ and a natural equivalence of functors $\alpha : Q^* \circ F \Rightarrow G \circ Q^*$ then there is a unique functor $F_\mathcal{B} : \mathcal{B} \to \mathcal{B}$ such that $Q^* \circ F = F_\mathcal{B} \circ Q^*$. In that case, we say that F is* **compatible with** Q^*.

Proof. Suppose $f : M \to N$ is a morphism in $\mathcal{A}$. Suppose that $Q^*(f)$ is invertible. Then $GQ^*(f) : GQ^*(M) \to GQ^*(N)$ also has some inverse s. The naturality of α and α^{-1} implies

$$\alpha_M^{-1} \circ s \circ \alpha_N \circ Q^*F(f) = \alpha_M^{-1} \circ s \circ GQ^*(f) \circ \alpha_M = \mathrm{id}_M,$$
$$Q^*F(f) \circ \alpha_M^{-1} \circ s \circ \alpha_N = \alpha_N^{-1} \circ GQ^*(f) \circ s \circ \alpha_N = \mathrm{id}_N,$$

hence $\alpha_M^{-1} \circ s \circ \alpha_N : Q^*F(N) \to Q^*F(M)$ is the inverse of $Q^*F(f)$. The conclusion is that for any f with $Q^*(f)$ invertible, $Q^*F(f)$ is invertible as well. In other words, (by the universal property of the localization), functor Q^*F factors through the quotient category $\mathcal{B}$, i.e. $\exists! F_\mathcal{B} : \mathcal{B} \to \mathcal{B}$ with $Q^* \circ F = F_\mathcal{B} \circ Q^*$. Q.E.D.

4 Ore localization for monoids

4.1 A **semigroup** is a set R with a binary *associative* operation. A semigroup with unit element $1 \in R$ is called a **monoid**. By definition, maps of semigroups are set maps which respect the multiplication, and maps of monoids should preserve unit element as well. Monoids and maps of monoids form a category $\mathfrak{Mon}$, which has arbitrary products. The notion of a submonoid is the obvious one.

A subset S of a monoid R is **multiplicative** if $1 \in S$ and whenever $s_1, s_2 \in S$ then $s_1 s_2 \in S$. For a set $S_1 \subset R$ there is a smallest multiplicative subset $S \subset R$ containing S_1, namely the set of all products $s_1 \cdots s_n$ where $s_i \in S_1$, including the product of the empty set of elements which equals 1 by definition. We say that S is **multiplicatively generated** by S_1.

4.2 A multiplicative subset $S \subset R$ is a **left Ore set** if

- $(\forall s \in S \ \forall r \in R \ \exists s' \in S \ \exists r' \in R) \, (r's = s'r)$ (left Ore condition);

- $(\forall n_1, n_2 \in R \ \forall s \in S) \, (n_1 s = n_2 s) \Rightarrow (\exists s' \in S, \, s' n_1 = s' n_2)$

  (left reversibility).

4.3 In traditional ring-theoretic terminology, S is a left **Ore set** if the first condition holds and S is a **left denominator set** if both conditions hold. We often say "left Ore set" for a left denominator set, as is increasingly common among geometers, and the notion of satisfying just the left Ore condition may be said simply "satisfying left Ore condition". By saying (plural:) "left Ore condition*s*" we subsume both the left Ore condition and the left reversibility.

4.4 A monoid R can be viewed as a small category $\mathrm{Cat}(R)$ with a single object R. Left multiplication by an element $a \in R$ is a morphism in $\mathrm{Cat} R$ denoted by L_a. We compose the morphisms by composing the maps. Any small category having one single object is clearly equivalent to $\mathrm{Cat}(R)$ for a suitable R.

This suggests a generalization of the notion of a denominator set (as well as its applications below) by replacing $\mathrm{Cat}(R)$ by an arbitrary category. A

multiplicative system in a category $\mathcal{A}$ is a class Σ of morphisms in $\mathcal{A}$ where all identity morphisms 1_A, where $A \in \mathrm{Ob}\,\mathcal{A}$, are in Σ, and for any two composable morphism $s, t \in \Sigma$ (i.e., the target (range) of t matches the source (domain) of s), also $s \circ t \in \Sigma$.

A multiplicative system Σ satisfies the left Ore conditions if it satisfies the ordinary left Ore condition with all quantifiers conditioned on the matching of the source and target maps appropriately.

More precisely, Σ satisfies the left Ore condition if

$$\forall(s : A \to B) \in \Sigma, \; \forall r : A' \to B, \; \exists(s' : D \to A') \in \Sigma, \; \exists r' : D \to A,$$

so that $r' \circ s = s' \circ r$. Σ satisfies left reversibility ('simplifiability') if

$$(\forall n_1, n_2 : A \to B, \; \forall(s : C \to A) \in \Sigma)\,(n_1 \circ s = n_2 \circ s)$$
$$\Rightarrow (\exists(s' : B \to D) \in \Sigma, \; s' \circ n_1 = s' \circ n_2).$$

We may picture the left simplifiability by the diagram

$$C \xrightarrow{\;s\;} A \underset{n_2}{\overset{n_1}{\rightrightarrows}} B \dashrightarrow^{s'} D$$

We say that Σ is a left denominator system, or equivalently, that the pair $(\mathcal{A}, \Sigma)$ forms a **left calculus of fractions** if the left Ore and left simplifiability condition hold. The book [104] has a good graphical treatment of that subject. See also [21, 35, 43, 103].

4.5 Lemma. *Let $f : R \to R'$ be a surjective map of monoids and $S \subset R$ left Ore. Then $f(S)$ is left Ore in R'.*

4.6 Let $\mathcal{D}$ be some category of monoids with additional structure, i.e. a category with a faithful functor $(-)_{\mathrm{mon}} : \mathcal{D} \to \mathfrak{Mon}$ preserving and reflecting finite equalizers. If R is an object in $\mathcal{D}$, a multiplicative set in R is by definition any multiplicative set $S \subset (R)_{\mathrm{mon}}$.

Definition. *Given a multiplicative set S in $R \in \mathcal{D}$ we introduce the category $\mathcal{C}_{\mathcal{D}}(R, S)$ as follows. The objects of $\mathcal{C}_{\mathcal{D}}(R, S)$ are all pairs (j, Y) where $Y \in \mathrm{Ob}\,\mathcal{D}$ and $j : R \to Y$ is a morphism in $\mathcal{D}$ satisfying*

- $(\forall s \in S)\,(\exists u \in Y)\,(uj(s) = j(s)u = 1)$ *in* $(Y)_{\mathrm{mon}}$;

The morphisms of pairs $\sigma : (j, Y) \to (j', Y')$ are precisely those morphisms $\sigma : Y \to Y'$ in $\mathcal{D}$ for which $\sigma \circ j = j'$.

In plain words, we consider those morphisms which invert all $s \in S$.

Now we would like the multiplication $j(s_1)^{-1}j(r_1) \cdot j(s_2)^{-1}j(r_2)$ to obtain again a 'left fraction' $j(s)^{-1}j(r)$. For this it is enough to be able to 'commute' the two middle terms in the sense $j(r_1)j(s_2)^{-1} = j(s')^{-1}j(r')$ as $j(s_1)^{-1}j(s')^{-1}) = j(s's_1)^{-1}$ and $j(r')j(r_2) = j(r'r_2)$ and we are done. This reasoning is the origin of the left Ore condition. Here is a formal statement:

4.7 Proposition. *(i) For* $(j, Y) \in \mathrm{Ob} \in \mathcal{C}_{\mathfrak{Mon}}(R, S)$*,* $j(S)$ *is left Ore in* $j(R)$ *iff*

$$j(S)^{-1}j(R) = \{j(s)^{-1}j(r) \mid s \in S, r \in R\} \subset Y$$

is a submonoid of Y*. In particular, if* S *is left Ore in* R*,* $j(S)^{-1}j(R)$ *is a submonoid of* Y *for each* $(j, Y) \in \mathrm{Ob}(\mathcal{C}(R, S))$*.*
(ii) If the equivalent conditions in (i) hold, then

$$\begin{array}{l} \forall(s, r) \in S \times R \\ j(s)^{-1}j(r) = j(s')^{-1}j(r') \end{array} \Leftrightarrow \left\{ \begin{array}{l} \exists \tilde{s} \in S, \exists \tilde{r} \in R, \\ j(\tilde{s})j(s') = j(\tilde{r})j(s) \\ j(\tilde{s})j(r') = j(\tilde{r})j(r). \end{array} \right. \tag{1}$$

Proof. (i) ($\Rightarrow$) Let $s_1, s_2 \in S$ $r_1, r_2 \in R$. By the left Ore condition $\exists s' \in S$ $\exists r' \in R$ with $j(s')j(r_1) = j(r')j(s_2)$. Hence the product $j(s_1)^{-1}j(r_1) \cdot j(s_2)^{-1}j(r_2) = j(s's_1)^{-1}j(r'r_2)$ belongs to Y.
($\Leftarrow$) If $j(S)^{-1}j(R)$ is a monoid then $j(r)j(s)^{-1} \in j(S)^{-1}j(R)$. In other words, $\exists s' \in S$ $\exists r' \in R$ such that $j(r)j(s)^{-1} = j(s')^{-1}j(r')$. Thus $j(s_1)j(r) = j(r_1)j(s)$.
(ii) By multiplying from the left by $j(s')$ one gets $j(s')j(s^{-1})j(r) = j(r')$. As S is left Ore, $\exists \tilde{s} \in S$ $\exists \tilde{r} \in R$ such that $\tilde{s}s' = \tilde{r}s$. This implies $j(\tilde{s})j(s') = j(\tilde{r})j(s)$ and, consequently, $j(s')j(s^{-1}) = j(\tilde{s})^{-1}j(\tilde{r})$; then $j(\tilde{s})^{-1}j(\tilde{r})j(r) = j(r')$ and, finally, $j(\tilde{r})j(r) = j(\tilde{s})j(r')$.

4.8 Proposition. *Let* S, R, Y, j *be as in* **4.7***, and let* R, Y *be each equipped with a binary operation, in both cases denoted by* $+_0$*, such that* $\cdot$ *is left distributive with respect to* $+_0$*. If* $j(S)^{-1}j(R)$ *is a submonoid of* Y*, then it is closed with respect to* $+_0$ *as well.*
Proof. The following calculation is valid in Y:

$$\begin{aligned} j(s_1)^{-1}j(r_1) +_0 j(s_2)^{-1}j(r_2) &= j(s_1)^{-1}(j(r_1) +_0 j(s_1)j(s_2)^{-1}j(r_2)) \\ &= j(s_1)^{-1}(j(r_1) +_0 j(\tilde{s})^{-1}j(\tilde{r})j(r_2)) \\ &= j(\tilde{s}s_1)^{-1}(j(\tilde{s})j(r_1) +_0 j(\tilde{r})j(r_2)) \in j(S)^{-1}j(R), \end{aligned}$$

where $j(\tilde{s})j(s_1) = j(\tilde{r})j(s_2)$ for some $\tilde{s}, \tilde{r}$ by the left Ore condition which holds due **4.7**.

4.9 Remark. We do not require $j(a +_0 b) = j(a) +_0 j(b)$ here.

4.9a Exercise. Generalize this to a family $\mathcal{F}$ of n-ary left distributive operations in place of $+_0$, i.e., of operations of the form $F : X^{\times n} \to X$, for various $n \geq 0$, such that $L_a \circ F = F \circ L_a^{\times n}$.

4.10 From now on we limit to the case where the category $\mathcal{D}$ above corresponds to a variety $\mathfrak{D}$ of algebras $(A, \mathcal{L}_A)$ (in the sense of universal algebra) of signature $\mathcal{L} = (\cdot, 1, \mathcal{F})$, where $\mathcal{F}_A$ is a family of left distributive operations on A on a $(A, \cdot_A, 1_A)$. The reader who does not care for this generality (suitable say for algebras with operators) can consider 3 basic cases: 1) $\mathcal{F} = \emptyset$ when $\mathcal{D} = \mathfrak{Mon}$; 2) $\mathcal{F} = \{+\}$ and algebras are unital rings; 3) $\mathcal{F} = \{+\}$ and algebras are associative unital **k**-algebras over a commutative ring **k**.

4.11 Denote by $\mathcal{C}_{l,\mathcal{D}}(R,S)$ the full subcategory of $\mathcal{C}_{\mathcal{D}}(R,S)$ consisting of those objects (j, Y) which satisfy

- $(\forall r, r' \in R)\, (j(r) = j(r') \Leftrightarrow \exists s \in S\, (sr = sr'))$.

- $j(S)^{-1} j(R)$ is a subring of Y

Similarly, $\mathcal{C}_{l,\mathcal{D}}^{-}(R,S)$ by definition consists of objects satisfying the first, but not necessarily the second property. Denote by $\mathcal{C}_{r,\mathcal{D}}(R,S)$ the full subcategory of $\mathcal{C}_{\mathcal{D}}(R,S)$ consisting of those objects (j, Y) which satisfy the symmetric conditions

- $(\forall r, r' \in R)\, (j(r) = j(r') \Leftrightarrow \exists s \in S\, (rs = r's))$.

- $j(R) j(S)^{-1}$ is a subring of Y

Finally, the objects in $\mathcal{C}_{r,\mathcal{D}}^{-}(R,S)$, by definition, satisfy the first, but not necessarily the second property. If there is a universal initial object in $\mathcal{C}_l(R,S)$ $(\mathcal{C}_r(R,S))$, we denote it by $(\iota, S^{-1}R)$ (resp.(ι, RS^{-1})) and we call the pair, or by abuse of language, also the ring $S^{-1}R$ $(RS^{-1}$ resp.$)$, the left (right) **Ore localization** of R at set S, and map ι the localization map. An alternative name for $S^{-1}R$ (RS^{-1}) is the left (right) **ring of fractions** (of ring R at set S).

4.12 Proposition. *If* $\forall (j, Y)$ *in* $\mathcal{C}_{l,\mathfrak{Mon}}^{-}(R,S)$ *the subset* $j(S)^{-1} j(R)$ *is a submonoid (i.e.*$\exists (j, Y) \in \mathcal{C}_{l,\mathfrak{Mon}}(R,S)$*), then it is so* $\forall (j, Y)$ *in* $\mathcal{C}_{l,\mathfrak{Mon}}^{-}(R,S)$*, i.e. the categories* $\mathcal{C}_{l,\mathfrak{Mon}}(R,S)$ *and* $\mathcal{C}_{l,\mathfrak{Mon}}^{-}(R,S)$ *coincide. In that case, S is a left denominator set in R.*

Proof. Let $j(S)^{-1} j(R)$ be a subring for some (j, Y). Then $j(S)^{-1} j(R)$ is Ore in $j(R)$ by the previous proposition. Thus for every $s \in S$, $r \in R$ $\exists s' \in S$ $\exists r' \in R$ such that $j(r) j(s)^{-1} = j(s')^{-1} j(r')$ and therefore $j(s') j(r) = j(r') j(s)$ which means $j(s'r) = j(r's)$. That implies $\exists s^{+} \in S$ with $s^{+} s'r = s^{+} r's)$. Therefore for any other (j', Y') in $\mathcal{C}_{l,\mathfrak{Mon}}(R,S)$ the subset $j'(S)^{-1} j'(R)$ is a subring. Moreover we have $s^{+} s' \in S$ and $s^{+} r'$ satisfy $(s^{+} s')r = (s^{+} r')s$. Since they were constructed for an arbitrary s and

r, S is left Ore in R.

Left reversibility: Let $r, r' \in R$, $s \in S$. Then $rs = r's \Rightarrow j(r)j(s) = j(r')j(s)$, so by invertibility of $j(s)$ also $j(r) = j(r')$. But (j, Y) is object in $C_{l,\mathfrak{Mon}}(R, S)$ so $j(r) = j(r') \Rightarrow \exists s' \in S$, $s'r = s'r'$.

4.13 Lemma. *(i) Let S be a left denominator set. Define the relation $\sim$ on $S \times R$ by*

$$(s, r) \sim (s', r') \quad \Leftrightarrow \quad (\exists \tilde{s} \in S \,\exists \tilde{r} \in R)\,(\tilde{s}s' = \tilde{r}s \text{ and } \tilde{s}r' = \tilde{r}r).$$

Then $\sim$ is an equivalence relation.

(ii) Let Σ be a system of left fractions in a category C. For any pair of objects X, Y in C let $(\Sigma \times C)(X, Y)$ be the class of all diagrams of the form $\left(X \xrightarrow{r} Z \xleftarrow{s} Y \right)$ in C. Define a relation $\sim$ on $(\Sigma \times C)(X, Y)$ by

$$\left(X \xrightarrow{r} Z \xleftarrow{s} Y \right) \sim \left(X \xrightarrow{r'} Z' \xleftarrow{s'} Y \right)$$
$$\Leftrightarrow \exists \left(X \xleftarrow{\tilde{r}} B \xrightarrow{\tilde{s}} Y \right), \begin{cases} \tilde{s} \circ s' = \tilde{r} \circ s : B \to Z \\ \tilde{s} \circ r' = \tilde{r} \circ r : B \to Z' \end{cases}.$$

The latter condition can be depicted by saying that the diagram

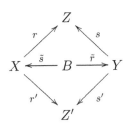

commutes. Conclusion: $\sim$ is an equivalence relation.

Here $\left(X \xleftarrow{\tilde{r}} B \xrightarrow{\tilde{s}} Y \right)$ is *not* a diagram in $(\Sigma^{\mathrm{op}} \times C)(X, Y)$.

Proof. Reflexivity is trivial.

Symmetry: By Ore $\exists r_1 \in R$, $s_1 \in S$ with $r_1 s = s_1 s'$. Also $\exists r_2 \in R$, $s_2 \in S$ with $r_1 s = s_1 s'$. Thus

$$r_2 \tilde{r} s' = r_2 \tilde{s} s = s_2 r_1 s = s_2 s_1 s'.$$

In other words $r_2 \tilde{r} - s_2 s_1 \in I_S$. Thus by the left reversibility, $\exists t \in S$ with $t(r_2 \tilde{r} - s_2 s_1) = 0$. Therefore $t(r_2 \tilde{r} - s_2 s_1)r' = 0$, hence $ts_2 s_1 r' = tr_2 \tilde{s} r$. Compare with $ts_2 s_1 s' = tr_2 \tilde{s} s$ to see that $(s', r') \sim (s, r)$.

Transitivity: Assume $(s, r) \sim (s', r')$ and $(s', r') \sim (s'', r'')$. This means

$$\exists \tilde{s}, \tilde{\tilde{s}} \in S \,\exists \tilde{r}, \tilde{\tilde{r}} \in R \quad \begin{cases} \tilde{s}s = \tilde{r}s' & \tilde{\tilde{s}}s' = \tilde{\tilde{r}}s'' \\ \tilde{s}r = \tilde{r}r' & \tilde{\tilde{s}}r' = \tilde{\tilde{r}}r'' \end{cases}$$

S is left Ore, hence $\exists s_* \in S\, \exists r_* \in R$ with $s_*\tilde{r} = r_*\tilde{s}$. Therefore

$$(s_*\tilde{s})s = s_*\tilde{r}s' = r_*\tilde{s}s' = (r_*\tilde{s})s''$$
$$(s_*\tilde{s})r = s_*\tilde{r}r' = r_*\tilde{s}r' = (r_*\tilde{s})r''$$

Hence $(s, r) \sim (s'', r'')$.

4.14 Simplifying observation. Consider a family of arrows $(s, r) \to ((ps)^{-1}, (pr))$ where $ps \in S$. Then $(s, r) \sim ((ps)^{-1}, (pr))$. If some property $\mathcal{P}$ of elements of $S \times R$ does not change along such arrows, then $\mathcal{P}$ is well-defined on classes $s^{-1}r := [s, r]/\sim \, \in S \times R/\sim$.

Proof. Clearly every $\sim$-arrow is a composition of one such arrow and the inverse of another such arrow.

4.15 Lemma. If $t \in S$ and $tr = tr'$ (by reversibility even better if $rt = r't$) then $(s, r_1rr_2) \sim (s, r_1r'r_2)$.

Proof. There are $t' \in S$, $r_1' \in R$ with $r_1't = t'r_1$. Then $(s, r_1rr_2) \sim (t's, t'r_1rr_2) \sim (t's, r_1'trr_2) \sim (t's, r_1'tr'r_2) \sim (t's, t'r_1r'r_2) \sim (s, r_1r'r_2)$.

4.16 Proposition. For (j, Y) in $\mathrm{Ob}\,\mathcal{C}_{l,\mathfrak{D}}(R, S)$ the statement

$$(\forall y \in Y\ \exists s \in S\ \exists r \in R)\,(y = j(s)^{-1}j(r)) \tag{2}$$

holds iff (j, Y) is a universal initial object in $\mathcal{C} = \mathcal{C}_{\mathfrak{D}}(R, S)$.

Proof. ($\Leftarrow$) Let $(j, Y) \in \mathcal{C}_l$ be universal in $\mathcal{C}$. Suppose $Y_0 := j(S)^{-1}j(R)$ is a proper subring of Y. We'll denote by j_0 the map from R to Y_0 agreeing with j elementwise. Then (j_0, Y_0) is an object in $\mathcal{C}_l$ and the inclusion $i : Y_0 \to Y$ is a morphism from (j_0, Y_0) into (j, Y). By universality of (j, Y) there is a morphism $i' : (j, Y) \to (j_0, Y_0)$. The composition of morphisms $i \circ i'$ is an automorphism of (j, Y) clearly different from the identity, contradicting the universality of (j, Y).

($\Rightarrow$) Let (j, Y) satisfy (2) and let (j', Y') be any object in $\mathcal{C}(R, S)$. We want to prove that there is unique map $i : Y \to Y'$ which satisfies $i(j(r)) = j'(r)\ \forall r \in R$. Note that $i(j(s)^{-1}j(s)) = i(j(s)^{-1})j'(s)$ implies $i(j(s)^{-1}) = j'(s)^{-1}$. Thus $i(j(s)^{-1}j(r)) = j'(s)^{-1}j'(r)$ so that the value of i is forced for all elements in Y proving the uniqueness.

This formula sets i independently of choice of s and r. Indeed, if $j(s)^{-1}j(r) = j(s')^{-1}j(r')$ then $j(r) = j(s)j(s')^{-1}j(r')$. As $j(S)$ is left Ore in $j(R)$, we can find $\tilde{s} \in S$ and $\tilde{r} \in R$ such that $j(\tilde{r})j(s') = j(\tilde{s})j(s)$ and therefore $j(s)j(s')^{-1} = j(\tilde{s})^{-1}j(\tilde{r})$. Thus $j(r) = j(\tilde{s})^{-1}j(\tilde{r})j(r')$ or $j(\tilde{s})j(r) = j(\tilde{r})j(r')$ and, finally, $j(\tilde{s}r) = j(\tilde{r}r')$. Thus $\exists s^+ \in S$, $s^+\tilde{s}r = s^+\tilde{r}r'$. Starting here and reversing the chain of arguments, but with j' instead of j, we get $j'(s)^{-1}j'(r) = j'(s')^{-1}j'(r')$.

4.17 Theorem. *If S is a left denominator set in R, then the universal object (j, Y) in $\mathcal{C}_{l,\mathfrak{mon}}(R, S)$ exists.*

Proof. We will construct a universal object $(j, Y) \equiv (\iota, S^{-1}R)$. <u>As a set,</u> $S^{-1}R := (S \times R)/ \sim$. Let $[s, r]$, and, by abuse of notation, let $s^{-1}r$ also denote the $\sim$-equivalence class of a pair $(s, r) \in S \times R$. Notice that $1^{-1}r = 1^{-1}r'$ may hold even for some $r \neq r'$, namely when $\exists s \in S$ and r, r' with $sr = sr'$. The equivalence relation is forced by (1).

<u>Multiplication</u> $\boxed{s_1^{-1}r_1 \cdot s_2^{-1}r_2 := (\tilde{s}s_1)^{-1}(\tilde{r}r_2)}$ where $\tilde{r} \in R$, $\tilde{s} \in S$ satisfy $\tilde{r}s_2 = \tilde{s}r_1$ (thus $\tilde{s}^{-1}\tilde{r} = r_1 s_2^{-1}$), as in the diagram:

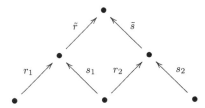

If we choose another pair of multipliers $\tilde{\tilde{r}} \in R, \tilde{\tilde{s}} \in S$ with $\tilde{\tilde{r}}s_2 = \tilde{\tilde{s}}r_1$ instead, then by the left Ore condition we can choose $r_* \in R, s_* \in S$ with $s_*\tilde{s} = r_*\tilde{\tilde{s}}$. Then

$$r_*\tilde{\tilde{r}}s_2 = r_*\tilde{\tilde{s}}r_1 = s_*\tilde{s}r_1 = s_*\tilde{r}s_2$$

and therefore $r_*\tilde{\tilde{r}} - s_*\tilde{r} \in I_S$.

In other words, $\exists s^+ \in S$ with $s^+ r_*\tilde{\tilde{r}} = s^+ s_*\tilde{r}$.

Therefore we have

$$s^+ s_*\tilde{r}r_2 = s^+ r_*\tilde{\tilde{r}}r_2$$
$$s^+ s_*\tilde{s}s_1 = s^+ r_*\tilde{\tilde{s}}s_1$$

which proves $(\tilde{s}s_1)^{-1}(\tilde{r}r_2) = (\tilde{\tilde{s}}s_1)^{-1}(\tilde{\tilde{r}}r_2)$. Thus multiplication is well defined as a map $\mu_1 : (S \times R) \times (S \times R) \to S^{-1}R$.

We have to show that μ_1 factors to $\mu : S^{-1}R \times S^{-1}R \to S^{-1}R$.

By **4.14**, it is sufficient to show that $a = \mu_1((s_1, r_1), (s_2, r_2))$ equals $b = \mu_1(((rs)_1, (rr_1)), ((ps)_2, (pr_2)))$ whenever $rs \in S$ and $ps \in S$.

$s_2'r_1 = r_1's_2$ for some $s_2' \in S$ and $r_1' \in R$. Then $a = (s_2's_1)^{-1}(r_1'r_2)$. As $ps_2 \in S$, $\exists p' \in R, s_* \in S$ with $p'(ps_2) = s_*r_1's_2 = s_*s_2'r_1$. Furthermore, $s_\sharp r = p_\sharp s_* s_2'$ for some $s_\sharp \in S$ and $p_\sharp \in R$. Putting these together, we infer $s_\sharp(rr_1) = p_\sharp s_* s_2'r_1 = p_\sharp p'ps_2$ and therefore $(rr_1)(ps_2)^{-1} \to s_\sharp^{-1}(p_\sharp p')$, i.e., by definition, that $b = (s_\sharp rs_1)^{-1}(p_\sharp p'pr_2)$, hence by above, $b = (p_\sharp s_* s_2's_1)^{-1}(p_\sharp p'pr_2)$. Now use lemma **4.15** and $(p'p)s_2 = (s_*r_1')s_2$ to conclude $b = (p_\sharp s_* s_2's_1)^{-1}(p_\sharp s_* r_1'r_2) = (s_2's_1)^{-1}(r_1'r_2) = a$.

Hence μ is well-defined. The unit element is clearly $1 = 1^{-1}1$. We need to show the associativity of μ. The product $s_1^{-1}r_1 \cdot s_2^{-1}r_2 \cdot s_3^{-1}r_3$ does not depend on the bracketing, essentially because one can complete the following commutative diagram of elements in R (the composition is the multiplication in R: any pair of straight-line (composed) arrows with the same target is identified with a pair in $S \times R$):

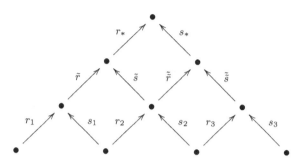

Finally, the construction gives the universal object because it clearly satisfies the equivalent condition in **4.16**.

5 Ore localization for rings

5.1 Exercise. The two left Ore conditions together immediately imply the **combined left Ore condition**:

If $n \in R$ is such that $ns = 0$ for some $s \in S$, then for every $r \in R$ there are $s' \in R$, $r' \in R$ such that $r'n = s'r$.

It is sometimes useful to quote this property in order to avoid introducing the additional variables needed for deriving it.

5.2 Lemma. *Let $f : R \to R'$ be a ring morphism and $S \subset R$ Ore. Then $f(S)$ is an Ore set in R.*

5.3 Notation. In this section we are concerned only with the category of unital rings. Thus $\mathcal{C}(R, S) := \mathcal{C}_{Rings}(R, S)$.

5.4 Notation. For any $S \subset R$ let $I_S := \{n \in R \,|\, \exists s \in S, sn = 0\}$. I_S is clearly a right ideal. If S is a left Ore set, then $sn = 0$ and the left Ore condition imply that $\forall r \in R \; \exists s_0 \in S, r_0 \in R$ with $r_0s = s_0r$, hence $s_0rn = r_0sn = 0$. Thus I_S is then a 2-sided <u>ideal</u>.

5.5 Corollary. If $S^{-1}R$ exists then $\forall(j, Y) \in \mathcal{C}(R, S)$, $\ker j \subset I_S$. In particular, an Ore localization of a domain is a domain.

5.6 Theorem. *If S is a left denominator set in R then the universal object (j, Y) in $\mathcal{C}_l(R, S)$ exists.*

Proof. In **4.17**, we have constructed a monoid structure on $Y = S \times R/\sim$. We exhibit an additive structure on Y such that j is a ring map and (j, Y) is indeed universal.

Addition: Suppose we are given two fractions with representatives (s_1, r_1) and (s_2, r_2). By the left Ore condition, $\exists \tilde{s} \in S$, $\exists \tilde{r} \in R$, such that $\tilde{s}s_1 = \tilde{r}s_2$. The sum is then defined as

$$\boxed{s_1^{-1}r_1 + s_2^{-1}r_2 := (\tilde{s}s_1)^{-1}(\tilde{s}r_1 + \tilde{r}r_2)}$$

Suppose we have chosen $(\tilde{\tilde{s}}, \tilde{\tilde{r}}') \in S \times R$ with $\tilde{\tilde{s}}s_1 = \tilde{\tilde{r}}s_2$, instead of $(\tilde{s}, \tilde{r})$. Then by left Ore we find $s_* \in S$ and $r_* \in R$ such that $s_*\tilde{s} = r_*\tilde{\tilde{s}}$. Then

$$r_*\tilde{\tilde{r}}s_2 = r_*\tilde{\tilde{s}}s_1 = s_*\tilde{s}s_1 = s_*\tilde{r}s_2$$

hence $(s_*\tilde{r} - r_*\tilde{\tilde{s}}) \in I_S$, i.e. $\exists s^\sharp \in S$ with

$$s^\sharp s_*\tilde{r} = s^\sharp r_*\tilde{\tilde{s}}$$

Then

$$(s^\sharp s_*)(\tilde{s}r_1 + \tilde{r}r_2) = (s^\sharp r_*)(\tilde{\tilde{s}}r_1 + \tilde{\tilde{s}}r_2)$$
$$(s^\sharp s_*)(\tilde{s}s_1) = (s^\sharp r_*)(\tilde{\tilde{s}}s_2)$$

Conclusion: $(\tilde{s}s, \tilde{s}r_1 + \tilde{r}r_2) \sim (\tilde{\tilde{s}}s, \tilde{\tilde{s}}r_1 + \tilde{\tilde{r}}r_2)$, as required.

Now let's check that the sum does not depend on the choice of the representative of the first summand. Suppose we are given two representatives of the first fraction $s_1^{-1}r_1 = s_1'^{-1}r_1'$. Then for some $(s^*, r^*) \in S \times R$ we have

$$s_*s_1 = r_*s_1' \quad \text{and} \quad s_*r_1 = r_*r_1'$$

The second fraction is $s_2^{-1}r_2$. Choose

$$(\tilde{s}, \tilde{r}) \in S \times R \quad \text{with} \quad \tilde{s}s_1' = \tilde{r}s_2.$$

Now choose $(s_\sharp, r_\sharp) \in S \times R$ such that $s_\sharp r_* = r_\sharp \tilde{s}$. Then $(r_\sharp \tilde{r})s_2 = r_\sharp \tilde{s}s_1' = s_\sharp r_* s_1' = (s_\sharp s_*)s_1$ and $(r_\sharp \tilde{s})r_1' = s_\sharp r_* r_1' = (s_\sharp s_*)r_1$. Therefore

$$
\begin{aligned}
s_1^{-1}r_1 + s_2^{-1}r_2 &= (s_\sharp s_* s_1)^{-1}(s_\sharp s_* r_1 + r_\sharp \tilde{r}r_2) \\
&= (r_\sharp \tilde{s}s_1')^{-1}(r_\sharp \tilde{s}r_1' + r_\sharp \tilde{r}r_2) \\
&= (\tilde{s}s_1')^{-1}(\tilde{s}r_1' + \tilde{r}r_2) \\
&= s_1'^{-1}r_1' + s_2^{-1}r_2
\end{aligned}
$$

We should also check that the sum does not depend on the second summand. This proof cannot be symmetric to the previous one as our definition of the sum is not symmetric. We shall choose an indirect proof. Denote the set-theoretic quotient map by $p : S \times R \to S^{-1}R$. By now we have completed the proof that addition as a map from

$$\tilde{+} : S^{-1}R \times (S \times R) \to S^{-1}R$$

is well defined. Now we prove that the map

$$\tilde{+}(p \times \mathrm{id})\tau : (S \times R) \times (S \times R) \to S^{-1}R$$

where τ is the transposition of factors coincides with $\tilde{+}(p \times \mathrm{id})$. Thus we have a well-defined addition as a map defined on $S^{-1}R \times S^{-1}R$ which is then automatically commutative. It is sufficient to prove that for any two pairs (s_1, r_1) and (s_2, r_2) and any

$$\tilde{s}, \tilde{\tilde{s}} \in S, \tilde{r}, \tilde{\tilde{r}} \in R \quad \text{with} \quad \tilde{s}s_1 = \tilde{r}s_2, \quad \tilde{\tilde{r}}s_1 = \tilde{\tilde{s}}s_2,$$

the classes

$$(\tilde{s}s_1)^{-1}(\tilde{s}r_1 + \tilde{r}r_2)$$
$$(\tilde{\tilde{s}}s_2)^{-1}(\tilde{\tilde{r}}r_1 + \tilde{\tilde{s}}r_2)$$

coincide in $S^{-1}R$. For that purpose, choose $s_\sharp \in S$ and $r_\sharp \in R$ such that $s_\sharp \tilde{r} = r_\sharp \tilde{\tilde{s}}$. Then

$$r_\sharp \tilde{s}s_1 = s_\sharp \tilde{r}s_2 = r_\sharp \tilde{\tilde{s}}s_2.$$

Next $r_\sharp \tilde{\tilde{r}}s_1 = r_\sharp \tilde{\tilde{s}}s_2 = s_\sharp \tilde{r}s_2 = s_\sharp \tilde{s}s_1$, and therefore $(r_\sharp \tilde{\tilde{r}} - s_\sharp \tilde{s}) \in I_S$ (**5.4**). Thus $\exists s^+ \in S$ with

$$s^+ r_\sharp \tilde{\tilde{r}} - s^+ s_\sharp \tilde{s} = 0.$$

In particular, $s^+ r_\sharp \tilde{\tilde{r}}r_1 = s^+ s_\sharp \tilde{s}r_1 = 0$. Thus

$$\begin{aligned}
(\tilde{s}s_1)^{-1}(\tilde{s}r_1 + \tilde{r}r_2) &= (s^+ s_\sharp \tilde{s}s_1)^{-1}(s^+ s_\sharp \tilde{s}r_1 + s^+ s_\sharp \tilde{r}r_2) \\
&= (s^+ r_\sharp \tilde{\tilde{s}}s_1)^{-1}(s^+ r_\sharp \tilde{\tilde{r}}r_1 + s^+ r_\sharp \tilde{\tilde{s}}r_2) \\
&= (\tilde{\tilde{s}}s_1)^{-1}(\tilde{\tilde{r}}r_1 + \tilde{\tilde{s}}r_2)
\end{aligned}$$

The associativity of addition is left to the reader.

The distributivity law follows by **4.8**.

The element $1^{-1}0$ in $S^{-1}R$ is the zero and thus $S^{-1}R$ is equipped with a natural unital ring structure.

Define $\iota : R \to S^{-1}R$ by $\iota(r) = [1, r] = 1^{-1}r$. Check that ι is a unital ring homomorphism. Check that $\iota(S)$ consists of units and that $\iota(S)^{-1}\iota(R) = Y$. Pair $(\iota, S^{-1}R)$ is a universal object in $\mathcal{C}_l(R, S)$, as it clearly satisfies the equivalent condition in **4.16**.

5.7 *Right* Ore conditions, and right Ore localizations with respect to $S \subset R$, are by definition the left Ore conditions and localizations with respect to $S \subset R^{\mathrm{op}}$. The *right* ring of fractions is denoted $RS^{-1} := (S^{-1}R^{\mathrm{op}})^{\mathrm{op}}$. It consists of certain equivalence pairs $rs^{-1} := [(r, s)]$, where $(r, s) \in R \times S$.

6   Practical criteria for Ore sets

This section is to be read only by those who want to test in practice whether they have an Ore set at hand.

6.1 Theorem. *(i) Let S and S' be multiplicative sets in ring R, where S is also left Ore in R. Assume*

1. for a map $j : R \to Y$ of unital rings, the image $j(S)$ consists of units in Y iff the image $j(S')$ consists of units in Y;

2. $sr = 0$ for some $s \in S$ iff $\exists s' \in S'$ with $s'r = 0$.

Then S' is left Ore as well and $S^{-1}R$ is canonically isomorphic to $S'^{-1}R$.

Proof. Under the assumptions the categories $\mathcal{C}_l(R, S)$ and $\mathcal{C}_l(R, S')$ are identical, so call them simply $\mathcal{C}$. The left Ore condition is equivalent to the existence of an initial object in $\mathcal{C}$; and the 2 localizations are just the 2 choices of an initial object, hence there is a unique isomorphism in $\mathcal{C}$ between them; its image under the forgetful functor $\mathcal{C} \to R - \text{Mod}$ into the category of unital rings, is the canonical isomorphism as required.

6.2 The left Ore condition is often checked inductively on a filtration, or an ordered set of generators. For this purpose we shall temporarily use some *nonstandard* notation which generalizes the left Ore condition. One fixes an (only) multiplicative set $S \subset R$. For *any* subset $A \subset R$, and any $(s, r) \in S \times A \subset S \times R$, introduce predicate

$$\mathbf{lOre}(s, r \uparrow A) := \mathbf{lOre}_{S,R}(s, r \uparrow A) \equiv (\exists s' \in S,\, \exists r' \in A,\, s'r = r's),$$

where S, R in subscripts may be skipped if known from context. Moreover if $A = R$ then $\uparrow A$ may be skipped from the notation. For example, $\mathbf{lOre}(s, r) = \mathbf{lOre}_{S,R}(s, r \uparrow R)$.

For any subsets $A_0 \subset A$ and $S_0 \subset S$, abbreviate

$$\mathbf{lOre}(S_0, A_0 \uparrow A) \equiv (\forall s \in S_0, \forall r \in R_0,\, \mathbf{lOre}_{S,R}(s, r \uparrow A)),$$

with rules for skipping $\uparrow A$ as before. For example, $\mathbf{lOre}(S, R)$ is simply the left Ore condition for $S \subset R$.

Finally,

$$\mathbf{slOre}(S_0, A) \equiv \mathbf{lOre}_{S,R}(S_0, A \uparrow A).$$

For an additive subgroup $A \subset R$ consider also the relative versions, e.g.

$$\mathbf{rel} - \mathbf{lOre}_{S,R}(s, r; I) \equiv (\exists s' \in S,\, \exists r' \in R,\, s'r - r's \in A).$$

If $A = I$ is an ideal, then this predicate is suitable for the study of some (non-Ore) generalizations (cf. [38] for such).

6.3 Extending the Ore property *Let $A, B \subset R$ be additive subgroups of R, $A \subset B \subset R$, and $S \subset R$ a multiplicative subset.*

(i) $(\mathbf{lOre}(S, A)$ and $\mathbf{rel} - \mathbf{lOre}(S, R; A)) \Rightarrow \mathbf{lOre}(S, R)$

(ii) $(\mathbf{lOre}(S, A \uparrow B)$ and $\mathbf{rel} - \mathbf{slOre}(S, B; A)) \Rightarrow \mathbf{lOre}(S, B)$

(iii) $(\mathbf{lOre}(S, A \uparrow B)$ and $\mathbf{rel} - \mathbf{slOre}(S, B; A)$
             and $SB \subset B) \Rightarrow \mathbf{slOre}(S, B)$

Proof. (i) is clearly the $B = R$ case of (ii). Let $b \in B$ and $s \in S$. Then $\mathbf{rel} - \mathbf{slOre}(S, B; A))$ means that $\exists s' \in S$, $\exists b' \in B$, $\exists a \in A$ such that $s'b - b's = a$. Now we compare s and a. There are $b_1 \in B$, $s_1 \in S$ such that $b_1 s = s_1 a$. Thus $s_1 s' b - s_1 b' s = s_1 a = b_1 s$, and finally, $(s_1 s')b = (s_1 b' + b_1)s$. S is multiplicative hence (ii), and if $SB \subset R$ then $s_1 b' + b_1 \in B$ hence (iii).

6.3a Remark. The above condition is usually checked for generators only. Also we can iterate the above criterion if we have a finite or denumerable family of nested subrings, for which the induction is convenient. One may also need to nest subsets of S, with refined criteria, like $\mathbf{lOre}(S_1 \uparrow S_2, A_1 \uparrow A_2)$, where the $\uparrow S_2$ means that s' may be chosen in S_2.

6.4 Lemma. If S_1 *multiplicatively* generates S, and $A \subset R$ then

$$\mathbf{lOre}(S_1, R) \Leftrightarrow \mathbf{lOre}(S, R),$$

$$\mathbf{slOre}(S_1, A) \Leftrightarrow \mathbf{slOre}(S, A).$$

Proof. The first statement is clearly a particular case of the second. Hence we prove the second statement; the nontrivial direction is $\Rightarrow$. By assumption, the set S can be written as a nested union $\cup_{n \geq 0} S_n$ where S_n consists of all those $s \in S$ which can be expressed as a product $\prod_{k=1}^{n'} s_k$ with $n' \leq n$ and $s_k \in S_k$; (hence S_1 is as the same as before). The assumption is $\mathbf{slOre}(S_1, A)$, hence by induction it is enough to prove that $\mathbf{slOre}(S_n, A) \Rightarrow \mathbf{slOre}(S_{n+1}, A)$ for all $n \geq 1$. Take $s = s_1 s_2 \cdots s_n$. Then $\mathbf{slOre}(S_n, A)$ means that for any $a \in A$ we have

$$\exists a' \in A \, \exists s' \in S \, (a' s_2 \cdots s_n = s'a),$$
$$\exists a'' \in A \, \exists s'' \in S \, (a'' s_1 = s''a')$$

and consequently

$$a'' s_1 s_2 \ldots s_n = s'' a' s_2 \ldots s_n = (s'' s')a,$$

with the desired conclusion by the multiplicative closedness of S.

6.5 Lemma. If $A_0^+, A^+ \subset R$ are the additive closures of A_0, A respectively, then (obviously)

$$\mathbf{lOre}(S, A_0 \uparrow A) \Rightarrow \mathbf{lOre}(S, A_0^+ \uparrow A^+).$$

6.6 Lemma. If A generates R as a ring, then

$$\mathbf{lOre}(S, A) \Rightarrow \mathbf{lOre}(S, R).$$

Proof. By **6.5** it is enough to check this multiplicativity:

$$(\forall i, \ \mathbf{lOre}(S, c_i)) \Rightarrow \mathbf{lOre}(S, c_n \cdots c_1),$$

when $c_i \in A$. However, this general statement holds for any choice of c_i whatsoever. Namely, if we do not require $c_i \in A$, we see, by induction on n, that it is enough to prove this statement for $n = 2$. For $s \in S$ and $c_1, c_2 \in R$ we can find $r_1', r_2' \in R$ and $s', s'' \in S$, so that $r_1's = s'c_1$ and $r_2's' = s''c_2$. The result is

$$r_2'r_1's = r_2's'c_1 = s''c_2c_1,$$

hence the lemma is proved.

6.7 Theorem. *If $A \subset B$ and the subring $\bar{A} \subset R$ is also contained in B, then for any S_1 multiplicatively generating a multiplicatively closed set S we have*

$$\mathbf{slOre}(S_1, A) \Rightarrow \mathbf{lOre}(S, \bar{A} \uparrow B),$$

$$\mathbf{lOre}(S, A) \Rightarrow \mathbf{lOre}(S, \bar{A}).$$

Proof. We know $\mathbf{slOre}(S_1, A) \Rightarrow \mathbf{slOre}(S, A)$. Hence the first assertion follows from the second by $\bar{A} \subset B$. We proved the second statement in the case $\bar{A} = R$. If $S \subset \bar{A}$, the statement clearly does not say anything more than it would say after replacing R by its subring $\bar{A}$. The proof of the general case is exactly the same, as $s \in R$ is never used, and our calculations and quantifiers may be taken over a bigger overring.

6.8 Warning-theorem. *If A generates R as a ring and S_1 generates S multiplicatively, then it is NOT necessarily true that*

$$\mathbf{lOre}(S_1, A) \Rightarrow \mathbf{lOre}(S, R), \tag{3}$$

even if S_1 has only one multiplicative generator. We know from **6.4** that (3) holds if we replace $\mathbf{lOre}$ by the stronger condition $\mathbf{slOre}$. Nevertheless, various intermediate conditions, standing between $\mathbf{lOre}$ and $\mathbf{slOre}$, often utilizing filtrations and combined arguments, are widely used in practice. However it is also common to use (3) without proper justification.

6.8a Counterexample. (proving the warning statement above)

Let R be the unital ring generated by 4 generators z_1, z_2, z_3, D modulo the following relations:

$$\begin{aligned} Dz_1 &= z_2 z_3 D, \\ D^2 z_2 &= z_3 z_1 D, \\ D^3 z_3 &= z_1 z_2 D. \end{aligned} \qquad (4)$$

Clearly the (4) are simply the identities needed to check **lOre**(S_1, A) where $S_1 = \{D\}$, and $A = \{z_1, z_2, z_3\}$. The powers of D on the left-hand side, which are $1, 2, 3$ may be replaced by $1, p, q$ respectively, where $p > 0$ and $q > 1$ are any integers, and the same proof applies, but the inequality $q > 1$ is indeed essential: if $p = q = 1$, this is not a counterexample at all.

We claim that for any nonnegative integer n, $D^n z_1 = PD^2$ does not have a solution for $P \in R$, hence the left Ore condition is *not* satisfied for the multiplicative set $S = \{1, D, D^2, \ldots\}$. The proof of the claim will be by contradiction, but we need first to study a convenient basis of ring R.

A basis of R consists of all ordered monomials where the right-hand side of any of the equations (4) does not appear as a factor. This is obtained using SHIRSHOV-BERGMAN's diamond lemma ([15, 17]) with the reduction system K having 3 reductions corresponding to the relations (4) with production arrows from right to left. This reduction system has clearly no ambiguities whatsoever and all reductions send monomials into monomials in generators z_1, z_2, z_3, D. It is direct to see using this basis that D is not a zero divisor.

Suppose then, that S is left Ore. Then there exist n such that

$$D^n z_1 = PD^2, \quad \text{for some } P \in R. \qquad (5)$$

We suppose that $n \geq 3$, and leave the remaining case to the reader. Equation (5) implies $D^{n-1} z_2 z_3 D = PD^2$. D is not a divisor, hence $D^{n-1} z_2 z_3 = PD$. Now write P as a sum of linearly independent K-reduced monomials P_i. Because D is not a zero divisor, monomials $P_i D$ are also linearly independent. Since the reductions in K send monomials to monomials, and $D^{n-1} z_2 z_3$ is a K-reduced monomial, we see that $D^{n-1} z_2 z_3$ can not be obtained as a sum of more than one of the linearly independent monomials $P_i D$, hence P has to be a monomial. The only way that PD in K-reduced form (which is $D^{n-1} z_2 z_3$) has z_3 as the most right-hand side factor is that $P = P' z_1 z_2$ for some P' in K-reduced form. Hence we obtain $D^{n-1} z_2 z_3 = P' z_1 z_2 D = P' D z_3$. Again using basis one can check directly that $Qz_3 = 0$ implies $Q = 0$, hence $D^{n-1} z_2 = P'D$. Now $D^{n-3} D^2 z_2 = D^{n-3} z_3 z_1 D = P'D$ implies $D^{n-3} z_3 z_1 = P'$. This substituted back in the expression for P and the equation (5) gives

$$D^n z_1 = P' z_1 z_2 D = D^{n-3} z_3 z_1 z_1 z_2 D^2 = D^{n-1} z_2 D^2 z_3 D.$$

This is a contradiction as the two sides differ even though they are K-reduced.

6.9 Proposition. *Let S and T be left Ore sets in some ring R. Then the set of all elements of the form st where $s \in S$ and $t \in T$ satisfies the left Ore condition in R (but it is not necessarily multiplicatively closed).*

Proof. Suppose the contrary, i.e. there is $st \in ST$ and $r \in R$, such that we can not find $s' \in S, t' \in T$ and $r' \in B$, so that $r'st = s't'r$. The set T is Ore, so there are $t' \in T$ and $r_1 \in R$ with $r_1 t = t'r$. Next we can find $s' \in S$ and $r_2 \in R$ so that $r_2 s = s'r_1$. Combining, we obtain

$$s't'r = s'r_1 t = r_2 st,$$

a contradiction.

6.10 Proposition. *Let S be a left Ore set in a k-algebra R. The set $1 \otimes S$ of all elements of $R \otimes_k R$ of the form $1 \otimes s$ where $s \in S$ satisfies the left Ore condition.*

Proof. $1 \otimes S$ is obviously multiplicatively closed.

If the Ore condition is not true, there is an element $y = \sum_{i=1}^{n} a_i \otimes b_i$ and an element $s \in S$ such that $(1 \otimes S)y \cap (R \otimes R)(1 \otimes s) = \emptyset$.

We use induction by n to find an element in the intersection. If $n = 1$ we simply use that S is left Ore to find $r' \in R$ and $s' \in S$ such that $r's = s'b_1$ and we have

$$(1 \otimes s')(a_1 \otimes b_1) = (a_1 \otimes r')(1 \otimes s),$$

which proves the basis of induction.

Suppose we found $s'_j \in S$ and $z = \sum_{i=1}^{j} a_i \otimes r_i'^j$ so that

$$(1 \otimes s'_j)\left(\sum_{i=1}^{j} a_i \otimes b_i\right) = \left(\sum_{i=1}^{j} a_i \otimes r_i'\right)(1 \otimes s).$$

Now we use again the property that S is left Ore to find $r_{j+1}'^{j+1} \in R$ and $\bar{s}_{j+1} \in S$ such that

$$r_{j+1}'^{j+1} s = \bar{s}_{j+1} s'_j b_{j+1}.$$

S is a multiplicatively closed set so $s'_{j+1} = \bar{s}_{j+1} s'_j$ is an element of S. Now we have

$$
\begin{aligned}
(1 \otimes s'_{j+1})(\textstyle\sum_{i=1}^{j+1} a_i \otimes b_i) = & \ (1 \otimes \bar{s}_{j+1})(1 \otimes s'_j)(\textstyle\sum_{i=1}^{j} a_i \otimes b_i) + \\
& + (1 \otimes \bar{s}_{j+1})(1 \otimes s'_j)(a_{j+1} \otimes b_{j+1}) \\
= & \ (\textstyle\sum_{i=1}^{j} a_i \otimes \bar{s}_{j+1} r_i'^j)(1 \otimes s) + (a_{j+1} \otimes r_{j+1}'^{j+1})(1 \otimes s) \\
= & \ (\textstyle\sum_{i=1}^{j+1} a_i \otimes r_i'^{j+1})(1 \otimes s)
\end{aligned}
$$

where we denoted $r_i'^{j+1} = \bar{s}_{j+1} r_i'^j$ for $i < j+1$ and $r_{j+1}'^{j+1}$ has already been defined.

7 Ore localization for modules

The modern point of view on Ore localization is to express it as a *localization functor* on some category of modules. The localization map $\iota : R \to S^{-1}R$ will be replaced by a localization functor Q_S^* from $R-$mod to $S^{-1}R-$mod.

7.1 Let S be a left Ore set in a ring R, and M a left R-module. Notice that $S^{-1}R$ may be viewed as an $S^{-1}R$-R-bimodule. The **module of fractions** $S^{-1}M$ of M with respect to Ore set S is the left $S^{-1}R$-module

$$
Q_S^*(M) = S^{-1}M := S^{-1}R \otimes_R M. \tag{6}
$$

For each morphism $f : M \to N$, set $Q_S^*(f) := 1 \otimes f : S^{-1}R \otimes_R M \to S^{-1}R \otimes_R N$. This defines a localization functor $Q_S^* : R - \mathrm{mod} \to S^{-1}R - \mathrm{mod}$ whose right adjoint is the obvious forgetful functor $Q_{S*} : S^{-1}R - \mathrm{mod} \to R - \mathrm{mod}$. If $\iota = \iota_R : R \to S^{-1}R$ is the localization map, then define the map of R-modules $\iota_M : M \to S^{-1}M$ via $\iota_M = \iota_R \otimes_R \mathrm{id}$ i.e. $m \mapsto 1 \otimes m$, also called the localization map. The maps ι_M define a natural transformation of functors, namely the adjunction $\iota : \mathrm{Id} \to Q_{S*}Q_S^*$.

7.1a Remark. If S is a right Ore set, and M a left R-module, then $M[S^{-1}] := R[S^{-1}] \otimes_R M$. If N is a right R-module then view RS^{-1} (or $S^{-1}R$) as a R-RS^{-1}- (resp. R-$S^{-1}R$)- bimodule and define $Q_S^*(N) := N \otimes_R R[S^{-1}]$ (resp. $N \otimes_R S^{-1}R$). We emphasize that the choice of right vs. left Ore sets is not correlated with the choice of right or left R-module categories, at least in the principle of the construction.

7.2 Universal property. For given R, M, S as above we define the category $\mathcal{M} = \mathcal{M}(R, M, S)$. The objects of $\mathcal{M}$ are pairs (N, h) where N is a left $S^{-1}R$-module and $h : M \to {}_R N$ a map of left R-modules. A morphism of pairs $\mu : (N, h) \to (N', h')$ is a map $\mu : N \to N'$ of $S^{-1}R$-modules such that $h' = \mu \circ h$.

Proposition. *The pair* $(S^{-1}M, \iota_M)$ *is an initial object in* $\mathcal{M}$.

Proof. For any pair $(N, h) \in \text{Ob}(\mathcal{M})$ there is a well-defined morphism

$$\alpha : (S^{-1}M, \iota_M) \to (N, h) \quad \text{by} \quad \alpha(s^{-1} \otimes_R m) = s^{-1}h(m). \tag{7}$$

Let now α' be any morphism from $(S^{-1}M, \iota)$ to (N, h). By $h = \alpha' \circ i$ we conclude

$$\alpha'(1 \otimes m) = h(m) = \alpha(1 \otimes m), \quad \forall m \in M.$$

The elements of the form $1 \otimes m$ generate $S^{-1}M$ as a module over $S^{-1}R$ and therefore $\alpha = \alpha'$.

7.3 Elements in the tensor product (6) are of the form $\sum_i s_i^{-1} \otimes m_i$ but such can be added up to a single term of that form, as the fractions can be always put to the common denominator. Namely, by the left Ore condition $\forall s, s' \in S \, \exists \tilde{s} \in S \, \exists \tilde{r} \in R, \, s's^{-1} = \tilde{s}^{-1}\tilde{r}$, and therefore

$$\begin{aligned} s^{-1} \otimes m + s'^{-1} \otimes m' &= s'^{-1}s's^{-1} \otimes m + s'^{-1}\tilde{s}^{-1}\tilde{s} \otimes m' \\ &= s'^{-1}\tilde{s}^{-1} \otimes (\tilde{r}m + \tilde{s}m'). \end{aligned} \tag{8}$$

Thus every element in $S^{-1}M$ may be written in the form $s^{-1} \otimes_R m$, hence there is a surjection of sets $\nu : S \times M \to S^{-1}M$. The set $S \times M$ may be viewed as a set retract $S \times \{1\} \times M$ of $S \times R \times M$ via the retraction $(s, r, m) \mapsto (s, rm)$. Clearly ν extends to $\nu' : S \times R \times M \to S^{-1}M$. By the universality of the free abelian group $\mathbb{Z}(S \times R \times M)$ with basis $S \times R \times M$, $\exists! \tilde{\nu}' : \mathbb{Z}(S \times R \times M) \to S^{-1}M$ which is additive and extends ν'. It is clear by checking on the basis elements (s, r, m) and uniqueness that the composition of the canonical projections $\mathbb{Z}(S \times R \times M) \to \mathbb{Z}(S^{-1}R \times M) \to S^{-1}R \otimes_R M$ equals $\tilde{\nu}'$.

For $r \in R$ for which $rs \in S$, $s^{-1} \otimes_R m = (rs)^{-1}r \otimes_R m = (rs)^{-1} \otimes_R rm$ implying that $\ker \nu' \subset \mathbb{Z}(S \times M)$ contains all differences $(s, m) - (s', m')$ in $\mathbb{Z}(S \times M)$ of pairs in $S \times M$ which are equivalent via

$$\boxed{(s, m) \sim (s', m') \Leftrightarrow \exists r, r' \in R \quad rs = r's' \in S \quad \text{and} \quad rm = r'm'.} \tag{9}$$

Lemma. *(i) On $(S \times M)/\sim$ there is a unique binary operation $+$ such that*

$$(s, m) + (s, m') \sim (s, m + m'). \tag{10}$$

(ii) $((S \times M)/\sim, +)$ is an abelian group. Hence by the universality of the free abelian group, the composition $S \times R \times M \to S \times M \to (S \times M)/\sim$ extends to a unique map $p : \mathbb{Z}(S \times R \times M) \to (S \times M)/\sim$ of abelian groups.
(iii) The map p factors through a map $p' : S^{-1}R \otimes_R M \to (S \times M)/\sim$.
(iv) p' is an inverse of ν, hence p' respects addition.

Proof. (i) *Uniqueness.* Suppose there are two such operations, $+_1, +_2$ and two classes (s_1, m_1) and (s_2, m_2) on which $+_1$ and $+_2$ disagree. By the left Ore condition choose $\tilde{s} \in S, \tilde{r} \in R$ with $\tilde{s}s_1 = \tilde{r}s_2$. Then $(s_1, m_1) +_i (s_2, m_2) \sim (\tilde{s}s_1, \tilde{s}m_1) +_i (\tilde{r}s_2, \tilde{r}m_2) \sim (\tilde{s}s_1, \tilde{s}m_1 + \tilde{r}m_2)$ which agree for $i = 1, 2$, giving a contradiction.

Existence. Take $(s_1, m_1) + (s_2, m_2) := (\tilde{r}_1 s_1, \tilde{r}_1 m_1 + \tilde{r}_2 m_2)$ with any choice of $\tilde{r}_1, \tilde{r}_2$ such that $\tilde{r}_1 s_1 = \tilde{r}_2 s_2 \in S$. We verify that the class of the result does not depend on the choices. If s_1, m_1 are replaced by $rs_1 \in S, rm_1 \in M$ we can by the combined Ore condition choose r_*, s_* with $s_* r = r_* \tilde{r}_1$, hence $S \ni s_* rs_1 = r_* \tilde{r}_1 s_1 = r_* \tilde{r}_2 s_2$. Hence the rule for the sum gives $(s_* rs_1, s_* rm_1 + r_* \tilde{r}_2 m_2) \sim (r_* \tilde{r}_1 s_1, r_* \tilde{r}_1 m_1 + r_* \tilde{r}_2 m_2) \sim (\tilde{r}_1 s_1, \tilde{r}_1 m_1 + \tilde{r}_2 m_2)$. By symmetry, we have the same independence for choice of (s_2, m_2). Finally, suppose that instead of choosing $\tilde{r}_1, \tilde{r}_2$ we choose $\tilde{b}_1, \tilde{b}_2$. As $\tilde{r}_1 s_1 \in S$ by the combined Ore condition we may choose $s_\sharp \in S, r_\sharp \in R$, such that $r_\sharp \tilde{r}_1 = s_\sharp \tilde{b}_1$ with $r_\sharp \tilde{r}_1 s_1 = s_\sharp (\tilde{b}_1 s_1) \in S$. Hence $(\tilde{r}_1 s_1, \tilde{r}_1 m_1 + \tilde{r}_2 m_2) \sim (r_\sharp \tilde{r}_1 s_1, r_\sharp \tilde{r}_1 m_1 + r_\sharp \tilde{r}_2 m_2) \sim (s_\sharp \tilde{b}_1 s_1, s_\sharp \tilde{b}_1 m_1 + r_\sharp \tilde{r}_2 m_2)$. Now $r_\sharp \tilde{r}_2 s_2 = r_\sharp \tilde{r}_1 s_1 = s_\sharp \tilde{b}_1 s_1 = s_\sharp \tilde{b}_2 s_2$, hence by left reversibility, there is s_2^* such that $s_2^* r_\sharp \tilde{r}_2 = s_2^* s_\sharp \tilde{b}_2$. Thus $(s_\sharp \tilde{b}_1 s_1, s_\sharp \tilde{b}_1 m_1 + r_\sharp \tilde{r}_2 m_2) \sim (s_2^* s_\sharp \tilde{b}_1 s_1, s_2^* s_\sharp \tilde{b}_1 m_1 + s_2^* s_\sharp \tilde{b}_2 m_2) \sim (\tilde{b}_1 s_1, \tilde{b}_1 m_1 + \tilde{b}_2 m_2)$, as required.

(ii) In the proof of existence in (i) we have seen that $+$ is commutative. Notice also that the class of $(s, 0)$ (independent on S) is the neutral element. For any pair, and hence for any triple of elements in $S \times M / \sim$, we can choose representatives such that all three are of the form (s, m) with the same s. For such triples the associativity directly follows by applying (10).

(iii) As p and the projection $S \times R \times M \to S \times M \to (S \times M)/\sim$ are additive it is sufficient to show that p sends the kernel of the projection to $0 \in (S \times M)/\sim$. The kernel of the projection is spanned by the elements of several obvious types, so we check on generators.

1. $(s, r, m) - (s', r', m)$ where $s^{-1}r = (s')^{-1}r'$. This means that for some $\tilde{s} \in S, \tilde{r} \in R$ we have $\tilde{s}s = \tilde{r}s'$ and $\tilde{s}r = \tilde{r}r'$. Compute $p(s, r, m) - p(s', r', m') = (s, rm) + (s', -r'm') = (\tilde{s}s, \tilde{s}rm) + (\tilde{r}s', -\tilde{r}r'm') = (\tilde{s}s, \tilde{s}rm - \tilde{r}r'm') = 0$.

2. Elements $(s, r + r', m) - (s, r, m) - (s, r + r', m)$, as well as $(s, rr', m) - (s, r, r'm)$ and $(s, r, m + m') - (s, r, m) - (s, r, m')$ go to 0 because, by (ii), in computing p one has to first act with the second component to the third.

7.4 Proposition. $(S \times M)/\sim$ *is additively canonically isomorphic to* $S^{-1}M$. *This isomorphism equips* $(S \times M)/\sim$ *with the canonical left* $S^{-1}R$-*module structure for which the following formulas can be taken as defining:*

$$t^{-1}r(s, m) = (s_* t, r_* m) \in (S \times M)/\sim$$

where $s_* \in S$ and $r_* \in R$ such that $s_* r = r_* s$.

Proof. By the lemma we have the first statement and hence we can view the class of (s, m) as $s^{-1} \otimes m$ and the formulas follow. They are defined because the formulas agree with the action transferred by an isomorphism, hence the existence, and by the left Ore condition for any $t^{-1} r$ and (s, m) there are s_*, r_* qualifying for the formulas, hence the uniqueness.

7.4a Corollary. *Let $S \subset R$ be left Ore and M a left R-module. Then $m = 0$ for some s iff $\exists s \in S$ and $sm = 0$.*

7.4b (long) **Exercise.** Let S, W be two left Ore sets in R, and $M = {}_R M$ a left R-module. The relation $\sim$ on $S \times W \times M$ given by

$$(s, w, m) \sim (s', w', m') \Leftrightarrow \begin{cases} \exists r, r', u, u', \tilde{r}, \tilde{r}' \in R, \exists \tilde{w}, \tilde{w}' \in W, \\ rs = r's' \in S, u\tilde{w} = u'\tilde{w}' \in W, \\ \tilde{w}r = \tilde{r}w, \tilde{w}'r' = \tilde{r}'w', z\tilde{r}m = z'\tilde{r}'m', \end{cases}$$

is an equivalence relation. The map $(S \times W \times M/\sim) \to S^{-1} W^{-1} M$ given by $(s, w, m) \mapsto s^{-1} \otimes w^{-1} \otimes m$ is well-defined and bijective.

8   Monads, comonads and gluing

8.1 A monoidal category is a category $\mathcal{C}$ equipped with a 'tensor' product bifunctor $\otimes : \mathcal{C} \times \mathcal{C} \to \mathcal{C}$; a distinguished object $1_{\mathcal{C}}$, a family of associativity isomorphisms $c_{ABC} : (A \otimes B) \otimes C \to A \otimes (B \otimes C)$, natural in objects A, B, C in $\mathcal{C}$; the left unit $l_A : 1_{\mathcal{C}} \otimes A \to A$ isomorphism and the right unit isomorphism $r_A : A \otimes 1_{\mathcal{C}} \to A$ are both indexed by and natural in objects A in $\mathcal{C}$; furthermore, require some standard coherence conditions (pentagon axiom for associativity coherence; left and right unit coherence conditions, cf. [21, 84]). A monoidal category $(\tilde{\mathcal{C}}, \otimes, 1_{\mathcal{C}}, c, r, l)$ is **strict** if c_{ABC}, l_A, r_A are actually all identity morphisms.

8.2 Monads and comonads. Given a diagram of categories $\mathcal{A}, \mathcal{B}, \mathcal{C}$, functors f_1, f_2, g_1, g_2 and natural transformations F, G as follows

$$\mathcal{A} \quad \overset{\overset{f_1}{\longrightarrow}}{\underset{\underset{f_2}{\longrightarrow}}{\Uparrow F}} \mathcal{B} \quad \overset{\overset{g_1}{\longrightarrow}}{\underset{\underset{g_2}{\longrightarrow}}{\Uparrow G}} \mathcal{C}, \tag{11}$$

one defines the natural transformation $G \star F : g_2 \circ f_2 \Rightarrow g_1 \circ f_1$ by

$$(G \star F)_A := G_{f_1(A)} \circ g_2(F_A) = g_1(F_A) \circ G_{f_2(A)} : g_2(f_2(A)) \to g_1(f_1(A)).$$

$(F, G) \mapsto F \star G$ is called the **Godement product** ('horizontal composition', cf. (11)). It is associative for triples for which $F \star (G \star H)$ is defined.

Given functors $f, g, h : \mathcal{A} \to \mathcal{B}$ and natural transformations $\alpha : f \Rightarrow g$, $\beta : g \Rightarrow h$, define their 'vertical' (or ordinary) composition $\beta \circ \alpha : f \Rightarrow h$ to be their composition taken objectwise: $(\beta \circ \alpha)_A := \beta_A \circ \alpha_A : f(A) \to h(A)$.

Note the **interchange law**: $(\alpha \circ \beta) \star (\gamma \circ \delta) = (\alpha \star \gamma) \circ (\beta \star \delta)$.

If T, T' are endofunctors in $\mathcal{A}$ and $\alpha : T \Rightarrow T'$, $\beta : T' \Rightarrow T$ natural transformations, one may also use the concatenation notation: $\alpha T : TT \Rightarrow T'T$ is given by $(\alpha T)_M := \alpha_{TM} : T(TM) \to T'(TM)$, i.e. $\alpha T \equiv \alpha \star 1_T$; similarly, $T\alpha$ equals $1_T \star \alpha$, or, $(T\alpha)_M = T(\alpha_M) : TTM \to TT'M$. This notation extends to the sequences with more functors but only one natural transformation – it is preferable to specify the product $\circ$ versus $\star$ among the words if each has some natural transformation mixed in. Here concatenation is higher binding than any of the composition signs. Notice that $1_T \star 1_S = 1_{T \circ S}$.

Given a strict monoidal category $\tilde{\mathcal{C}} := (\mathcal{C}, \otimes, 1_{\tilde{c}})$ a **monoid** in $\mathcal{C}$ is a pair (X, μ) of an object X and a multiplication morphism $\mu : X \otimes X \to X$ which is associative and there is a 'unit' map $\eta : 1_{\tilde{c}} \to X$ such that $\mu \circ (\eta \otimes \mathrm{id}) = \mu \circ (\mathrm{id} \otimes \eta) \cong \mathrm{id}$ (here the identification $1_{\tilde{c}} \otimes X \cong X$ is assumed). As this characterizes the unit map uniquely, one may or may not include η in the data, writing triples (X, μ, η) when convenient.

For a fixed small category $\mathcal{A}$, the category $\mathrm{End}_\mathcal{A}$ of endofunctors (as objects) and natural transformations (as morphisms) is a strict monoidal category: the product of endofunctors is the composition, the product of natural transformations is the Godement product, and the unit is $\mathrm{Id}_\mathcal{A}$.

A **monad** (T, μ, η) in an arbitrary category $\mathcal{A}$ is a monoid in $\mathrm{End}_\mathcal{A}$, and a **comonad** $(\perp, \delta, \eta)$ in $\mathcal{A}$ is a monad in $\mathcal{A}^{\mathrm{op}}$. The natural transformations $\delta :\perp \to \perp \circ \perp$ and $\epsilon :\perp \to \mathrm{Id}_\mathcal{A}$ are also called the coproduct and the counit of the comonad respectively.

An **action** of a monad $\mathbf{T} = (T, \mu, \eta)$ on an object M in $\mathcal{A}$ is a morphism $\nu : T(M) \to M$ such that the diagram

$$
\begin{array}{ccc}
TT(M) & \overset{\mu_M}{\to} & T(M) \\
T(\nu) \downarrow & & \downarrow \nu \\
T(M) & \overset{\nu}{\to} & M
\end{array}
$$

commutes and $\nu \circ \eta_M = \mathrm{Id}_M$. We say that (M, μ) is a **module** (older terminology: algebra) over $\mathbf{T}$. A map $(M, \mu) \to (N, \nu)$ is a morphism $f : M \to N$ in $\mathcal{A}$ intertwining the actions in the sense that $f \circ \nu_M = \nu_N \circ T(f) : T(M) \to N$. For a fixed $\mathbf{T}$, modules and their maps constitute a category $\mathcal{A}^\mathbf{T} \equiv \mathbf{T} - \mathrm{Mod}$, called the **Eilenberg-Moore** category of $\mathbf{T}$ ([37]). The natural forgetful functor $U^\mathbf{T} : \mathcal{A}^\mathbf{T} \to \mathcal{A}$, $(M, \nu) \mapsto M$ is faithful, reflects isomorphisms and has a left adjoint $F : M \mapsto (M, \mu_M)$. The unit of adjunction $\eta : \mathrm{Id}_\mathcal{A} \Rightarrow U^\mathbf{T} F = T$ coincides with the unit of $\mathbf{T}$, and the counit

$\epsilon : FU^{\mathbf{T}} \to \mathrm{Id}_{A^{\mathbf{T}}}$ is given by $\epsilon_{(M,\nu)} = \nu$. The essential image of F, is a full and replete subcategory $A_{\mathbf{T}} \subset A^{\mathbf{T}}$ and it is called the **Kleisli category** of **T**, while its objects are called **free T-modules**.

Dually, for a comonad $\mathbf{G} = (G, \delta, \epsilon)$ in A, a **G-comodule** is an object in the category $\mathbf{G} - \mathrm{Comod} := ((A^{\mathrm{op}})^{\mathbf{G}})^{\mathrm{op}} = (\mathbf{G} - \mathrm{Mod})^{\mathrm{op}}$; equivalently it is a pair (M, σ) where M is an object in A and $\sigma : M \to G(M)$ is a morphism in A such that

$$
\begin{array}{ccc}
M & \xrightarrow{\sigma} & G(M) \\
\sigma \downarrow & & \downarrow G(\sigma) \\
G(M) & \xrightarrow{\delta} & GG(M)
\end{array}
$$

and $\epsilon_M \circ \sigma = \mathrm{Id}_M$. A map $(M, \rho_M) \to (N, \nu_N)$ is a morphism $f : M \to N$ in A intertwining the coactions in the sense that $\sigma_N \circ f = T(f) \circ \sigma_M :$ $M \to T(N)$. The forgetful functor $\mathbf{G} - \mathrm{Mod} \to A^{\mathrm{op}}$ may be interpreted as a functor $U_{\mathbf{G}} : \mathbf{G} - \mathrm{Comod} \to A$ which thus has a *right* adjoint $H :$ $M \mapsto (G(M), \delta)$ whose essential image by definition consists of **cofree G-comodules**. The counit of the adjunction agrees with the counit of the comonad $\epsilon : G = U_{\mathbf{G}} H \Rightarrow \mathrm{Id}_A$ and the unit $\eta : \mathrm{Id}_{\mathbf{G}-\mathrm{Comod}} \Rightarrow HU_{\mathbf{G}}$ is given by $\eta_{(M,\sigma)} = \sigma : (M, \sigma) \to (G(M), \delta)$.

8.3 An archetypal example of a monad is constructed from a pair of adjoint functors $Q^* \dashv Q_*$ where $Q_* : B \to A$. In other words there are natural transformations $\eta : \mathrm{Id}_A \Rightarrow Q_*Q^*$ called the unit and $\epsilon : Q^*Q_* \Rightarrow \mathrm{Id}_B$, called the counit of the adjunction, such that the composites in the two diagrams

$$
Q_* \xrightarrow{\eta Q_*} Q_*Q^*Q_* \xrightarrow{Q_*(\epsilon)} Q_*, \quad Q^* \xrightarrow{Q^*(\eta)} Q^*Q_*Q^* \xrightarrow{\epsilon Q_*} Q^*,
$$

are the identity transformations. Then $\mathbf{T} := (Q_*Q^*, 1_{Q_*} \star \epsilon \star 1_{Q^*}, \eta)$ is a monad in A. In other words, the multiplication is given by

$$
\mu_M = Q_*(\epsilon_{Q^*(M)}) : Q_*Q^*Q_*Q^*(M) = TT(M) \to Q_*Q^*(M) = T(M).
$$

The **comparison functor** $K^{\mathbf{T}} : B \to A^{\mathbf{T}}$ is defined by

$$
M \mapsto (Q_*(M), Q_*(\epsilon_M)), \quad F \mapsto Q_*(f).
$$

It is full and Q^* factorizes as $B \xrightarrow{K^{\mathbf{T}}} A^{\mathbf{T}} \xrightarrow{U^{\mathbf{T}}} A$. More than one adjunction (varying B) may generate the same monad in A in this vein.

Dually, $\mathbf{G} := (Q^*Q_*, Q^*\eta Q_*, \epsilon)$ is a comonad, i.e. a monad in B^{op}. The comparison functor $K^{\mathbf{G}} : A^{\mathrm{op}} \to (B^{\mathrm{op}})^{\mathbf{T}}$ is usually identified with a 'comparison functor' $K_{\mathbf{G}} : A \to ((B^{\mathrm{op}})^{\mathbf{T}})^{\mathrm{op}} \equiv \mathbf{G} - \mathrm{Comod}$ which is hence given by $N \mapsto (Q^*(N), \eta_{Q^*(N)})$. $K_{\mathbf{G}}$ is full and Q_* factorizes as $\mathcal{A} \overset{K_{\mathbf{G}}}{\Rightarrow}$ $\mathbf{G} - \mathrm{Comod} \overset{U_{\mathbf{G}}}{\Rightarrow} \mathcal{B}$.

8.4 A map of monoids $f : (A, \mu, \eta) \rightarrow (A', \mu', \eta')$ in a monoidal category $(\mathcal{A}, \otimes, 1_{\mathcal{A}}, a, l, r)$ is a morphism $f : A \rightarrow A'$ in $\mathcal{A}$, commuting with multiplication: $\mu \circ (f \otimes f) = f \circ \mu$; and with the unit map: $\eta' \circ \mu = \eta \otimes \eta$, where on the left the application of one of the isomorphisms $l_{1_{\mathcal{A}}}, r_{1_{\mathcal{A}}} : 1_{\mathcal{A}} \otimes 1_{\mathcal{A}} \rightarrow 1_{\mathcal{A}}$ is assumed. In particular, the morphism $\phi : (T, \mu, \eta) \rightarrow (T', \mu', \eta')$ of monads in $\mathcal{A}$ is a natural transformation $\phi : T \rightarrow T'$ such that $\mu \circ (\phi \star \phi) = \phi \circ \mu :$ $TT \Rightarrow T'$ and $\eta' \circ \mu = \eta \star \eta : TT \Rightarrow \mathrm{Id}_{\mathcal{A}}$. If M is an object in $\mathcal{A}$ and ν a T'-action on M, then $\nu' \circ \phi_M : TM \Rightarrow M$ is a T-action on M. More precisely, a natural transformation $\phi : T \Rightarrow T'$ and rules $\mathcal{A}^\phi(M, \nu) = (M, \nu' \circ \phi_M)$ and $\mathcal{A}^\phi(f) = f$ define a functor

$$\mathcal{A}^\phi : \mathcal{A}^{\mathbf{T'}} \rightarrow \mathcal{A}^{\mathbf{T}}$$

iff ϕ is a morphism of monads and every functor $\mathcal{A}^{\mathbf{T'}} \rightarrow \mathcal{A}^{\mathbf{T}}$ inducing the identity on $\mathcal{A}$ is of that form.

8.5 Let Δ be the 'simplicial' category: its objects are nonnegative integers viewed as finite ordered sets $\mathbf{n} := \{0 < 1 < \ldots < n\}$ and its morphisms are nondecreasing monotone functions. Given a category $\mathcal{A}$, denote by $\mathrm{Sim}\mathcal{A}$ the category of **simplicial objects** in $\mathcal{A}$, i.e. functors $F : \Delta^{\mathrm{op}} \rightarrow \mathcal{A}$. Represent F in $\mathrm{Sim}\mathcal{A}$ as a sequence $F_n := F(\mathbf{n})$ of objects, together with the face maps $\partial_i^n : F_n \rightarrow F_{n-1}$ and the degeneracy maps $\sigma_i^n : F_n \rightarrow F_{n+1}$ for $i \in \mathbf{n}$ satisfying the familiar simplicial identities ([145, 84]). The notation $F_\bullet$ for this data is standard.

Given a comonad $\mathbf{G}$ in $\mathcal{A}$ one defines the sequence $\mathbf{G}_\bullet$ of endofunctors $\mathbb{Z}_{\geq 0} \ni n \mapsto \mathbf{G}_n := G^{n+1} := G \circ G \circ \ldots \circ G$, together with natural transformations $\partial_i^n : G^i \epsilon G^{n-i} : G^{n+1} \rightarrow G^n$ and $\sigma_i^n : G^i \delta G^{n-i} : G^{n+1} \rightarrow G^{n+2}$, satisfying the simplicial identities. Use $\epsilon G \circ \delta = G\epsilon \circ \delta = \mathrm{Id}_{\mathcal{A}}$ in the proof. Hence any comonad $\mathbf{G}$ canonically induces a simplicial endofunctor, i.e. a functor $\mathbf{G}_\bullet : \Delta^{\mathrm{op}} \rightarrow \mathrm{End}\mathcal{A}$, or equivalently, a functor $\mathbf{G}_\bullet : \mathcal{A} \rightarrow \mathrm{Sim}\mathcal{A}$. The counit ϵ of the comonad $\mathbf{G}$ satisfies $\epsilon \circ \partial_0^1 = \epsilon \circ \partial_1^1$, hence $\epsilon : \mathbf{G}_\bullet \rightarrow \mathrm{Id}_{\mathcal{A}}$ is in fact an augmented simplicial endofunctor.

This fact is widely used in homological algebra ([8, 84, 145]), and now also in the cohomological study of noncommutative spaces ([109]).

8.6 Barr-Beck lemma. ([9, 85]) Let $Q^* \dashv Q_*$ be an adjoint pair $\mathbf{T}$ its associated monad, and $\mathbf{G}$ its associated comonad (as in **8.3**). Recall the notions of preserving and reflecting (co)limits from **1.6**.

If Q_* preserves and reflects coequalizers of all parallel pairs in $\mathcal{A}$ (for which coequalizers exists) and if any parallel pair mapped by Q_* into a pair *having a coequalizer in $\mathcal{B}$ has a coequalizer in $\mathcal{A}$ then the comparison functor $K : \mathcal{B} \to \mathcal{A}^{\mathbf{T}}$ is an equivalence of categories.*

If Q^* preserves and reflects equalizers of all parallel pairs in $\mathcal{B}$ (for which equalizers exists) and if any parallel pair mapped by Q^* into a pair having an equalizer in $\mathcal{A}$ has an equalizer in $\mathcal{B}$ then the comparison functor K' : $\mathcal{A} \to \mathbf{G} - \mathrm{Comod}$ is an equivalence of categories.

Left (right) exact functors by definition preserve finite limits (colimits) and faithful functors clearly reflect both. In particular, this holds for (co)equalizers of parallel pairs. In abelian categories such (co)equalizers always exist. Hence

Corollary. *Consider an adjoint pair $Q^* \dashv Q_*$ of additive functors between abelian categories. If Q_* is faithful and exact, then the comparison functor for the associated comonad is an equivalence. If Q^* is faithful and exact, then the comparison functor for the associated monad is an equivalence.*

Given a functor $U : \tilde{\mathcal{M}} \to \mathcal{M}$ one may ask when there is a monad $\mathbf{T}$ in $\mathcal{M}$ and an equivalence $H : \tilde{\mathcal{M}} \to \mathcal{M}^{\mathbf{T}}$ such that $U^T H = U$. The conditions are given by the *Beck monadicity (=tripleability) theorem(s)* ([9, 11, 84, 85]). If we already know that U has left adjoint, this may be rephrased by asking if the comparison functor for the *associated* monad is an equivalence. The Barr-Beck lemma gives only sufficient conditions for this case, it is easier to use, and is widely applicable.

8.7 ([109]) **A comonad associated to a family of continuous functors.** Let $\{Q^*_\lambda : \mathcal{A} \to \mathcal{B}_\lambda\}_{\lambda \in \Lambda}$ be a small family of continuous (= having a right adjoint) functors. The categories $\mathcal{B}_\lambda$ are *not* necessarily constructed from $\mathcal{A}$ by a localization.

One may consider the category $\mathcal{B}_\Lambda := \prod_{\lambda \in \Lambda} \mathcal{B}_\lambda$ whose objects are families $\prod_{\lambda \in \Lambda} M^\lambda$ of objects M^λ in $\mathcal{B}_\lambda$ and morphisms are families $\prod_{\lambda \in \Lambda} f_\lambda$: $\prod_{\lambda \in \Lambda} M^\lambda \to \prod_{\lambda \in \Lambda} N^\lambda$ where $f_\lambda : M^\lambda \to N^\lambda$ is a morphism in $\mathcal{B}_\lambda$, with componentwise composition. This makes sense as the family of objects is literally a function from Λ to the disjoint union $\coprod_\lambda \mathrm{Ob}\, \mathcal{B}_\lambda$ which is in the same Grothendieck universe.

The family of adjoint pairs $Q^*_\lambda \dashv Q_{\lambda *}$ defines an inverse image functor $\mathbf{Q}^* = \prod Q^*_\lambda : \mathcal{A} \to \mathcal{A}_\Lambda$ by $\mathbf{Q}^*(M) := \prod_{\lambda \in \Lambda} Q^*_\lambda(M)$ on objects and $\mathbf{Q}^*(f) := \prod_\lambda Q^*_\lambda(f)$ on morphisms. However, a direct image functor may not exist. We may naturally try $\mathbf{Q}_* : \prod'_\lambda M^\lambda \mapsto \prod'_\lambda Q_*(M^\lambda)$ where $\prod'$ is now the symbol for the *Cartesian* product in $\mathcal{A}$ which may not always exist. For *finite* families, with $\mathcal{A}$ abelian, these trivially exist. Let $\mathcal{A}^\Lambda = \prod_{\lambda \in \Lambda} \mathcal{A}$ be the power category. Assume a fixed choice of the Cartesian product for all

Λ-tuples in $\mathcal{A}$. Then $\{M^\lambda\}_\lambda \mapsto \prod'_{\lambda \in \Lambda} M^\lambda$ extends to a functor $\mathcal{A}^\Lambda \to \mathcal{A}$, and the universality of products implies that the projections $p'_{\nu\{M^\lambda\}_\lambda} : \prod_\lambda M^\lambda \to M^\nu$ form a natural transformation of functors $p'_\nu : \prod'_\lambda \mathrm{Id}_\mathcal{A} \Rightarrow p_\nu$ where $p_\nu : \mathcal{A}^\Lambda \to \mathcal{A}$ is the ν-th formal projection $\prod_\lambda M^\lambda \to M^\nu$. The unique liftings $\eta_M : M \to \mathbf{Q}_*\mathbf{Q}^*(M)$ of morphisms $\eta_{\nu M} : M \to Q_{\nu *}Q^*_\nu(M)$ in the sense that $(\forall \nu)\, \eta_{\nu M} = p'_{\nu M} \circ \eta_M$ hence form a natural transformation $\eta : \mathrm{Id}_\mathcal{A} \Rightarrow \mathbf{Q}_*\mathbf{Q}^*$.

Define $\epsilon \equiv \prod_\lambda \epsilon_\lambda : \prod_\lambda Q^*_\lambda Q_{\lambda *} \Rightarrow \prod_\lambda \mathrm{Id}_{\mathcal{B}_\lambda} = \mathrm{Id}_{\mathcal{B}_\Lambda}$ componentwise. This way we obtain an adjunction $\mathbf{Q}^* \dashv \mathbf{Q}_*$. If $Q_{\lambda *}$ is faithful and exact for every λ then $\mathbf{Q}_*$ is as well.

Consider the comonad $\mathbf{G}$ in $\mathcal{B}_\Lambda$ associated to $\mathbf{Q}^* \dashv \mathbf{Q}_*$. We are interested in situation when the comparison functor $K_\mathbf{G}$ is an equivalence of categories. That type of a situation arises in practice in two different ways:

- 1) All categories $\mathcal{A}, \mathcal{B}_\lambda$ and flat localization functors $Q^*_\lambda, Q_{\lambda *}$ are given at start and the construction is such that we know the faithfulness of $\mathbf{Q}_*$.

- 2) Only categories $\mathcal{B}_\lambda$ are given (not $\mathcal{A}$) but equipped with **gluing morphisms** i.e. the family Φ of flat functors (not necessarily localizations) $\phi^*_{\lambda,\lambda'} : \mathcal{B}_\lambda \to \mathcal{B}_{\lambda'}$ for each pair λ, λ', where Φ satisfies some cocycle condition.

8.7a In 1), to ensure the faithfulness of $\mathbf{Q}_*$ we require that the family $\{Q^*_\lambda\}_{\lambda \in \Lambda}$ is a flat **cover** of $\mathcal{A}$. That means that this is a small *flat* family of functors with domain $\mathcal{A}$ which is **conservative**, i.e. a morphism $f \in \mathcal{A}$ is invertible iff $Q^*_\lambda(f)$ is invertible for each $\lambda \in \Lambda$. A flat map whose direct image functor is conservative is called **almost affine**. In particular, this is true for adjoint triples $f^* \dashv f_* \dashv f^!$ coming from a map $f : R \to S$ of rings. Adjoint triples where the direct image functor is conservative are called **affine morphisms**.

8.7b In 2), we *a posteriori* construct $\mathcal{A}$ to be $\mathcal{B}_\Lambda$ as before but equipped with functors $Q^*_\lambda \dashv Q_{\lambda *}$ constructed from ϕ-functors. The cocycle condition for gluing morphism is equivalent to the associativity of the associated comonad ([21]). The remaining requirements are made to ensure that the comparison functor is an equivalence and the other original data may be reconstructed as well. The Eilenberg-Moore category of the associated monad may be constructed directly from gluing morphisms, and it appears to be just a reformulation of the descent category. In a generalization, the category $\mathcal{B}_{\lambda\lambda'}$ which is the essential image of $\phi_{\lambda\lambda'}$ in $\mathcal{B}_{\lambda'}$ may be replaced by any 'external' category $\mathcal{B}_{\lambda\lambda'}$, but then, instead of $\phi^*_{\lambda\lambda'}$ one requires not only flat functors $\phi^*_{\lambda\lambda'} =: \phi^{\lambda*}_{\lambda\lambda'} : \mathcal{B}_\lambda \to \mathcal{B}_{\lambda\lambda'}$, but also flat functors $\tilde{\phi}^*_{\lambda\lambda'} := \phi^{\lambda'*}_{\lambda\lambda'} : \mathcal{B}_{\lambda'} \to \mathcal{B}_{\lambda\lambda'}$. This generalization is essentially more general only if we allow the direct image functors (of the second type, i.e. $\phi^{\lambda'}_{\lambda\lambda'*}$), to be not necessarily fully faithful (hence $\mathcal{B}_{\lambda\lambda'}$ may not be viewed as a full subcategory of $\mathcal{B}_{\lambda'}$). Another generalization of this descent situation, which can be phrased as having a pseudofunctor from a finite poset Λ (viewed as a 2-category with only identity 2-cells) to the 2-category of categories has been studied by V. LUNTS ([78]). This analogue of the descent category is called a *configuration category*.

8.7c The usual formalism of descent is via fibred categories, cf. [143]. For the correspondence between the two formalisms see, e.g. [110].

8.8 Globalization lemma. (Version for Gabriel filters: in [112] p. 103) *Suppose* $\{Q^*_\lambda : R - \mathrm{Mod} \to \mathcal{M}_\lambda\}_{\lambda\in\Lambda}$ *is a finite cover of* $R - \mathrm{Mod}$ *by flat localization functors (e.g. a conservative family of Ore localizations* $\{S^{-1}_\lambda R\}_{\lambda\in\Lambda}$*). Denote* $Q_\lambda := Q_{\lambda*}Q^*_\lambda$ *where* $Q_{\lambda*}$, *is the right adjoint to* Q^*_λ. *Then for every left* R-*module* M *the sequence*

$$0 \to M \xrightarrow{\iota_{\Lambda,M}} \prod_{\lambda\in\Lambda} Q_\lambda M \xrightarrow{\iota_{\Lambda\Lambda,M}} \prod_{(\mu,\nu)\in\Lambda\times\Lambda} Q_\mu Q_\nu M$$

is exact, where $\iota_{\Lambda M} : m \mapsto \prod \iota_{\lambda,M}(m)$, *and*

$$\iota_{\Lambda\Lambda M} := \prod_\lambda m_\lambda \mapsto \prod_{(\mu,\nu)} (\iota^\mu_{\mu,\nu,M}(m_\mu) - \iota^\nu_{\mu,\nu,M}(m_\nu)).$$

Here the order matters: pairs with $\mu = \nu$ may be (trivially) skipped, but, unlike in the commutative case, **we can not confine to the pairs of indices with** $\mu < \nu$ **only**. *Nota bene!*

Proof. A direct corollary of Barr-Beck lemma. For proofs in terms of Gabriel filters and torsion see [113], pp. 23–25, and [62, 138, 139].

8.9 A monad $\mathbf{T} = (T, \mu, \eta)$ in $\mathcal{A}$ is **idempotent** if the multiplication $\mu : TT \Rightarrow T$ is an equivalence of endofunctors. As μ_M is the left inverse of η_{TM}, and of $T(\eta_M)$, then μ_M is invertible iff any of them is, hence both, and then $\eta_{TM} = T(\eta_M) = \mu^{-1}_M$.

If $\nu : TM \to M$ is a $\mathbf{T}$-action, then by naturality $\eta_M \circ \nu = T(\nu) \circ \eta_{TM} = T(\nu) \circ T(\eta_M) = T(\nu \circ \eta_M) = \mathrm{Id}_{TM}$, hence η_M is 2-sided inverse of ν in $\mathcal{A}$, hence *every* $\mathbf{T}$-*action is an isomorphism*. Conversely, If every $\mathbf{T}$-action is an isomorphism, μ_M is in particular, and $\mathbf{T}$ is idempotent. Moreover, if every action $\nu : TM \to M$ is an isomorphism, then its right inverse η_M must be the 2-sided inverse, hence there may be *at most* one action on a given object M in $\mathcal{A}$. By naturality of η, its inverse ν which is well-defined on the full subcategory of $\mathcal{A}$ generated by objects in the image of $U^{\mathbf{T}} : \mathcal{A}^{\mathbf{T}} \to \mathcal{A}$, is also natural, i.e. it intertwines the actions, hence it is in fact a morphism in $\mathcal{A}^{\mathbf{T}}$, hence the forgetful functor $U^{\mathbf{T}} : \mathcal{A}^{\mathbf{T}} \to \mathcal{A}$ is not only faithful but also full. Its image is strictly full (full and closed under isomorphisms) as the existence of **T**-actions on M depends only on the isomorphism class of M in $\mathcal{A}$. To summarize, $\mathcal{A}^{\mathbf{T}}$ includes via $U^{\mathbf{T}}$ as a strictly full subcategory into $\mathcal{A}$ and the inclusion has a left adjoint $F^{\mathbf{T}}$.

In general, a **(co)reflective subcategory** $\mathcal{B} \hookrightarrow \mathcal{A}$ is a strictly full subcategory, such that the inclusion $U : \mathcal{B} \hookrightarrow \mathcal{A}$ has a (right) left adjoint, say F. As $F^{\mathbf{T}} \dashv U^{\mathbf{T}}$, we have just proved that $\mathcal{A}^{\mathbf{T}}$ is canonically isomorphic to a reflective subcategory of $\mathcal{A}$ via inclusion $U^{\mathbf{T}}$ if **T** is idempotent. On the other hand, it may be shown that for any reflective subcategory $U : \mathcal{B} \hookrightarrow \mathcal{A}$ the corresponding monad $(UF, U(\epsilon_F), \eta)$ is idempotent and the comparison functor $\mathcal{K} : \mathcal{B} \cong \mathcal{A}^{\mathbf{T}}$ is an isomorphism. Similarly, coreflective subcategories are in a natural correspondence with idempotent comonads.

9 Distributive laws and compatibility

9.1 A distributive law from a monad $\mathbf{T} = (T, \mu^T, \eta^T)$ to an endofunctor P is a natural transformation $l : TP \Rightarrow PT$ such that

$$l \circ (\eta^T)_P = P(\eta^T), \qquad l \circ (\mu^T)_P = P(\mu^T) \circ l_T \circ T(l). \qquad (12)$$

Then P **lifts** to a unique endofunctor $P^{\mathbf{T}}$ in $\mathcal{A}^{\mathbf{T}}$, in the sense that $U^{\mathbf{T}} P^{\mathbf{T}} = PU^{\mathbf{T}}$. Indeed, the endofunctor $P^{\mathbf{T}}$ is given by $(M, \nu) \mapsto (PM, P(\nu) \circ l_M)$.

9.1a A distributive law from a monad $\mathbf{T} = (T, \mu^T, \eta^T)$ to a monad $\mathbf{P} = (P, \mu^P, \eta^P)$ in $\mathcal{A}$ ([10]) (or "of **T** over **P**") is a distributive law from **T** to the endofunctor P, compatible with μ^P, η^P in the sense that

$$l \circ T(\eta^P) = (\eta^P)_T, \qquad l \circ T(\mu^P) = (\mu^P)_T \circ P(l) \circ l_P.$$

For clarity, we show the commutative diagram for one of the relations.

$$
\begin{array}{ccccc}
TPP & \xrightarrow{l_P} & PTP & \xrightarrow{P(l)} & PPT \\
{\scriptstyle T(\mu^P)}\downarrow & & & & \downarrow{\scriptstyle P(\mu^T)} \\
TP & & \xrightarrow{\quad l \quad} & & PT
\end{array}
$$

Then **P** lifts to a unique monad $\mathbf{P}^{\mathbf{T}} = (P^{\mathbf{T}}, \tilde{\mu}, \tilde{\eta})$ in $\mathcal{A}^{\mathbf{T}}$, such that $P^{\mathbf{T}}$ lifts P, and for all $N \in \mathcal{A}^T$ we have $U^{\mathbf{T}}(\tilde{\eta}_N) = (\eta^P)_{U^{\mathbf{T}}N}$ and $U^{\mathbf{T}}(\tilde{\mu}_N) = (\mu^P)_{U^{\mathbf{T}}N}$.

Indeed, such a lifting is defined by the formulas $P^{\mathbf{T}}(M,\nu) := (PM, P(\nu) \circ l_M)$, $\mu^P_{(M,\nu)} = \mu_M$, $\eta^P_{(M,\nu)} = \eta_M$. On the other hand, if $\mathbf{P}^{\mathbf{T}}$ is a lifting of $\mathbf{P}$ then a distributive law $l = \{l_M\}$ is defined, namely l_M is the composition

$$TPM \overset{TP(\eta_M)}{\longrightarrow} TPTM \overset{U^{\mathbf{T}}(\epsilon_{PF^{\mathbf{T}}M})}{\longrightarrow} PTM.$$

where $F^{\mathbf{T}} : \mathcal{A} \to \mathcal{A}^{\mathbf{T}}$ is the free **T**-module functor from **8.2**.

Distributive laws from **T** to **P** are in a canonical bijective correspondence with those monads in $\mathcal{A}$ whose underlying functor is PT, whose unit is $\eta_P \star \eta_T$, such that $P\eta^T : P \to PT$, and $\eta^P_T : T \to PT$ are triple maps and which satisfy the **middle unitary law** $\mu \circ (P(\eta^T) \star \eta^P_T) = \mathrm{id} : PT \to PT$ (cf. [10]). In this correspondence, $\mu^{PT}_M : PTPTM \to PTM$ is obtained by $\mu^{PT}_M = (\mu^P \star \mu^T) \circ P(l_{TM})$, and conversely, l_M by composition $TPM \overset{TP(\eta_M)}{\longrightarrow}$
$TPTM \overset{\eta_{TPTM}}{\longrightarrow} PTPTM \overset{\mu^{PT}_M}{\longrightarrow} PTM.$

9.1b Distributive law from a comonad $\mathbf{G} = (G, \delta^G, \epsilon^G)$ **to a comonad** $\mathbf{F} = (F, \delta^F, \epsilon^F)$ is a natural transformation $l : F \circ G \Rightarrow G \circ F$ such that

$$G(\epsilon^F) \circ l = (\epsilon^F)_G, \quad (\delta^G)_F \circ l = G(l) \circ l_G \circ F(\delta^G),$$
$$(\epsilon^G)_F \circ l = F(\epsilon^G), \quad G(\delta^F) \circ l = l_F \circ F(l) \circ (\delta^F)_G.$$

9.1c Mixed distributive law from a monad T to a comonad G: a natural transformation $l : TG \Rightarrow GT$ such that

$$\epsilon^T \circ l = T(\epsilon), \quad l \circ \mu_G = G(\mu) \circ l_T \circ T(l),$$
$$l \circ \eta_G = G(\eta), \quad l \circ T(\delta) = \delta_T \circ G(l) \circ l_G.$$

Such an l corresponds to a lifting of the comonad **G** to a comonad $\mathbf{G}^{\mathbf{T}}$ in $\mathcal{A}^{\mathbf{T}}$, where $G^{\mathbf{T}}(M,\nu) = (MG, G(\nu) \circ l_M)$.

9.1d Mixed distributive law from a comonad G to a monad T: a natural transformation $l : GT \Rightarrow TG$ such that

$$T(\epsilon) \circ l = \epsilon_T, \quad l \circ G(\mu) = \mu_G \circ T(l) \circ l_T$$
$$l \circ \eta_G = G(\eta), \quad T(\delta) \circ l = l_G \circ G(l) \circ \delta_T.$$

Such distributive laws are in a correspondence with liftings of a monad **T** to a monad $\mathbf{T}^{\mathbf{G}}$ in $\mathbf{G} - \mathrm{Comod}$.

9.2 Examples are abundant ([21, 119]; [131],II). In a common scenario, the objects in a category of interest are in fact objects in a simpler 'base' category, together with multiple extra structures, satisfying "compatibility" conditions between the structures, which correspond to a fixed choice of distributive laws. Consider, left R-modules and right S-modules, where the base category is the category **Ab** of abelian groups. The rebracketing map $(R \otimes ?) \otimes S \mapsto R \otimes (? \otimes S)$ gives rise to a distributive law from $R \otimes ?$ and $? \otimes S$ in **Ab**. Thus, it induces a monad **V** with underlying functor $V = (S \otimes ?) \circ (T \otimes ?)$. **V**-modules over U are precisely $R - S$-bimodules. Similarly, for a group (or Hopf algebra) G, one can describe G-equivariant versions of many standard categories of sheaves or modules with extra structure, by considering one (co)monad for the underlying structure and another expressing the G-action.

9.3 A monad (T, μ, η) in arbitrary 2-category (even bicategory) $\mathcal{C}$ has been studied ([131]): $T : X \to X$ is now a 1-cell, where X is a fixed 0-cell, and 2-cells μ and η should satisfy analogous axioms as in the usual case, which corresponds to $\mathcal{C} = \text{Cat}$. On the other hand, if $\mathcal{C}$ is a bicategory with a single object X, it may be identified with a monoidal 1-category. The distributive laws in that case supply a notion of compatibility of monoids and comonoids in an arbitrary monoidal category. The distributive laws between monoids and comonoids in Vec_k are called *entwining structures* ([24]).

9.4 Let $\mathbf{T} := (T, \mu, \eta)$ be a monad, and $Q^* : \mathcal{A} \to \mathcal{B}$ a localization functor. The monad $\mathbf{T}$ is **compatible** with the localization if its underlying endofunctor T is compatible with the localization, i.e. there is a functor $T_{\mathcal{B}} : \mathcal{B} \to \mathcal{B}$ with $Q^*T = T_{\mathcal{B}}Q^*$, cf. **3.5**. In that case, $T_{\mathcal{B}}$ is the underlying endofunctor of a unique monad $\mathbf{T}_{\mathcal{B}} := (T_{\mathcal{B}}, \mu^{\mathcal{B}}, \eta^{\mathcal{B}})$ in $\mathcal{B}$ such that $(\mu^{\mathcal{B}})_{Q^*N} = Q^*(\mu_N)$ for every N in $\text{Ob}\,\mathcal{A}$.

Proof. Let $f : N \to N'$ be a morphism in $\mathcal{A}$, and $g : Q^*N' \to M$ an isomorphism in $\mathcal{B}$. Consider the diagram

$$
\begin{array}{ccccccc}
T_{\mathcal{B}}T_{\mathcal{B}}Q^*N & \xrightarrow{=} & Q^*TTN & \xrightarrow{Q^*\mu_N} & Q^*TN & \xrightarrow{=} & T_{\mathcal{B}}Q^*N \\
\downarrow{\scriptstyle T_{\mathcal{B}}T_{\mathcal{B}}Q^*f} & & \downarrow{\scriptstyle Q^*TTf} & & \downarrow{\scriptstyle Q^*Tf} & & \downarrow{\scriptstyle T_{\mathcal{B}}Q^*f} \\
T_{\mathcal{B}}T_{\mathcal{B}}Q^*N' & \xrightarrow{=} & Q^*TTN' & \xrightarrow{Q^*\mu_{N'}} & Q^*TN' & \xrightarrow{=} & T_{\mathcal{B}}Q^*N' \\
\downarrow{\scriptstyle T_{\mathcal{B}}T_{\mathcal{B}}g} & & & & & & \downarrow{\scriptstyle T_{\mathcal{B}}g} \\
T_{\mathcal{B}}T_{\mathcal{B}}M & & \xrightarrow{\hspace{2cm}(\mu^{\mathcal{B}})_M\hspace{2cm}} & & & & T_{\mathcal{B}}M.
\end{array}
$$

The upper part of the diagram clearly commutes. In particular, if Q^*f is identity, then $Q^*\mu_N = Q^*\mu_{N'}$. The vertical arrows in the bottom part are isomorphisms, so there is a map $(\mu^{\mathcal{B}})_M$ filling the bottom line. One has to show that this map does not depend on the choices and that such maps form a natural transformation. The localization functor is a composition of a quotient functor onto the quotient category and an equivalence. We may assume that Q^* is the functor onto the quotient category. Then, by the construction of the quotient category, every morphism g is of the zig-zag form as a composition of the maps of the form Q^*f and formal inverses of such maps, and if g is an isomorphism, both kinds of ingredients are separately invertible in $\mathcal{B}$. To show that $(\mu^B)_M = (T_{\mathcal{B}}g) \circ Q^*(\mu_{N'}) \circ (T_{\mathcal{B}}T_{\mathcal{B}}g)^{-1}$ for every isomorphism $g : Q^*N' \to M$ is the consistent choice, we use the upper part of the diagram repeatedly (induction by the length of zig-zag) for the zig-zag isomorphism $h = (g_1)^{-1}g_2 : Q^*N_2 \to Q^*N_1$ where $g_i : Q^*N_i \to M$. One obtains $T_{\mathcal{B}}h \circ Q^*\mu_{N_2} \circ (T_{\mathcal{B}}T_{\mathcal{B}}h)^{-1} = Q^*\mu_{N_1}$, hence $(T_{\mathcal{B}}g_2) \circ Q^*(\mu_{N_2}) \circ (T_{\mathcal{B}}T_{\mathcal{B}}g_2)^{-1} = (T_{\mathcal{B}}g_1) \circ Q^*(\mu_{N_1}) \circ (T_{\mathcal{B}}T_{\mathcal{B}}g_1)^{-1}$.

Upper part of the diagram also shows the naturality for μ^B with respect to each arrow of the form Q^*f and with respect to formal inverses of such. For any morphism $h : M \to M'$ in $\mathcal{B}$, using its zig-zag representation, we extend this to the naturality diagram $(\mu^B)_{M'} \circ T_{\mathcal{B}}T_{\mathcal{B}}h = T_{\mathcal{B}}T_{\mathcal{B}}h \circ (\mu^B)_M$. Uniqueness of μ^B is clear by the requirement $(\mu^B)_{Q^*N} = Q^*(\mu_N)$ and the naturality. The unit morphism $\eta : 1_{\mathcal{B}} \to T_{\mathcal{B}}$ satisfies $g \circ \eta^B_M = T_{\mathcal{B}}(g) \circ Q^*(\eta_N)$ for every isomorphism $g : Q^*N \to M$ in $\mathcal{B}$ such that $N \in \text{Ob}\,\mathcal{A}$. In particular, $(\eta^B)_{Q^*N} = Q^*(\eta_N)$. The very axioms of a monad may be checked in a similar vein.

9.4b If $\mathbf{G} = (G, \delta, \epsilon)$ is a comonad and the endofunctor G is compatible with each localization in family $\{Q^*_\lambda : \mathcal{A} \to \mathcal{B}_\lambda\}_{\lambda \in \Lambda}$, then there is a unique family of comonads $\{\mathbf{G}_\lambda = (G_\lambda, \delta_\lambda, \epsilon_\lambda)\}_{\lambda \in \Lambda}$ such that $Q^*_\lambda G = G_\lambda Q^*_\lambda$ for each λ. We have then $(\delta_\lambda)_{Q^*_\lambda M} = Q^*_\lambda(\delta_M)$ and $(\epsilon_\lambda)_{Q^*_\lambda M} = Q^*_\lambda(\epsilon_M)$ for every $M \in \text{Ob}\,\mathcal{A}$.

9.5 If $Q^* : \mathcal{A} \to \mathcal{B}$ is a localization with right adjoint Q_*, and T' is an endofunctor in $\mathcal{B}$, then $Q_*T'Q^*$ is compatible with Q^*. Indeed ϵ is an isomorphism by **3.4**, hence $\epsilon_{T'Q^*} : Q^*Q_*T'Q^* \Rightarrow T'Q^*$ is an isomorphism, and the assertion follows by **3.5**.

9.6 Example from Hopf algebra theory. Let B be a **k**-bialgebra and (E, ρ) a right B-comodule together with a multiplication $\mu : E \otimes_{\mathbf{k}} E \to E$ making it a B-comodule algebra, i.e. an algebra in the category of right B-comodules. The coaction $\rho : E \to E \otimes H$ is **compatible** with a fixed Ore localization $\iota : E \to S^{-1}E$ if there is a coaction $\rho_S : S^{-1}E \to S^{-1}E \otimes H$ which is an algebra map and such that $\rho_S \circ \iota = (\iota \otimes \text{id}_B) \circ \rho$. B induces a natural comonad $T = T(B, E)$ in $E - \text{Mod}$, such that $(E - \text{Mod})^T$ is a category of so-called (E, B)-Hopf modules. The compatibility above ensures that the localization lifts to a localization of $(E - \text{Mod})^T$ ([122], 8.5). Hence, T is compatible with the localization in the usual sense, with numerous applications of this type of situation ([118, 120, 121, 122]).

9.7 The compatibility of certain localizations of noncommutative spaces with differential functors is central in the treatment [79, 80, 81, 82] of $\mathcal{D}$-modules on noncommutative spaces.

9.8 Distributive laws become a much simpler issue when both monads in question are idempotent in the sense of **8.9**. In the localization literature this is roughly the situation treated under the name of **mutual compatibility of localizations**.

Let S, W be two Ore sets. Then the set SW of products $\{sw | s \in S, w \in W\}$ is not necessarily multiplicatively closed.

Suppose SW is multiplicatively closed. This means that $\forall s \in S, \forall w \in W$ if the product ws is in SW then $\exists w' \in W, \exists s' \in S$ such that $ws = s'w'$. Suppose now $M := {}_R M \in R-\text{Mod}$. Each element in $S^{-1}R \otimes_R W^{-1}R \otimes_R M$ is of the form $s^{-1} \otimes w^{-1} \otimes m$ with $s \in S$, $w \in W$ and $m \in M$. By a symmetric argument, $ws \in SW$, and thus $\exists w' \in W, \exists s' \in S$ such that $ws = s'w'$. Choosing s', w' by this rule we obtain an assignment $s^{-1} \otimes w^{-1} \otimes m \mapsto (w')^{-1} \otimes (s')^{-1} \otimes m$. We claim that this assignment is well-defined and a map of left R-modules $S^{-1}R \otimes_R W^{-1}R \otimes_R M \to W^{-1}R \otimes_R S^{-1}R \otimes_R M$. As M runs through $R - \text{Mod}$, such maps form a natural transformation $Q_S Q_W \to Q_W Q_S$ of functors, which is clearly an isomorphism.

In fact, this natural transformation is a distributive law. Although the compatibility of Q_S and Q_W is symmetric, the converse does not hold: compatibility does not mean that SW is multiplicatively closed. Indeed, let R be a $\mathbb{C}$-algebra with two generators a, b and relation $ab = qba$ where $q \neq \pm 1, 0$. Then the set multiplicatively generated by A and set multiplicatively generated by B are 2-sided Ore sets, and the corresponding localization functors are compatible; however AB is not multiplicatively closed.

If S, W are left Ore in R, that does not mean that $\iota_W(S)$ is left Ore in $W^{-1}R$. Namely, the left Ore condition for (the image of) S in $W^{-1}R$ includes the following: $\forall s \in S, \forall t \in T, \exists s' \in S, \exists (w')^{-1}r' \in W^{-1}R, (w')^{-1}r's = s'w^{-1}$. If R is a domain, this means that $r'sw = w's'$. This is almost the same condition as that SW is multiplicatively closed (above), except that one can choose extra r'. In the same away as in the former case, we derive the compatibility of Q_S and Q_W. If we change left Ore sets to right Ore sets, or S being Ore in $W^{-1}R$ to W being Ore in $S^{-1}R$ we get similar "Ore" equations $swr = w's'$, $wsr = s'w'$ etc. From the abstract point of view (say torsion theories) these compatibilities are indistinguishable.

The compatibility implies that the localization at the smallest multiplicative set generated by S and W is isomorphic to the consecutive localization by S and then by W. This simplifies the formalism of localization (cf. semiseparated schemes and Čech resolutions of [109], cf. **8.8**, [62] etc.).

10 Commutative localization

Here we describe specifics in the commutative case, and further motivation from commutative algebraic geometry, and its abstractions.

10.1 Suppose R is a unital associative ring, $Z(R)$ its center, and $S \subset Z(R)$ a multiplicative subset. Obviously, S is automatically a left and right Ore subset in R, with simpler proofs for the construction and usage of the Ore localization. We say that $S^{-1}R = S \times R/ \sim$ is the **commutative localization** of R at S. *The equivalence relation* $\sim$ *(4.13) simplifies to*

$$\boxed{s^{-1}r \sim s'^{-1}r' \Leftrightarrow \exists \tilde{s} \in S, \ \tilde{s}(sr' - s'r) = 0.} \tag{13}$$

Proof. By the definition, $\exists \tilde{s} \in S$, $\exists \tilde{r} \in R$, such that $\tilde{s}r = \tilde{r}r'$ and $\tilde{s}s = \tilde{r}s'$. Therefore,

$$\tilde{s}sr' = \tilde{r}s'r' = \tilde{r}r's' = \tilde{s}rs' = \tilde{s}s'r.$$

Unlike sometimes (mis)stated in the literature (e.g. [139],p.14), the commutative formula (13) (and variants of it) is inappropriate even for mildly noncommutative rings and even 2-sided Ore sets which are not in center. E.g. take the unital $\mathbb{C}$-algebra generated by two elements b and d with $bd = qdb$, where $\mathbb{C} \ni q \neq 1$. That algebra has no zero divisors. Let S be the 2-sided Ore set multiplicatively generated by b and d. Formula $b^{-1} = (db)^{-1}d$, and the criterion above would imply that $db = bd$ with contradiction.

For general R and S, formula (13) is actually not even an equivalence relation on $R \times S$.

10.2 General requirements on scheme-like theories. One wants to mimic several major points from the classical case. We first decide which geometric objects constitute the category $\mathcal{C}$ of affine schemes; then find a suitable larger geometric category $\mathfrak{Esp}$ of spaces, in the sense that it is equipped with a fully faithful functor $\mathcal{C} \hookrightarrow \mathfrak{Esp}$, where the objects in the image will be called here *geometric affine schemes*; finally there is a gluing procedure which assigns to a collection $\{\mathfrak{C}_h\}_{h \in H}$ of geometric affine schemes and some additional 'gluing' data Z, a space which may be symbolically denoted by $(\coprod_h \mathfrak{C}_h)/Z \in \mathfrak{Esp}$, together with canonical morphisms $\mathfrak{C}_h \to (\coprod_h \mathfrak{C}_h)/Z$ in $\mathfrak{Esp}$. For a fixed (type of) gluing procedure $\mathcal{G}$, and including all the isomorphic objects, one constructs this way a subcategory of **locally affine spaces** of type $(\mathcal{C}, \mathcal{G})$ in $\mathfrak{Esp}$.

Additional requirements and typical choices are in place.

10.2a Abstract affine schemes. Most often one deals with some monoidal category $\tilde{A} = (\mathcal{A}, \otimes, 1_{\tilde{A}})$. Then $\mathcal{C} = \mathrm{Aff}(\tilde{A}) := (\mathrm{Alg}(\tilde{A}))^{\mathrm{opp}}$ is the **category of affine schemes in** $\mathcal{A}$ i.e. the *opposite to the category of algebras (monoids) in* $\tilde{A}$. The basic example is the monoidal category of R-bimodules, where R is a **k**-algebra over a commutative ring **k**. The affine schemes in this category are given by **k**-algebra maps $f : R \to R'$ (making such R' an R-bimodule).

If $\mathcal{A}$ is *symmetric* via a symmetry τ where $\tau_{AB} : A \otimes B \to B \otimes A$, then one may consider only τ-commutative algebras (A, μ, η), i.e. for which $\mu \circ \tau = \mu$. τ-**commutative affine schemes** in $\mathcal{A}$ are the objects of the opposite of the category of τ-commutative algebras $\mathcal{C} = \mathrm{cAff}(\tilde{A}, \tau) := (\mathrm{cAlg}(\tilde{A}, \tau))^{\mathrm{opp}}$ (one often skips τ in the notation). Examples are (super)commutative affine schemes in a $\otimes$-category of **k**-modules and also the opposite to the category $\underline{\mathrm{cdga}}_{\mathbf{k}}$ of commutative differential graded **k**-algebras, which is important in recent 'derived algebraic geometry' program ([133, 134]).

10.2b Gluing for ringed topological spaces (and a version for local (l.) rings). A (locally) ringed space is a pair (X, F) consisting of a topological space X, and a ('structure') sheaf F of (l.) rings on X. A morphism of (l.) ringed spaces is a continuous map $f : X \to X'$ with a *comorphism* i.e. a map of sheaves of (l.) rings $F' \to f^* X'$ over X'. We obtain a category $\mathfrak{rSp}$ ($\mathfrak{lSp}$). Given a full subcategory $\mathcal{A}$ of $\mathfrak{rSp}$, considered as a category of (heuristic term here) "local models" we may consider all (l.) ringed space X for which there is a cover (in usual sense) $X = \cup_{\alpha} X_{\alpha}$ of underlying topological spaces and for each α an isomorphism $X_{\alpha}^0 \cong X_{\alpha}$ in $\mathfrak{rSp}$ ($\mathfrak{lSp}$) where X_{α}^0 is in $\mathcal{A}$. More abstractly, but equivalently, consider all families of morphisms $\{i_{\alpha} : X_{\alpha} \to X\}_{\alpha}$, which are 'covers by embeddings': topologically covers of X by families of monomorphisms (continuous, open and injective), and sheaf-wise isomorphisms on stalks. The intersections $X_{\alpha} \cap X_{\beta}$ represent fibred products $X_{\alpha} \times_X X_{\beta}$ in $\mathfrak{rSp}$. Spaces glued from objects in $\mathcal{A}$ are nothing but the colimits of the diagrams of the type $\coprod_{\alpha\beta} Y_{\alpha\beta} \rightrightarrows \coprod Y_{\alpha}$ where each of the morphisms $Y_{\alpha\beta} \to Y_{\alpha}$ and $Y_{\alpha\beta} \to Y_{\beta}$ are embeddings, and $Y_{\beta} \in \mathrm{Ob}\,\mathcal{A}$. There is a natural condition on $\mathcal{A}$: each X in $\mathcal{A}$ as a topological space has a basis of topology made out of (spaces of) some family of objects in $\mathcal{A}$ (or isomorphic to them); and this family may be chosen so that the restrictions of the structure sheaf agree. In the theory of schemes, affine schemes are such models: affine subschemes make a basis of topology, but not every open subset is affine; nor their intersections. Still the intersections and colimits exist from the start, in our ambient category of ringed spaces which is big enough.

10.2c Grothendieck (pre)topologies (G.(p)t.) and gluing in commutative algebraic geometry ([85, 143]). Schemes are glued from affine schemes in the Zariski topology (which may be considered both as an ordinary topology and a G.(p)t.); useful generalizations (e.g. algebraic spaces) in flat and étale G.(p)t. etc. A **sieve** is an assignment of a collection $J(R)$ of morphisms in $\mathcal{C}$ ([85]) with target R, such that if the target of f is $J(R)$ and $g : R \to R'$ then $g \circ f \in J(R')$. A G.t. is a collection of sieves $J(R)$ with target R for each R in $\mathrm{Ob}\,\mathcal{C}$, such that if $f \in J(R)$ and $g : R \to R'$ then $g \circ f \in J(R')$), satisfying some axioms ([85]). A G.pt. in a category $\mathcal{C}$ with fibred products is a class $\mathcal{T}$ of families $\{U_\alpha \to U\}_\alpha$ of morphisms (with one target per family), such that $\{\mathrm{Id} : U \to U\} \in \mathcal{T}$; if $\{f_\alpha : U_\alpha \to U\}_\alpha \in \mathcal{T}$ and $\forall \alpha \ \{g_{\alpha\beta} : U_{\alpha\beta} \to U_\alpha\}_\beta \in \mathcal{T}$ then $\{f_\alpha \circ g_{\alpha\beta}\}_{\alpha,\beta} \in \mathcal{T}$; and finally if $\{f_\alpha : U_\alpha \to U\}_\alpha \in \mathcal{T}$ and $g : V \to U$ is a morphism then $\{g^*(f_\alpha) : V \times_U U_\alpha \to V\} \in \mathcal{T}$. Elements of $\mathcal{T}$ are called **covers** (in $\mathcal{T}$), and the pair $(\mathcal{C}, \mathcal{T})$ a site. To any (ordinary) topological space X one associates a "small" site $\mathfrak{Duv}_X$: objects are open subsets in X; morphisms are inclusions; $\{U_\alpha \hookrightarrow U\}_\alpha$ is a cover if $\cup_\alpha U_\alpha = U$.

A presheaf F of sets on X is a functor $F : (\mathfrak{Duv}_X)^{\mathrm{op}} \to \mathrm{Sets}$; a presheaf F on any site $(\mathcal{C}, \mathcal{T})$ with values in a category $\mathcal{D}$ with products is a functor $F : \mathcal{C}^{\mathrm{op}} \to \mathcal{D}$. Given a cover $\{U_\alpha \to U\}_\alpha \in \mathcal{T}$, there are two obvious embeddings $\prod U_\alpha \times_U U_\beta \rightrightarrows \prod_\alpha U_\alpha$. A presheaf is a sheaf on $(\mathcal{C}, \mathcal{T})$ if $\{U_\alpha \to U\}_\alpha \in \mathcal{T}$ if for every such cover the induced diagram $F(U) \longrightarrow \prod_\alpha F(U_\alpha) \rightrightarrows \prod_{\alpha\beta} F(U_\alpha \times_U U_\beta)$ is an equalizer diagram.

For gluing, one again needs some bigger ambient category (or, instead, some universal construction). Our local models are now (commutative) affine schemes $\mathrm{Aff} := \mathrm{Aff}(\mathrm{Ab})$ with a G.pt. $\mathcal{T}$. The Yoneda embedding $X \mapsto X\check{} := \mathrm{Aff}(?, X)$ is a fully faithful functor from Aff into the category $\mathrm{PFas}\,(\mathrm{Aff})$ of presheaves of sets on Aff. One typically deals with *subcanonical* G.t. which means that the presheaves in the Yoneda image (representable functors) are sheaves. As in the case of ordinary topologies, to construct the global locally $\mathcal{T}$-affine spaces, one needs colimits of certain diagrams of the form $\prod_{\alpha\beta} V_{\alpha\beta} \rightrightarrows \prod_\alpha U_\alpha$, where, in the addition, the colimit cone $\prod U_\alpha \to U$ corresponds to a cover in $\mathcal{T}$. As for example, the nonseparated schemes in Zariski topology, some locally $\mathcal{T}$-affine spaces may not be produced this way with $V_{\alpha\beta}$ being in Aff. Similar problems for biflat covers by localizations in noncommutative geometry are known (lack of compatibility of localizations; nonsemiseparated covers). Furthermore, one needs to extend the notion of $\mathcal{T}$-covers to the target category of sheaves. We hope that the reader sees at this point the meaning of this abstract machinery. We won't proceed with the full construction. Namely, in the commutative case, it is usually replaced by equivalent constructions. For example, to construct the algebraic spaces, one usually does not glue affine schemes "over inter- sections" but rather starts with an equivalence relation in the category of all separated schemes. In the noncommutative case, G.t.'s are elaborated in [69, 110] and partly in [115]. There is also an approach to G.t.'s and quasicoherent sheaves for noncommutative affine schemes by ORLOV ([99]), utilizing ringed sites and sheaves of groupoids, but implicit application to the construction of noncommutative locally affine spaces is not given.

10.3 Though we assume that the reader has been exposed to the com- mutative scheme theory, we here sketch the basic construction of an affine scheme, as a part of a widely-known easy generalization to pairs of the form (noncommutative ring, central subring), cf. [106].

A left ideal $\mathfrak{p} \subset R$ is **prime** if for any two left ideals I, J, if $IJ \subset \mathfrak{p}$ then $I \subset \mathfrak{p}$ or $J \subset \mathfrak{p}$. A left ideal $\mathfrak{p} \subset R$ is **completely prime** if $fg \in \mathfrak{p}$ then $f \in \mathfrak{p}$ or $g \in \mathfrak{p}$; equivalently $R/\mathfrak{p}$ is a domain; or $R\backslash\mathfrak{p}$ is multiplicative. Each completely prime ideal is prime: otherwise one could find $f \in I\backslash\mathfrak{p}$ and $g \in J\backslash\mathfrak{p}$, such that $fg \in IJ \subset \mathfrak{p}$ with contradiction. If R is commutative the converse holds, as one can see by specializing the definitions to the principal left ideals $I = Rf$, $J = Rg$, $IJ = RfRg = Rfg$.

Consider the category $\mathcal{R}$, whose objects are pairs (R, C), of a unital ring R and a central subring $C \subset Z(R)$; and morphisms $(R, C) \to (R', C')$ are maps of rings $\phi : R \to R'$ such that $\phi(C) \subset C'$.

$\operatorname{Spec} C$ is the set of all prime ideals $\mathfrak{p}$ of C. For any ideal $I \subset C$, define $V(I) \subset \operatorname{Spec} C$ as the set of all $\mathfrak{p} \subset C$, such that $I \subset \mathfrak{p}$. Sets of the form $V(I)$ depend only (contravariantly with respect to inclusions) on the radical $\sqrt{I}$ (the intersection of all prime ideals containing I) and satisfy the axioms of antitopology. Complements of such sets hence form a topology on $\operatorname{Spec} C$, called **Zariski topology**. **Principal open sets** are the sets of the form $U_f = V((f))$, where (f) is the (principal) ideal generated by $f \in C$. They make a basis of Zariski topology, i.e., any Zariski open set is a union of sets of that form. Open sets and inclusions form category $\mathfrak{Ouv}_C$. Principal open sets and inclusions form its full subcategory $\mathfrak{Ouv}\mathfrak{P}_C$.

Define $\mathcal{O}'_C(U_f) := C[f^{-1}]$ and $\mathcal{O}'_{R,C}(U_f) := R[f^{-1}]$. Every inclusion $U_f \hookrightarrow U_g$ induces the unital ring maps $\phi_{f,g,i} : \mathcal{O}'_i(U_g) \to \mathcal{O}'_i(U_f)$, $i \in \{C, (R, C)\}$. Hence we have contravariant functors $\mathcal{O}'_i : \mathfrak{Ouv}\mathfrak{P}_C \to \underline{Rings}$. Natural inclusions $\mathbf{in}_f : C[f^{-1}] \to R[f^{-1}]$ form a natural transformation, i.e., $\mathcal{O}'_C$ is a subfunctor of $\mathcal{O}'_{C,R}$. Functors $\mathcal{O}'_i$ extend naturally to functors $\mathcal{O}_i : \mathfrak{Ouv}_C \to \underline{Rings}$ which are sheaves, and this requirement fixes sheaves

$\mathcal{O}_i$ uniquely up to isomorphism of sheaves. Namely, represent any open set U as the union $\cup U_f$ of a family $\mathcal{U}$ of (some or all) $U_f \subset U$. Define a diagram $\Delta(\mathcal{U})$ in *Rings* as follows. Vertices are of the form $\mathcal{O}_i(U_f)$ and of the form $\mathcal{O}_i(U_{fg}) = \mathcal{O}_i(U_f \cap U_g)$ where $U_f, U_g \in \mathcal{U}$, and the arrows are $\phi_{f,fg,i}$. Then $\mathcal{O}_i(U)$ has to be isomorphic to the inverse limit of the diagram $\Delta(\mathcal{U})$ and may be consistently set so. Moreover the natural transformation **in** extends making $\mathcal{O}_C$ a subfunctor of $\mathcal{O}_{C,R}$. In fact, $\mathcal{O}_{C,R}$ is an algebra in the monoidal category of (2-sided) $\mathcal{O}_C$-modules. All the stalks $(\mathcal{O}_C)_{\mathfrak{p}}$ of $\mathcal{O}_C$ are local rings, namely the localizations $C_{\mathfrak{p}} := C[\{f^{-1}\}_{f \notin \mathfrak{p}}]$ 'at prime ideal $\mathfrak{p}$'.

Let $\psi : R \to R'$ be a map of unital rings. The inverse image $\psi^{-1}(\mathfrak{p})$ of a completely prime ideal is completely prime. Let $\psi : (R, C) \to (R', C')$ be a morphism in $\mathcal{R}$, and let map $\phi : C \to C'$ agrees with ψ on C. One has map $\phi^* : \operatorname{Spec} C' \to \operatorname{Spec} C$ given by $\phi^* : \mathfrak{p} \mapsto \phi^{-1}(\mathfrak{p})$. If $U \subset \operatorname{Spec} C'$ is open, then $\phi^{-1}(U) \subset \operatorname{Spec} C$ is open as well, because $\phi^*(V(I)) = V(\phi^{-1}(I))$ for each ideal $I \subset C'$. Hence, ϕ^* is continuous. If $g \notin \phi^{-1}(\mathfrak{p})$ then $\phi(g) \notin \mathfrak{p}$. Thus all elements $g \in C$, newly inverted in $C_{\phi^{-1}(\mathfrak{p})}$ (and $R_{\phi^{-1}(\mathfrak{p})}$) are also invertible in $C'_{\mathfrak{p}}$ (and $R'_{\mathfrak{p}}$). Hence, by the universality of localization, one has a unique map $\psi_{\mathfrak{p}} : (R_{\phi^{-1}(\mathfrak{p})}, C_{\phi^{-1}(\mathfrak{p})}) \to (R'_{\mathfrak{p}}, C'_{\mathfrak{p}})$ such that $\psi_{\mathfrak{p}} \circ \iota_{\phi^{-1}(\mathfrak{p})} = \iota_{\mathfrak{p}} \circ \psi$. Define $\phi_U : \mathcal{O}_{R,C}(\phi^* U) \to \mathcal{O}_{R',C'}(U)$ by $(\phi_U(r))_{\mathfrak{p}} = \psi_{\mathfrak{p}}(r_{\phi^{-1}(\mathfrak{p})})$. One has to check that $\phi_U(r)$ is indeed in $\mathcal{O}_{R',C'}(U)$, i.e., that $\mathfrak{p} \mapsto \psi_{\mathfrak{p}}(r_{\phi^{-1}(\mathfrak{p})})$ is indeed a section. For this consider $U_f \subset U$ affine, i.e., $U_f = \{\mathfrak{p} \mid f \notin \mathfrak{p}\}$ and $\phi^* U_f := \{\phi^{-1}(\mathfrak{p}), \mathfrak{p} \not\ni f\}$. An argument as above gives map $\psi_f : \mathcal{O}_{R,C}(\phi^* U_f) \to \mathcal{O}_{R',C'}(U_f)$ satisfying $\psi_f \circ \iota_{\phi^{-1}(U_f)} = \iota_{U_f} \circ \psi$. It is easy to check then that ψ_f induces $\psi_{\mathfrak{p}}$ in stalk over $\mathfrak{p} \in U_f$. As a result, we obtain a map $\psi^{\sharp} : \phi_* \mathcal{O}_{R,C} \to \mathcal{O}_{R',C'}$ of sheaves over $\operatorname{Spec} C'$.

10.3a Let $\mathrm{l}\mathfrak{Sp}_2$ be a category whose object are locally ringed spaces $(X, \mathcal{O})$ in $\mathrm{l}\mathfrak{Sp}$ together with a sheaf $\mathcal{O}^{nc}$ of noncommutative algebras in the category of $\mathcal{O}$-modules. A morphism $\psi : (X, \mathcal{O}_X, \mathcal{O}_X^{nc}) \to (Y, \mathcal{O}_Y, \mathcal{O}_Y^{nc})$ in $\mathrm{l}\mathfrak{Sp}_2$ is a morphism $\psi^c : (X, \mathcal{O}_X) \to (Y, \mathcal{O}_Y)$ in $\mathrm{l}\mathfrak{Sp}$, together with a map of sheaves of $\mathcal{O}_X$-modules $\psi^{\sharp} : \mathcal{O}_Y^{nc} \to \psi^* \mathcal{O}_X^{nc}$.

We have above constructed a functor $\operatorname{Spec}_2 : \mathcal{R} \to \mathrm{l}\mathfrak{Sp}_2$.

10.3b In fact, $\mathcal{O}^{nc}$ is in a smaller category of *quasicoherent sheaves of $\mathcal{O}$-modules* (shortly: quasicoherent modules). A presheaf F of $\mathcal{O}$-modules on a ringed space $(X, \mathcal{O})$ is quasicoherent (EGA 0.5.1.1) if $\forall x \in X \ \exists U^{\text{open}} \ni x$ with an exact sequence $\mathcal{O}^I \to \mathcal{O}^J \to F \to 0$ where $\mathcal{O}^I$, $\mathcal{O}^J$ are free $\mathcal{O}$-modules (of possibly infinite rank). If the ringed space is locally $\mathcal{T}$-affine for some Grothendieck topology $\mathcal{T}$ on the category of commutative affine schemes, then one may take U affine, and an equivalent definition of quasicoherence is that for any pair of affine open subsets $W \subset V \subset X$,

$$F(W) = \mathcal{O}(W) \otimes_{\mathcal{O}(V)} F(V). \qquad (14)$$

The noncommutative structure sheaf $\mathcal{O}^{nc}$ of $\mathrm{Spec}_2(R, C)$ is a quasicoherent $\mathcal{O}$-module. For presheaves of $\mathcal{O}^{nc}$ left modules one may use the same formula (14). For bimodules one distinguishes left and right quasicoherence [67, 78, 148] (in the right hand version the tensoring with $\mathcal{O}(W)$ in formula (14) is from the RHS instead). Formula (14) is nothing else but the formula for a localization functor $Q^* : F(V) \mapsto F(W)$ from $\mathcal{O}(V)$-modules to $\mathcal{O}(W)$-modules. In this case, both Q^* and its right adjoint Q_* are exact functors. Ring theorists call such localizations of full module categories **perfect localizations** ([62]). Equivalently, the canonical forgetful functor from the localized category to the category of modules over the localized ring is an equivalence of categories.

F. van Oystaeyen ([139]) defines quasicoherent presheaves on the lattice of hereditary torsion theories (= localizations where Q^* is exact and the torsion subcategory is coreflective) over a $\mathbb{Z}_{\geq 0}$-graded rings, by using the appropriate localization instead of tensoring. For affine case see [113], I 6.0.3 (b) and I 6.2. One has theorems on gluing of such modules using Barr-Beck lemma.

D. Orlov ([99]) defines quasicoherent (pre)sheaves on X where X is a presheaf of sets (in particular, by Yoneda, on any object) on a given ringed site $(\mathcal{C}, \mathcal{T})$.

10.4 Now we will quote two theorems. A theorem of Deligne shows that in the context of commutative affine schemes, a formula which can be recognized as a localization at a Gabriel filter (see below), describes the behaviour of the category of quasicoherent sheaves under passage to a not necessarily affine subset. Hence the "noncommutative localization" is already seen there! The theorem may be proved directly, and we suggest to the reader to at least convince oneself that the RHS formula is indeed an A-module. "Deligne theorem" ([59]) in fact, cf. ([60], Appendix)[2], has been originally inferred from the Gabriel theorem below – the general statement that quasicompact open subsets of (nonaffine) schemes correspond always to exact localizations of abelian categories; combined with the Gabriel's formulas on how such localizations look like ([42]).

10.4a Deligne's theorem. Let $X = \mathbf{Spec}\,A$ be an affine Noetherian scheme, i.e. A is a commutative Noetherian ring. Let U be a Zariski open set (not necessarily affine!), and I an ideal such that $V(I) = X \backslash U$. Let

---

[2] I thank Prof. Hartshorne for an email on the history of the result.

M be an A-module, and $\tilde{M}$ the corresponding quasicoherent $\mathcal{O}_X$-module. Denote by $\tilde{M}|_U$ its restriction to U. Then

$$\Gamma(\tilde{M}|_U) \cong \lim_{n\to\infty} \operatorname{Hom}_A(I^n, M).$$

This isomorphism of A-modules is natural in M.

10.4b (P. Gabriel [42], VI.3) Let $(X, \mathcal{O}_X)$ be a scheme, and U an open subset of X, such that the canonical injection $j : U \hookrightarrow X$ is quasicompact. Let $Q_U^* : \mathfrak{Qcoh}_X \to \mathfrak{Qcoh}_U$ be a functor which associates to every quasicoherent $\mathcal{O}_X$-module M the restriction $M|_U$ of M to subscheme $(U, \mathcal{O}_X|_U)$. Functor Q_U^* canonically decomposes as $\mathfrak{Qcoh}_X \to \mathfrak{Qcoh}_X/\operatorname{Ker} Q_U^* \overset{\cong}{\to} \mathfrak{Qcoh}_U$ into the canonical projection onto the quotient category by $\operatorname{Ker} Q_U^*$ and an isomorphism.

11 Ring maps vs. module functors

11.1 As usual, we write $_RM$ when we want to emphasize that M is (understood as) a left R-module; $R - \operatorname{Mod}$ is the category of left R-modules. Let $f : R \to S$ be any map of (not necessarily unital) rings. We have the following functors induces by map f:

- **extension of scalars** $f^* : R - \operatorname{Mod} \to S - \operatorname{Mod}$, $M \mapsto S \otimes_R M$;

- **restriction of scalars** (forgetful functor) $f_* : S - \operatorname{Mod} \to R - \operatorname{Mod}$, $_SM \mapsto {_RM}$;

- $f^! : R - \operatorname{Mod} \to S - \operatorname{Mod}$, $M \mapsto \operatorname{Hom}_R({_RS}, M)$.

Denote $F \dashv G$ when functor F is left adjoint to functor G. Easy fact: $f^* \dashv f_* \dashv f^!$. Hence f^* is left exact, $f^!$ right exact and f_* exact.

11.2 It is of uttermost importance to have in mind the geometrical picture of this situation in the case when R and S are commutative and unital. Denote by $\mathfrak{lSp}$ the category of locally ringed spaces. An object $(X, \mathcal{O}_X) \in \mathfrak{lSp}$ is a pair of a topological space X and a sheaf of commutative local rings $\mathcal{O}_X$ over X; and a morphism is a pair $(f^o, f^\sharp)$ where $f^o : X \to Y$ is a map of topological spaces and a 'comorphism' $f^\sharp : f^\bullet \mathcal{O}_Y \to \mathcal{O}_X$ is a map of sheaves of local rings over X, and $f^\bullet$ a sheaf-theoretic pullback functor. There is a contravariant functor Spec : $\underline{CommRings} \to \mathfrak{lSp}$ assigning spectrum to a ring. A map $f : R \to S$ is replaced by a map of locally ringed spaces

$$\operatorname{Spec} f = (f^0, f^\sharp) : \operatorname{Spec} S \to \operatorname{Spec} R.$$

The comorphism $f^\sharp$ is important: e.g. if $f^o : X \hookrightarrow Y$ is an inclusion, the difference between a subvariety $(X, \mathcal{O}_X)$ and, say, its n-th infinitesimal neighborhood in $(Y, \mathcal{O}_Y)$, may be expressed by a proper choice of $f^\sharp$.

11.3 After geometrizing rings, one may proceed to geometrize modules. The basic fact here is the affine SERRE's theorem establishing a correspondence $M \leftrightarrow \tilde{M}$ between the R-modules and quasicoherent sheaves of $\mathcal{O}_X$-modules, for $X = \operatorname{Spec} R$. This correspondence is an equivalence of categories $R - \operatorname{Mod} \leftrightarrow \mathfrak{Qcoh} X$. Using this equivalence of categories, functors f^* and f_* may be rephrased as additive functors

$$f^* : \mathfrak{Qcoh} X \to \mathfrak{Qcoh} Y, \quad f_* : \mathfrak{Qcoh} Y \to \mathfrak{Qcoh} X;$$

and moreover, these can be defined for any morphism $f = (f^o, f^\sharp)$ between locally ringed spaces. In this wider context, functor f^* is called the **inverse image functor** of f, given by $f^*\mathcal{F} := \mathcal{O}_Y \otimes_{\mathcal{O}_X} f^\bullet \mathcal{F}$ where $f^\bullet \mathcal{F}$ is usual, sheaf theoretic, pullback of sheaf $\mathcal{F}$ via f^o. The restriction of scalars functor then generalizes to the **direct image functor** for sheaves which is on presheaf level given by

$$f_*\mathcal{F}(U) := \mathcal{F}(f^{-1}(U)),$$

and which sends $\mathcal{O}_X$-modules to $\mathcal{O}_Y$-modules via $\mathcal{O}_Y$-action given by the composition of $f^\sharp \times \operatorname{id}$ and the $\mathcal{O}_X$-action. Functors f^* and f_* are naturally defined between the bigger categories, $\mathcal{O}_X - \operatorname{Mod}$ and $\mathcal{O}_Y - \operatorname{Mod}$, where they simply preserve the quasicoherence.

11.4 Functor f_* (in all settings above) is a right adjoint to f^*, hence it is left exact and the inverse image functor f^* is right exact. This suggests that a pair of adjoint additive functors between abelian categories may be viewed as (coming from) a morphism in geometrical sense. Actually this point of view appears fruitful. On the other hand, surely the choice of a functor in its equivalence class is not essential; and the emphasis on the inverse image vs. direct image functor is the matter of choice as well.

Given two abelian categories $\mathcal{A}$, $\mathcal{B}$, (equivalent to small categories) a **morphism** $f : \mathcal{B} \to \mathcal{A}$ is an equivalence class of right exact additive functors from $\mathcal{A}$ to $\mathcal{B}$. An **inverse image functor** $f^* : \mathcal{A} \to \mathcal{B}$ of f is simply a representative of f, which is usually assumed to be made.

An additive functor $f^* : \mathcal{B} \to \mathcal{A}$ between abelian categories is ([109])

- **continuous** if it has a right adjoint, say f_*;
- **flat** if it is continuous and exact;
- **almost affine** if it is continuous and its right adjoint f_* is faithful and exact;

• **affine** if it is almost affine, and its right adjoint f_* has its own right adjoint, say $f^!$, cf. **11.1**.

Morphism f is continuous (flat, almost affine, affine) if its inverse image functor f^* is. Some authors say that a functor F is continuous if it has a left adjoint instead, which means that they view F as a direct image functor f_* of some continuous morphism f. On the other hand, a continuous morphism is **coflat** if its direct image is exact, and **biflat** if it is flat and coflat. Usually one equips the categories with distinguished objects ("structure sheaves"); then the morphisms respect the distinguished object.

11.4a ROSENBERG ([109]) introduced an abstract notion of a **quasi-compact relative noncommutative scheme** $(\mathcal{A}, \mathcal{O})$ over a category $\mathcal{C}$ as an abelian category $\mathcal{A}$ with a distinguished object $\mathcal{O}$, finite biflat affine cover by localizations $Q_\lambda^* : \mathcal{A} \to \mathcal{B}_\Lambda$, with a continuous morphism g from $\mathcal{A}$ to $\mathcal{C}$ (think of it as a direct image of a morphism $X \to \operatorname{Spec} \mathbf{k}$) such that each $g_* \circ Q_{\lambda*} : \mathcal{B}_\lambda \to \mathcal{C}$ is affine. This guarantees application of many usual geometric procedure (for basic cohomological needs one does not need $f^!$). Such 'schemes' can be related to some spectra and some Grothendieck topologies on Aff(Ab) ([115]). Quotient spaces for comodule algebras over Hopf algebras may be sometimes constructed as nonaffine noncommutative schemes [120].

11.5 If R and S are rings and ${}_S B_R$ a $S - R$-bimodule, then the functor $f^B : M \mapsto {}_S B_R \otimes_R M$ is a right exact functor from $R - \operatorname{Mod}$ to $S - \operatorname{Mod}$. If $S = \mathbb{Z}$ then B_R is called flat right R-module if $f^B : M \mapsto B_R \otimes_R M$ is exact.

Proposition. (WATTS [144], EILENBERG [36]) *(i) Let R be a (not necessarily unital) ring, and f^* a flat endofunctor in the category of nonunital left R-modules. Then f^* is equivalent to the functor*

$$M \mapsto f^*(R_1) \otimes_R M,$$

where R_1 is the corresponding unital ring. In particular, $f^*(R_1)$ is flat as a left R-module.

(ii) Let R be a unital ring, and f^* a flat endofunctor in the category $R - \operatorname{Mod}$ of unital left R-modules. Then f^* is equivalent to the functor $M \mapsto f^*(R) \otimes_R M$.

(iii) Let $f^* : R - \operatorname{Mod} \to S - \operatorname{Mod}$ be a continuous functor. Then there is a $S - R$-bimodule ${}_S B_R$ such that f^* is equivalent to the functor $M \mapsto B \otimes_R M$.

Notice that when applied to localizations word "flat" is here used in the sense that Q^* is flat (weaker), and in some other works means that

$Q := Q_*Q^*$ is flat. The latter case, for $R - \mathrm{Mod}$, is the situation of Watts theorem, and it is known under the name of **perfect localization**. Equivalently ([103]), the canonical forgetful functor from the localized category $(R - \mathrm{Mod})/\mathrm{Ker}Q$ to the modules over the localized ring $(QR) - \mathrm{Mod}$ is an equivalence of categories.

11.6 A **bicategory** (= weak 2-category) $\mathcal{A}$ consists of

(1) a class $\mathrm{Ob}\,\mathcal{A}$ of *objects* ('1-cells');

(2) for each pair of objects A, B a small category $\mathcal{A}(A, B)$; The objects of $\mathcal{A}(A, B)$ are called *arrows* (or 'morphisms' or '1-cells'), and the morphisms in $\mathcal{A}(A, B)$ are called *2-cells*;

(3) for each triple of objects A, B, C a bifunctor ('composition map')

$$\Phi_{A,B,C} : \mathcal{A}(A, B) \times \mathcal{A}(B, C) \to \mathcal{A}(A, C),$$

(4) for each object $A \in \mathcal{A}$, an arrow $1_A \in \mathcal{A}(A, A)$ ('identity arrow'); together with natural equivalences

$$a_{ABCD} : \Phi_{ACD}(\Phi_{ABC} \times \mathrm{Id}) \Rightarrow \Phi_{ABD}(\mathrm{Id} \times \Phi_{BCD}),$$
$$\lambda_{ABC} : \mathrm{Id} \times \Phi_{ABC} \Rightarrow \Phi_{ABC}, \qquad \rho_{ABC} : \Phi_{ABC} \times \mathrm{Id} \Rightarrow \Phi_{ABC},$$

satisfying some natural conditions. If $a_{ABCD}, \rho_{ABC}, \lambda_{ABC}$ are all identities, then the bicategory is called (strict) 2-category.

We omit further details in the definition, and sketch just the most important example to us: the bicategory of rings and bimodules $\mathfrak{Bim}$. Beforehand, notice that from any monoidal category $\tilde{\mathcal{C}} = (\mathcal{C}, \otimes, 1_{\mathcal{C}}, a, l, r)$ we can tautologically form a bicategory $\Sigma\tilde{\mathcal{C}}$ with one object $0 := \mathcal{C}$, and with $\Sigma\tilde{\mathcal{C}}(0, 0) := \tilde{\mathcal{C}}$ and with the composition $M \circ_{\Sigma\tilde{\mathcal{C}}} N := M \otimes N$; one further defines $a(M, N, P) := a_{0,0,0}(M, N, P)$, $l_A := \lambda_{0,0}(A)$ and $r_A := \rho_{0,0}(A)$. The definitions may be reversed to form a monoidal category out of any bicategory with a distinguished object 0 (forgetting the other objects). Hence monoidal categories may be viewed as bicategories with a single object; the notion of being 'strict' in both senses agrees as well.

The objects of $\mathfrak{Bim}$ are unital rings. For any $R, S \in \mathrm{Ob}\,\mathfrak{Bim}$ let $\mathfrak{Bim}(R, S)$ be the category of R-S-bimodules and (R, S)-bilinear mappings between them. The composition

$$\mathfrak{Bim}(R, S) \times \mathfrak{Bim}(S, T) \to \mathfrak{Bim}(R, T)$$

is given by the tensor product bifunctor $(M, N) \mapsto {}_RM_S \otimes_S {}_SN_T$, and the rest of the data is obvious. These data indeed define a bicategory.

The invertible 1-cells of 𝔅im are called MORITA equivalences. It has been observed in various applications of noncommutative geometry, for instance in physics, that Morita invariance is a common feature of natural constructions.

The Eilenberg-Watts' theorem identifies bimodules with flat functors. As a pair of adjoint functors, they resemble geometric morphisms among topoi. M. VAN DEN BERGH ([136]) defines a (generalized) bimodule to be any pair of adjoint functors between Grothendieck categories. He also considers right exact functors as so-to-say weak (versions of) bimodules. Some situations, for example the duality for coherent sheaves, involve functors for which the right and left adjoints ([60]). coincide. They are known as Frobenius functors ([25]). In the spirit of Van den Bergh's terminology, PAPPACENA calls *Frobenius bimodules* those adjoint pairs $F \dashv G$ where F is simultaneously left and right adjoint of G. In the abstract homotopy theoretic setting, the existence of two-sided adjoints is studied with appropriate (Bousfield-type) localization arguments ([90, 96]).

One of motivations for this ([96]) is to extend the GROTHENDIECK duality theory for coherent sheaves on varieties to $\mathcal{D}$-modules. This may be viewed as an example of noncommutative geometry. Namely, the role of the structure sheaf $\mathcal{O}$ is played by the sheaf $\mathcal{D}$ of regular differential operators which is a sheaf of noncommutative $\mathcal{O}$-algebras (cf. [149, 102] for the viewpoint of noncommutative geometry at $\mathcal{D}$-modules). In triangulated categories, the Serre-Grothendieck duality is axiomatized as an existence of so-called SERRE functor ([18, 19, 20, 100]), with applications at the borderline between the commutative and noncommutative geometry.

It is a remarkable observation ([73]), that the noncommutative geometry via operator algebras, could be also organized around similar bicategories. Roughly speaking, operator algebras (C^*-algebras; von Neumann algebras respectively) are 0-cells, appropriate bimodules as 1-cells (Hilbert bimodules; correspondences), and bimodule morphisms as 2-cells; while the monoidal products of 1-cells are specific tensor products which became prominent earlier in noncommutative geometry a la Connes, and related K-theories (Rieffel interior tensor product of Hilbert bimodules; Connes fusion product). Invertible 1-cells are called **Morita equivalences** in all these cases. There are also analogues concerning regular bibundles over Lie groupoids, and also analogues in symplectic and Poisson geometry. The latter may be viewed as a quasiclassical limit of noncommutative geometry. For Morita equivalence of Poisson manifolds and a similar notion of symplectic dual pairs see [74, 117].

12 Ore localization in filtered rings

After prerequisites on filtrations, we prove some general lemmas on localizations in filtered rings, mainly easy generalizations of some statements quoted without proof in [81] and in the manuscript [82][3].

We focus on *'negative'* filtrations. The main application in mind is the noncommutative deformation of commutative objects. Such filtrations arise from expanding the algebra operations in power series in the deformation parameter $q - 1$ ([4, 67, 78]). There is a more refined technique in algebraic analysis, *algebraic microlocalization*, see [137, 139] and the references in [137]. *'Positive'* filtrations involve different techniques than ours. In the study of noncommutative projective algebraic geometry there is a (negative) filtration related to deformation, but also a positive grading corresponding to the study of projective schemes. The latter grading may be refined to $\mathbb{Z}^{\times n}$-grading or grading with respect to the weight lattice P, as in the study of the quantum flag varieties ([81, 82]). If the latter, positive, grading complicates the picture, one restricts attention to homogeneous Ore sets only.

12.1 Given a (typically abelian) category $\mathcal{C}$, e.g. the category of modules over a ring $\mathbf{k}$, a $\mathbb{Z}$-**filtration** on an object M in $\mathcal{C}$ is a nested (unbounded) sequence of subobjects $F_*M = \{\cdots \subset F_{n-1}M \subset F_nM \subset F_{n+1}M \cdots \subset M\}$. A $\mathbb{Z}$-filtered object in $\mathcal{C}$ an object with a $\mathbb{Z}$-filtration on it. All filtrations in this article are assumed to be **exhaustive** i.e. the supremum subobject $\sup_{n\in\mathbb{Z}} F_nM$ exists and equals M (e.g. for modules $M = \cup_n F_nM$).

Let $M = \cup_{n\in\mathbb{Z}} F_nM$ be a filtered $\mathbf{k}$-module. The **degree** $d(m)$ of an element $m \in M$ is the smallest integer n, if it exists (otherwise $-\infty$), such that $m \in F_nM$ and $m \notin F_{n-1}M$. Conversely, if $d : M \to \{-\infty\} \cup \mathbb{Z}$ is subadditive $d(a + b) \leq d(a) + d(b)$, and $d(0) = -\infty$, then d is the degree function of a unique exhaustive filtration on M, indeed the one where $a \in F_nM$ iff $d(a) \leq n$. An (exhaustive) filtration is **separated** if $\cap_n F_nM = 0$. Then $d(m)$ is finite for all $m \neq 0$. *This will be our assumption from now on.*

12.1a A $\mathbb{Z}$-**filtered k-algebra** is a $\mathbf{k}$-algebra (E, μ) with a filtration F_*E on E as a $\mathbf{k}$-module where the multiplication μ restricted to $F_nE \times F_mE$ takes values within $F_{n+m}E$, for all n, m. This obviously generalizes to algebras in any monoidal category $\tilde{\mathcal{C}} := (\mathcal{C}, \otimes, 1_{\mathcal{C}})$ (notice that the resulting notion is different than if we consider these algebras as auxiliary *objects* in an abstract category $\mathcal{C}'$ of algebras, when **12.1a** applies, rather than as *algebras* in a *monoidal* category $\tilde{\mathcal{C}}$). For $\mathbf{k} = \mathbb{Z}$ we talk about $\mathbb{Z}$-**filtered rings**.

Given a filtered $\mathbf{k}$-algebra (F_*E, μ) a $\mathbb{Z}$-**filtered F_*E-module** is a $\mathbb{Z}$-

---

[3]I thank VALERY LUNTS for introducing me to this subject and sharing his notes.

filtered **k**-module F_*M such that F_nE acting on F_mM takes values within $F_{n+m}M$ for all n and m. In particular, F_*E is a $\mathbb{Z}$-filtered module over itself.

Given a filtered **k**-algebra $E = \cup_n F_nE$, an **associated graded algebra** is the $\mathbb{Z}$-graded **k**-module $\operatorname{gr} E = \oplus_n (\operatorname{gr} E)_n := \oplus_n F_nE/F_{n-1}E$ with the multiplication defined as follows. The **symbol map** $^-: E \to \operatorname{gr} E$, $e \mapsto \bar{e}$, maps e to the class $\bar{e}$ of e in $F_{d(e)}E/F_{d(e)-1}E$. An element $c \in \operatorname{gr} E$ is in the image of the symbol map iff c is homogeneous. For $c = \bar{e}, c' = \bar{e}'$, the formula $cc' := \bar{e}\bar{e}' := \overline{ee}'$ does not depend on the choice of e, e'. Therefore it defines a binary operation. The additive extension of this operation is the associative multiplication on $\operatorname{gr} E$.

It is always true $d(e+e') \leq \max\{d(e), d(e')\}$, with equality if $d(e) \neq d(e')$. On the other hand, if $d(e) = d(e')$ then $d(e+e')$ may be in general anything less or equal $d(e)$, as e and e' may cancel in several of the top degrees. Consequently the symbol map is *not* additive. However...

12.2 *...if $\operatorname{gr} E$ is a domain, then $d(ab) = d(a)d(b)$, hence the symbol map $E \to \operatorname{gr} E$ is multiplicative.*

12.3 For any subset $S \subset E$ not containing 0, we can always define a filtration on the set $S \times E$, by formula $d(s, e) := d(e) - d(s)$. If S is left Ore, the localized ring $S^{-1}E$ may be constructed as in **5.6**, as certain quotient of $S \times E$. Hence we have a filtration on $S^{-1}E$ as a set with degree function $d(s^{-1}e) = \inf_{s'^{-1}e' = s^{-1}e} d(s', e')$. Recall that $(s, e) \sim (s', e')$ means $\exists \tilde{s} \in S$, $\exists \tilde{e} \in E$, $\tilde{s}s = \tilde{s}s'$ and $\tilde{s}e = \tilde{e}e'$.

If the degree function is multiplicative, e.g. E is a domain, then

$$
\begin{aligned}
d(s, e) &= d(e) - d(s) \\
&= d(e) - d(\tilde{e}) - (d(e) - d(\tilde{s})) \\
&= d(\tilde{s}r) - d(\tilde{s}s) \\
&= d(\tilde{e}e') - d(\tilde{e}s') \\
&= d(e') - d(s') \\
&= d(s', e'),
\end{aligned}
$$

hence taking the infimum in the expression for $d(s^{-1}e)$ is superfluous, as all the representatives of $s^{-1}e$ give the same result. Therefore the degree is well-defined by $d(s^{-1}e) := d(e) - d(s)$.

The symbol image of a set $S \subset E$ is denoted by $\overline{S}$. If S is left Ore in E, and $\operatorname{gr} E$ is a domain, then $\overline{S}$ is clearly left Ore in $\operatorname{gr} E$.

12.4 Lemma. *If the symbol map $E \to \operatorname{gr} E$ is multiplicative, then the induced degree function on $S^{-1}E$ is multiplicative as well.*

12.5 Proposition. ([81], II 3.2) *(i) We have a well-defined map $\theta :$ $\overline{S}^{-1}\operatorname{gr} E \to \operatorname{gr} S^{-1}E$ given by $(\bar{s})^{-1}\bar{e} \mapsto s^{-1}e$.*

(ii) This map is an isomorphism of graded rings.

Proof. (i) Let $s_1, s_2 \in S$ with $\bar{s}_1 = \bar{s}_2$. By $s_1^{-1}s_2 = 1 + s_1^{-1}(s_2 - s_1)$ we get $s_1^{-1}e = s_2^{-1}e + s_1^{-1}(s_2 - s_1)s_2^{-1}e$. Using **12.4** and $d(s_2 - s_1) < d(s_1)$, we see that for each $0 \neq e \in E$, $s_1^{-1}(s_2 - s_1)s_2^{-1}e$ is lower degree than $s_2^{-1}e$. Thus $\overline{s_1^{-1}e} = \overline{s_2^{-1}e} \in \overline{S^{-1}E}$. In the same vein, but easier, we see that $\overline{s^{-1}e}$ does not depend on the choice of $e \in \bar{e}$.

Finally, choose different classes $\bar{t}$ and $\bar{f}$ with $(\bar{t})^{-1}\bar{f} = (\bar{s})^{-1}\bar{e}$. That is $\exists \bar{s}_* \in \bar{S}, \exists \bar{e}_* \in \operatorname{gr} E$ with $\bar{s}_*\bar{t} = \bar{e}_*\bar{s} \in \bar{S}$ and $\bar{s}_*\bar{f} = \bar{e}_*\bar{e}$. Then $\overline{s_*t} = \overline{e_*s}$ and $\overline{s_*f} = \overline{e_*e}$ for any choice of representatives e_*, s_* of $\bar{e}_*, \bar{s}_*$. Hence there are r_1, r_2 of lower degrees than e_*s, e_*e respectively, such that $s_*t = e_*s + r_1$ and $s_*f = e_*e + r_2$. Then $t^{-1}f = (s_*t)^{-1}s_*f = (e_*s + r_1)^{-1}(e_*e + r_2)$ which by the above equals $(e_*s)^{-1}(e_*e) = s^{-1}e$ up to elements of lower order, *provided that $e_*s \in S$*. As $e_*s + r_1 \in S$, this is always true if S is *saturated*, see below. However, the conclusion follows without that assumption. Indeed, by the left Ore condition, choose $s^\sharp, e^\sharp$ with $S \ni s^\sharp(e_*s + r_1) = e^\sharp s$. Then $e^\sharp e = e^\sharp ss^{-1}e = s^\sharp e_*e + s^\sharp r_1 s^{-1}e$, where, by the multiplicativity, $s^\sharp r_1 s^{-1}e$ is of lower order. Consequently, $t^{-1}f = [s^\sharp(e_*s + r_1)]^{-1}s^\sharp(e_*e + r_2) = (e^\sharp s)^{-1}e^\sharp e +$ lower order $= s^{-1}e +$ lower order, as required.

(ii) Since both the degree of $(\bar{s})^{-1}\bar{e}$ and of $\overline{s^{-1}e}$ are $d(e) - d(s)$, this map respects the grading. The obvious candidate $\overline{s^{-1}e} \mapsto (\bar{s})^{-1}\bar{e}$ for the inverse is well-defined by more straightforward reasons than the map θ. Namely, if $\overline{t^{-1}f} = \overline{s^{-1}e}$ then $\exists h$ of lower order with $t^{-1}f = s^{-1}e + h = s^{-1}(e + sh)$. As $\bar{e} = \overline{e + sh}$ it is enough to check the case $h = 0$. For some $s_* \in S$, $r_* \in r$ we have $s_*t = r_*s \in S$ and $s_*f = r_*e$. Then $\bar{s}_*\bar{t} = \bar{r}_*\bar{s} \in \bar{S}$ and $\bar{s}_*\bar{f} = \bar{r}_*\bar{e}$, hence $\bar{t}^{-1}\bar{f} = \bar{s}^{-1}\bar{e}$, as required.

12.6 Let $N = \cup_{k \in \mathbb{Z}}F_k N$, be a right and $M = \cup_{k \in \mathbb{Z}}F_k M$ a left filtered E-module, then $N \otimes_{\mathbf{k}} M$ is filtered with respect to the unique degree function additively extending formulas $d(n \otimes_{\mathbf{k}} m) = d(n) + d(m)$. The canonical quotient map $p_E : N \otimes_{\mathbf{k}} M \to N \otimes_E M$ induces the filtration $F_k(N \otimes_E M) := p_E(F_k(N \otimes_{\mathbf{k}} M))$. If N is a filtered $E' - E$-bimodule, one obtains a filtration of $N \otimes_E M$ as a left E'-module. In particular, *given $E = \cup_k F_k E$, where* $\operatorname{gr} E$ *is a domain, and given a filtered left E-module $M = \cup_k F_k M$, any Ore localization $S^{-1}M = S^{-1}E \otimes_E M$ is a filtered left $S^{-1}E$-module with the degree function*

$$d(s^{-1}m) = d(s^{-1} \otimes_E m) = d(s^{-1}1_E) + d(m) = d(m) - d(s).$$

12.7 Lemma. *If the symbol map $E \to \operatorname{gr} E$ is multiplicative, and M a filtered left E-module, then the degree functions are compatible with action in the sense that $d_M(e.m) = d_E(e)d_M(m)$. Furthermore, for any left Ore* set $S \subset E$,

$$
\begin{aligned}
d_{S^{-1}M}(s^{-1}e.t^{-1}m) &= d_{S^{-1}E}(s^{-1}e)d_{S^{-1}M}(t^{-1}m) \\
&= d_E(e) + d_M(m) - d_E(s) - d_E(t).
\end{aligned}
$$

12.8 Proposition. *(i) For a filtered ring E, for which gr E is a domain, and any filtered E-module M, we have a well-defined map $\theta_M : \bar{S}^{-1}\text{gr } M \to$ gr $S^{-1}M$ given by $(\bar{s})^{-1}\bar{m} \mapsto \overline{s^{-1}m}$.*

(ii) θ_M is an isomorphism of graded gr $S^{-1}E = \bar{S}^{-1}$ gr E-modules.

The proof is by the same techniques as **12.5**. The compatibility with the action **12.7** replaces the multiplicativity, and the formula (9) for the equivalence relation $\sim$ on $S \times M$ (with $(S \times M/\sim) \cong S^{-1}M$) replaces the equivalence relation $\sim$ from **4.13** on $S \times R$ in that proof.

12.9 Ore conditions recursively. *(i) Let S be a multiplicative set in a ring E with an exhaustive filtration*

$$
F_*E = \{\ldots \subset F_{-r}E \subset F_{-r+1}E \subset \ldots \subset F_{-1}E \subset F_0E \subset F_1E \subset \cdots \subset E\}.
$$

*Let S satisfy the **bounded below filtered-relative left Ore condition** in F_*E:*

$\exists r, \infty > r \geq -n, \forall s \in S, \forall k, -r \leq k \leq n, \forall e \in F_kE, \exists s' \in S, \exists e' \in E$ *such that $s'e - e's \in F_{k-1}E$ if $k > -r$, and $s'e - e's = 0$ if $k = -r$.*

Then S satisfies the left Ore condition for S in E.

*(ii) Assume that S is **bounded filtered left reversible** in F_*E:*

$\exists r < \infty, \forall e_k \in F_kE$, *if $\exists s \in S$ with $e_k s \in F_{k-1}E$ then $\exists s' \in S$ such that $s'e_k \in F_{k-1}E$ if $k > -r$, and $s'e_k = 0$ if $k = -r$.*

Then S is left reversible in E.

Proof. (i) Let $s \in S$ and $e = e_n \in F_nE$. By induction, we can complete sequences $e_n, \ldots, e_{-r}, e'_n, \ldots, e'_{-r}$ (here $e_k, e'_k \in F_kE$) and $s'_n, \ldots, s'_{-r} \in S$, with $e'_k s = s'_k e_k - e_{k-1}$ for all k with $e_{-r-1} := 0$. By descending induction on k,

$$
(e'_k + s'_k e'_{k+1} + \ldots + s'_k \cdots s'_{n-1} e'_n) s = s'_k \cdots s'_n e_n - e_{k-1},
$$

for each $k > -r$, and finally,

$$
(e'_{-r} + s'_{-r} e'_{-r+1} + \ldots + s'_{-r} s'_{-r+1} \cdots s'_{n-1} e'_n) s = s'_{-r} \cdots s'_n e_n.
$$

(ii) Suppose $e \in F_kE = E$ and $es = 0$ for some $s \in S$. It is sufficient to inductively choose a descending sequence and $s'_{k+1} = 1, s'_k, s'_{k-1}, \ldots, s'_{-r} \in S$, with requirements $s'_j e \in F_{j-1}E$ for all $j > -r$ and $s'_{-r}e = 0$. Suppose we have chosen $s_k, \ldots, s_{j+1}$. Then $(s'_{j+1}e)s = s'_{j+1}(es) = 0$ with $s'_{j+1}e \in F_j E$, hence by the assumption there exist some $\tilde{s} \in S$ such that $\tilde{s}s'_{j+1}e \in F_{j-1}E$. Set therefore $s'_j := \tilde{s}s'_{j+1} \in S$.

12.9a Let $F_* E$ be an exhaustive filtration of E with $F_{-r}E = 0$ for some finite r, and $S \subset E$ be a multiplicative set. If its image $\bar{S}$ under the symbol map satisfies the left Ore condition in gr E, then the conditions in **12.9** hold. Hence S satisfies the left Ore condition in E as well.

12.9b Let $t \in E$ be a regular element $(tE = Et)$ in a ring E. Then for each $n > 0$ the ideal $t^n E$ is 2-sided, hence $E_n := E/(t^n E)$ is a quotient ring in which the element t is nilpotent of order less or equal n. Rule $F_{-k}E_n = (t^k E)/(t^n E) \subset E/(t^n E) \equiv E_n$ defines a bounded 'negative' filtration

$$F_* E_n = \{0 = F_{-n}E_n \subset \ldots \subset F_{-1}E_n \subset F_0 E_n = E_n\}$$

in which (the image of) t is of degree -1. If gr E is a domain then *both* the symbol map $E \to \mathrm{gr}\, E$, and its truncation $E_n \to \mathrm{gr}\, E_n$ are multiplicative.

12.10 Theorem. *Let S be a left Ore set in some ring $E_n = F_0 E_n$ with a bounded negative filtration $F_\bullet E_n$. Suppose $S' \subset E_n$ is a multiplicative set such that $s' \in S' \cap F_j E$ iff $\exists b \in F_{j-1}E$ such that $s' = s - b$. Then S' is left Ore as well and $S^{-1}E_n = (S')^{-1}E_n$ as graded rings.*

Proof. Since the left Ore localization is a universal object in the category $\mathcal{C}_l(E_n, S)$ (cf. Chap.4) it is enough to see that a map of rings $j : E_n \to Y$ is in it iff it is in $\mathcal{C}_l(E_n, S')$. If $j(s)$ is invertible in Y, let $c = j(s)^{-1}j(b)$. The mapping j induces a (non-separated in general) filtration on Y such that j is a map of filtered rings, by taking the degree to be the infimum of expressions $d(e) - d(t)$ for elements which can be represented in the form $j(t)^{-1}j(e)$ and $-\infty$ otherwise. With our numerical constraints on the degree, for nonvanishing $e \in E_n$ this difference can not be less than $-n$. As $d(c) < 0$ we obtain $d(c^n) < n - 1$, hence $c^n = 0$. Thus we can invert $j(s)^{-1}(j(s - b)) = 1 - c$ to obtain the geometrical progression $\sum_{j=0}^{n-1} c^j$. Then $\sum_{j=0}^{n-1} c^j j(s)^{-1}j(s-b) = 1$ hence $j(s - b)$ is invertible in Y.

It remains to check that $se = 0$ for some $s \in S$ iff $\exists s'' \in S'$ with $s''e = 0$. We proceed by induction on the degree j of e starting at $-n$ where $s'e - se \in F_{-n-1} = 0$ for $s' = s - b$ with the degree of b smaller than of s' hence negative. For any j, $s'e = (s - b)e = -be$ has the degree at most $j - 1$. On the other hand, by the left Ore condition, we can find $s_* \in S$, and $b_* \in E$ with $s_*be = b_*se = 0$, hence $s_*(s'e) = 0$. Set $e' := s'e$. Since $s_*e' = 0$ with $d(e') < d(e)$, by the inductive assumption there exists $s'_* \in S'$ with $s'_*s'e = s'_*e' = 0$. Set $s'' := s'_*s'$.

The induced grading on the two localized rings is the same after the identification, because the symbol maps evaluate to the same element on s

and $s' = s + b$ (or, alternatively, after the identification, the gradings on the localization are induced by the *same* ring map).

Definition. *A multiplicative subset $S \subset E_n$ is*
- **admissible** *if $\forall s \in S$, $0 \neq \bar{s} \in \operatorname{gr} E_n$;*
- **saturated** *if $S = \{s \in E_n \mid \bar{s} \in \bar{S}\}$.*

12.11 Corollary. *Let $E, t, E_n, F_\bullet E_n$ be as in* **12.9**. *Suppose* $\operatorname{gr} E$ *is a commutative domain. Let S be a multiplicative subset in E_n. Then*

a) S is left and right Ore.

b) $S^{-1}E_n \neq 0$ iff S is admissible.

c) $S^{-1}E_n$ depends only on $\bar{S} \subset \bar{E}$.

d) $S^{-1}E_n$ is filtered by powers of t and $(S^{-1}E_n)/\langle t \rangle \cong \bar{S}^{-1}E_n$.

e) Any two saturated Ore sets S, T are compatible, i.e. $S^{-1}T^{-1}E_n \cong T^{-1}S^{-1}E_n$ and $ST = \{st \mid s \in S, t \in T\}$ is also saturated.

f) Let S be admissible. Then $\operatorname{gr}(S^{-1}E_n) \cong \bar{S}^{-1}\operatorname{gr} E_n$. In particular, $\overline{S^{-1}E_n} \cong \bar{S}^{-1}\overline{E}_n$.

Sketch of the proof. a) follows as a simple case of **12.9**; b) is trivial; c) follows by **12.10**; d) is evident; f) follows from **12.8** after truncating both sides from E to the quotient filtered ring E_n (it is not a special case of **12.8**, though, as E_n is *not* a domain); e) Because T is saturated, $(T^{-1}E)_n \cong T^{-1}E_n$. In $T^{-1}E_n$, set S is still multiplicative, hence by a) applied to $(T^{-1}E)_n$ it is left Ore. This is equivalent to compatibility (cf. Sec. 10).

13   Differential Ore condition

An extensive literature is dedicated to differential structures of various kind associated to objects of noncommutative geometry: derivations and rings of regular differential operators on NC rings, 1st and higher order differential calculi, with and without (bi)covariance conditions, NC connections and de Rham complexes etc.

13.1 *Let $\partial : R \to R$ be an R-valued derivation on R and S a left Ore set in R. Then the formula*

$$\bar{\partial}(s^{-1}r) = s^{-1}\partial(r) - s^{-1}\partial(s)s^{-1}r, \quad s \in S, r \in R, \tag{15}$$

defines a derivation $\bar{\partial} : S^{-1}R \to S^{-1}R$.

The same conclusion if we started with $\partial : R \to S^{-1}R$ instead.

Proof. 1. $\bar{\partial}$ IS WELL DEFINED.
Suppose $s^{-1}r = t^{-1}r'$ for some $r, r' \in R$, $s, t \in S$. Then

$$\exists \tilde{s} \in S, \exists \tilde{r} \in R, \quad \tilde{s}t = \tilde{r}s, \quad \tilde{s}r' = \tilde{r}r.$$

$$s^{-1} = t^{-1}\tilde{s}^{-1}\tilde{r}$$

$$
\begin{aligned}
t^{-1}\partial(r') &= t^{-1}\tilde{s}^{-1}\tilde{s}\,\partial(r') \\
&= t^{-1}\tilde{s}^{-1}[\partial(\tilde{s}r') - \partial(\tilde{s})r'] \\
&= t^{-1}\tilde{s}^{-1}[\partial(\tilde{r}r) - \partial(\tilde{s})r'] \\
t^{-1}\partial(t) &= t^{-1}\tilde{s}^{-1}\tilde{s}\,\partial(t) \\
&= t^{-1}\tilde{s}^{-1}[\partial(\tilde{s}t) - \partial(\tilde{s})t] \\
&= t^{-1}\tilde{s}^{-1}[\partial(\tilde{r}s) - \partial(\tilde{s})t]
\end{aligned}
$$

$$
\begin{aligned}
\bar{\partial}(t^{-1}r') &= t^{-1}\partial(r') - t^{-1}\partial(t)t^{-1}r' \\
&= t^{-1}\tilde{s}^{-1}[\partial(\tilde{r}r) - \partial(\tilde{s})r'] - t^{-1}\tilde{s}^{-1}[\partial(\tilde{r}s) - \partial(\tilde{s})t]t^{-1}r' \\
&= t^{-1}\tilde{s}^{-1}\partial(\tilde{r}r) - t^{-1}\tilde{s}^{-1}\partial(\tilde{r}s)t^{-1}r' \\
&= t^{-1}\tilde{s}^{-1}\partial(\tilde{r}r) - t^{-1}\tilde{s}^{-1}\partial(\tilde{r}s)s^{-1}r \\
&= t^{-1}\tilde{s}^{-1}\partial(\tilde{r})r + t^{-1}\tilde{s}^{-1}\tilde{r}\partial(r) \\
&\quad - t^{-1}\tilde{s}^{-1}\partial(\tilde{r})ss^{-1}r - t^{-1}\tilde{s}^{-1}\tilde{r}\partial(s)s^{-1}r \\
&= s^{-1}\partial(r) - s^{-1}\partial(s)s^{-1}r \\
&= \bar{\partial}(s^{-1}r)
\end{aligned}
$$

2. $\bar{\partial}$ IS A DERIVATION. We have to prove that for all $s, t \in S$ and $r, r' \in R$

$$\bar{\partial}(s^{-1}rt^{-1}r') = \bar{\partial}(s^{-1}r)\,t^{-1}r' + s^{-1}r\bar{\partial}(t^{-1}r'). \tag{16}$$

The argument of $\bar{\partial}$ on the left hand side has to be first changed into a left fraction form before we can apply the definition of $\bar{\partial}$. By the left Ore condition, we can find $r_* \in R$, $s_* \in S$ such that $r_*t = s_*r$ i.e. $rt^{-1} = s_*^{-1}r_*$.

We first prove identity (16) in the case $s = r' = 1$ i.e.

$$\bar{\partial}(rt^{-1}) = \partial(r)\,t^{-1} + r\bar{\partial}(t^{-1}). \tag{17}$$

The left-hand side of (17) is

$$
\begin{aligned}
\bar{\partial}(rt^{-1}) &= \bar{\partial}(s_*^{-1}r_*) \\
&= s_*^{-1}\partial(r_*) - s_*^{-1}\partial(s_*)s_*^{-1}r_* \\
&= s_*^{-1}\partial(r_*) + \bar{\partial}(s_*^{-1})r_*.
\end{aligned}
$$

The right-hand side of (17) is

$$
\begin{aligned}
\partial(r)t^{-1} - rt^{-1}\partial(t)t^{-1} &= \partial(r)t^{-1} - s_*^{-1}r_*\partial(t)t^{-1} \\
&= \partial(r)t^{-1} - s_*^{-1}\partial(r_*t)t^{-1} - s_*^{-1}\partial(r_*)tt^{-1} \\
&= \partial(r)t^{-1} - \bar{\partial}(s_*^{-1}r_*t)t^{-1} + \bar{\partial}(s_*^{-1})r_* - s_*^{-1}\partial(r_*) \\
&= \partial(r)t^{-1} - \partial(r)t^{-1} - \bar{\partial}(s_*^{-1})r_* - s_*^{-1}\partial(r_*) \\
&= \bar{\partial}(s_*^{-1})r_* - s_*^{-1}\partial(r_*).
\end{aligned}
$$

Hence (17) follows. Using (17), we prove (16) directly:

$$
\begin{aligned}
\bar{\partial}(s^{-1}rt^{-1}r') &= \bar{\partial}((s_*s)^{-1}r_*r') \\
&= (s_*s)^{-1}\partial(r_*r') - (s_*s)^{-1}\partial(s_*s)(s_*s)^{-1}r_*r' \\
&= s^{-1}s_*^{-1}\partial(r_*)r' + s^{-1}s_*^{-1}r_*\partial(r') \\
&\quad -s^{-1}s_*^{-1}\partial(s_*)s_*^{-1}r_*r' - s^{-1}\partial(s)s^{-1}s_*^{-1}r_*r' \\
&= s^{-1}s_*^{-1}\partial(r_*)r' + s^{-1}t^{-1}r\partial(r') + s^{-1}\bar{\partial}(s_*^{-1})r_*r' + \bar{\partial}(s^{-1})s_*^{-1}r_*r' \\
&= s^{-1}\bar{\partial}(s_*^{-1}r_*)r' - s^{-1}\bar{\partial}(s_*^{-1})r_*r + s^{-1}rt^{-1}\partial(r') + \bar{\partial}(s^{-1})rt^{-1}r' \\
&= s^{-1}\bar{\partial}(rt^{-1}) - s^{-1}\bar{\partial}(s_*^{-1})r_*r + s^{-1}rt^{-1}\partial(r') + \bar{\partial}(s^{-1})rt^{-1}r' \\
&\overset{(17)}{=} s^{-1}\partial(r)t^{-1}r' + s^{-1}r\bar{\partial}(t^{-1})r' + s^{-1}rt^{-1}\partial(r') + \bar{\partial}(s^{-1})rt^{-1}r' \\
&= \bar{\partial}(s^{-1}r)t^{-1}r' + s^{-1}r\bar{\partial}(t^{-1}r').
\end{aligned}
$$

Standard textbooks have incomplete proofs of **13.1**, e.g. [32, 116].

13.2 Definition. *A **Poisson bracket** on a unital associative* **k**-*algebra is an antisymmetric bilinear operation* $\{,\} : A \otimes A \to A$ *satisfying the Jacobi identity* $\{f, \{g, h\}\} + \{h, \{f, g\}\} + \{g, \{h, f\}\} = 0$ *for all* $f, g, h \in A$ *and such that for each* f, **k**-*linear map* $X_f : g \mapsto \{f, g\}$ *is a* **k**-*derivation of* A. *A **Poisson algebra** is a* commutative *algebra with a Poisson bracket.*

Proposition. *Let* A *be a* **k**-*algebra with a Poisson bracket* $\{,\}$, *and* $S \subset A \backslash \{0\}$ *a central multiplicative set. Then*

(i) $S^{-1}A$ *posses a bilinear bracket* $\{,\} = \{,\}^S$ *such that the localization map* $\iota_S : A \to S^{-1}A$ *intertwines the brackets:* $\{,\}^S \circ (\iota_S \otimes_\mathbf{k} \iota_S) = \iota_S \circ \{,\}$.

(ii) If either $\{s, t\} \in \mathrm{Ker}\, \iota_S$ *for all* $s, t \in S$, *or if* A *is commutative, then there is a unique such bracket* $\{,\}^S$ *which is, in addition, skew-symmetric.*

(iii) If A *is commutative then this unique* $\{,\}^S$ *is a Poisson bracket.*

Proof. (i) Each X_b by **13.1** induces a unique derivation $X_b^S = \bar{\partial}$ on $S^{-1}A$ by (15) for $\partial = X_b$. The map $b \mapsto X_b^S$ is **k**-linear by uniqueness as $X_b^S + X_c^S$ is a derivation extending X_{b+c} as well. For each $s^{-1}a \in S^{-1}A$ define **k**-linear map $Y_{s^{-1}a} : A \to S^{-1}A$ by

$$
Y_{s^{-1}a} : b \mapsto -X_b^S(s^{-1}a) = -s^{-1}\{b, a\} + s^{-1}\{b, s\}s^{-1}a.
$$

Because s *is central,* $Y_{s^{-1}a}$ *is a* **k**-*linear derivation. Namely,*

$$
\begin{aligned}
Y_{s^{-1}a}(bc) &= -s^{-1}\{bc, a\} + s^{-1}\{bc, s\}s^{-1} \\
&= -s^{-1}\{b, a\}c - s^{-1}b\{c, a\} + \\
&\quad + s^{-1}b\{c, s\}s^{-1}a + s^{-1}\{b, s\}cs^{-1}a,
\end{aligned}
$$

and, on the other hand,

$$
\begin{aligned}
Y_{s^{-1}a}(b)c + bY_{s^{-1}a}(c) &= -s^{-1}\{b, a\}c + s^{-1}\{b, s\}s^{-1}ac - \\
&\quad - bs^{-1}\{c, a\} + bs^{-1}\{c, s\}s^{-1}a.
\end{aligned}
$$

Hence $Y_{s^{-1}a}$ extends to a derivation $Y_{s^{-1}a}^S$ on $S^{-1}A \to S^{-1}A$ by formula (15) as well. Define $\{s^{-1}a, t^{-1}b\} := Y_{s^{-1}a}^S(t^{-1}b)$.

(ii) To show the skew-symmetry, we calculate,

$$
\begin{aligned}
Y_{s^{-1}a}^S(t^{-1}b) &= t^{-1}Y_{s^{-1}a}(b) - t^{-1}Y_{s^{-1}a}t^{-1}b \\
&= -t^{-1}X_b^S(s^{-1}a) + t^{-1}X_t^S(s^{-1}a)t^{-1}b \\
&= -t^{-1}s^{-1}X_b(a) + t^{-1}s^{-1}X_b(s)s^{-1}a \\
&\quad + t^{-1}s^{-1}X_t(a)t^{-1}b - t^{-1}s^{-1}X_t(s)s^{-1}at^{-1}b.
\end{aligned}
$$

$$
\begin{aligned}
Y_{t^{-1}b}^S(s^{-1}a) &= s^{-1}Y_{t^{-1}b}(a) - s^{-1}Y_{t^{-1}b}(s)s^{-1}a \\
&= -s^{-1}X_a^S(t^{-1}b) + s^{-1}X_s^S(t^{-1}b)s^{-1}a \\
&= -s^{-1}t^{-1}X_a(b) + s^{-1}t^{-1}X_a(t)t^{-1}b \\
&\quad + s^{-1}t^{-1}X_s(b)s^{-1}a - s^{-1}t^{-1}X_s(t)t^{-1}bs^{-1}a.
\end{aligned}
$$

Using $X_b(a) = -X_a(b)$ etc. and centrality of s, t we see that the first 3 terms in $Y_{s^{-1}a}^S(t^{-1}b)$ match with negative sign the first 3 terms (in order 1,3,2) in expression for $Y_{t^{-1}b}^S(s^{-1}a)$. If a and b mutually commute, the 4th term agrees the same way, and if they don't but $X_s(t) = 0$ in the localization $S^{-1}A$, then they are simply 0, implying skew-symmetry $\{s^{-1}a, t^{-1}b\} + \{t^{-1}b, s^{-1}a\} = 0$.

Uniqueness: $Z_{s^{-1}a}(t^{-1}b) := \{s^{-1}a, t^{-1}b\}$ defines a derivation $Z_{s^{-1}a}$ on $S^{-1}A$, which restricts to a derivation $Z_{s^{-1}a}| : A \to S^{-1}A$. On the other hand, $s^{-1}a \mapsto Z_{s^{-1}a}(b)$ is $-X_b^S$ by its definition. Hence the value of $Z_{s^{-1}a}|$ is determined at every b, and by **13.1** this fixes $Z_{s^{-1}a}$.

(iii) We'll prove that if the Jacobi rule holds for given (a, b, c) and (s, b, c) in $S^{-1}A^{\times 3}$, then it follows for $(s^{-1}a, b, c)$ provided s is invertible. By symmetry of the Jacobi rule and by renaming $s^{-1}a \mapsto a$ we infer that it follows for $(s^{-1}a, t^{-1}b, c)$, as well, and finally for the general case by one more application of this reasoning. Thus we only need to show that Jacobi(a, b, c) implies Jacobi$(s^{-1}a, b, c)$. For commutative $S^{-1}A$ this is a straightforward calculation, using the Jacobi identity, lemma above and skew-symmetry. We name the summands:

$$
\begin{aligned}
\{s^{-1}a, \{b, c\}\} &= s^{-1}\{a, \{b, c\}\} - s^{-2}\{s, \{b, c\}\}a =: (A1) + (A2) \\
\{b, \{c, s^{-1}a\}\} &= s^{-1}\{b, \{c, a\}\} - s^{-2}\{b, s\}\{c, a\} - s^{-2}\{b, \{c, s\}\}a \\
&\quad - s^{-2}\{c, s\}\{b, s\}a - s^{-1}\{c, s\}\{b, a\} \\
&=: (B1) + (B2) + (B3) + (B4) + (B5) \\
\{c, \{s^{-1}a, b\}\} &= s^{-1}\{c, \{a, b\}\} - s^{-2}\{c, s\}\{a, b\} - s^{-2}\{c, \{s, b\}\}a \\
&\quad + s^{-3}\{s, b\}\{c, s\}a - s^{-2}\{s, b\}\{c, a\} \\
&=: (C1) + (C2) + (C3) + (C4) + (C5).
\end{aligned}
$$

Then $(A1) + (B1) + (C1) = 0$ and $(A2) + (B3) + (C3) = 0$ by Jacobi for (a, b, c), and (b, c, s) respectively. By skew-symmetry $(B2) + (C5) = 0$, $(B5) + (C2) = 0$ and $(B4) + (C4) = 0$ which finishes the proof.

This fact for A (super)commutative is used for example in the theory of integrable systems, sometimes in connection to 'quantization' which is a rich source of examples in noncommutative geometry.

13.3 Let $(R, \cdot, +)$ be a ring (**k**-algebra), not necessarily unital. A *first order differential calculus* (FODC) is a $R - R$-bimodule $\Omega^1(R)$ together with an additive (**k**-linear) map $d : R \to \Omega^1(R)$ satisfying Leibnitz identity

$$d(ab) = d(a)b + ad(b), \quad a, b \in R,$$

and such that $\Omega^1(R)$ is generated by differentials dr, $r \in R$ as a *left module*. Define a category $\mathfrak{Fodc}$: objects are pairs of a ring R and a FODC $(\Omega^1(R), d)$ on R. A morphisms is a pair $(f, e) : (R, \Omega^1(R), d) \to (R', \Omega^1(R'), d')$ of a ring map $f : R \to R'$ and a map $e : \Omega^1(R) \to \Omega^1(R')$ of $R - R$-bimodules such that $e \circ d = d' \circ f$. Fixing R and allowing only morphisms of the form (Id_R, d) we obtain a (non-full) subcategory $\mathfrak{Fodc}_R$ of $\mathfrak{Fodc}$. If R is unital, then $(\mathrm{Ker}(R \otimes_{\mathbf{k}} R \overset{\cdot}{\to} R, d)$ where $da = 1 \otimes a - a \otimes 1$, and the R-bimodule structure is $_R R \otimes_{\mathbf{k}} R_R$, is an initial object of that category.

Two objects $c_R = (R, \Omega^1(R), d)$, $c_{R'} = (R', \Omega^1(R'), d')$ in $\mathfrak{Fodc}$ are **compatible along** $f : R \to R'$ if there is an e such that $(f, e) \in \mathfrak{Fodc}(c_R, c_{R'})$.

Differential calculi restrict: Given $c_{R'} \in \mathfrak{Fodc}_{R'}$ and f as above, define $f_\Omega^1 \Omega^1(R')$ to be the smallest additive subgroup of $\Omega^1(R')$ containing all the elements of the form $f(a)\partial'(f(b))$, $a, b \in R$. It appears to be an $R - R$-bimodule. Define $f^\sharp(c_{R'}) := (R, f_\Omega^1 \Omega^1(R'), d' \circ f)$. Then $f^\sharp(c_{R'}) \in \mathfrak{Fodc}_R$ because $\partial(b).c = \partial'(f(b))f(c) = \partial'(f(bc)) - f(b)\partial'(f(c)) = \partial(bc) - b.\partial(c) \in f^\sharp \Omega^1(R')$, where $\partial = \partial' \circ f : R \to f^\sharp \Omega^1(R')$ is the restricted differential. Note the decomposition of $(f, e) : c_R \to c_{R'}$ into $(f, e) : c_R \to f^\sharp c_{R'}$ and $(\mathrm{id}_{R'}, \mathrm{incl}) : f^\sharp c_{R'} \to c_{R'}$, where $\mathrm{incl} : f_\Omega^1 \Omega^1(R') \to \Omega^1(R')$ is the inclusion of R'-bimodules.

Unlike restricting, there is *no general recipe for extending* the calculus along ring maps $f : R \to R'$, except for the special case when $R' = S^{-1}R$ and $\Omega^1 R = _R R_R$, treated in **13.1**. That case is of central importance in the study of the regular differential operators and D-modules over noncommutative spaces ([79, 80, 81]). We'll just mention a slight generalization.

13.4 Theorem. *Let $S \subset R$ be a left Ore set in a ring R, and suppose $\{x \in \Omega^1(R) \mid \exists t \in S, xt = 0\} = 0$.*

The following are then equivalent:

(i) The $S^{-1}R$-R-bimodule structure on $S^{-1}\Omega^1(R) \equiv S^{-1}R \otimes_R \Omega^1(R)$ extends to an (actually unique) $S^{-1}R$-bimodule structure which may carry a differential $d_S : S^{-1}R \to S^{-1}\Omega^1(R)$ such that the pair of localization maps

$(\iota_S, \iota_{S,\Omega^1(R)})$ *is a morphism in* $\mathfrak{Foc}$ *(i.e. 'the calculi are compatible along the localization').*

(ii) *The* **differential Ore condition** *is satisfied:*

$$\forall t \in S, \forall r \in R, \exists s \in S, \exists \omega \in \Omega^1(R), \quad s\,dr = \omega t.$$

Proof. (i) $\Rightarrow$ (ii). If $S^{-1}\Omega^1(R) \equiv S^{-1}R \otimes_R \Omega^1(R)$ is a $S^{-1}R$-bimodule then $(dr)t^{-1} \in S^{-1}\Omega^1(R)$ for $t \in S$, $r \in R$. All the elements in $S^{-1}\Omega^1(R)$ are of the form $s^{-1}\omega$ where $s \in S$ and $\Omega \in \Omega^1(R)$. Hence $\exists s \in S$, $\exists \omega \in \Omega^1(R)$ such that $s\,dr = \omega t$ in the localization. By **7.4** this means $s\,dr = \omega t + \omega'$ in $\Omega^1(R)$, where $s'\omega' = 0$ for some $s' \in S$. Pre-multiplying by s' we obtain $(s's)dr = (s'\omega)t$, of the required form.

(ii) $\Rightarrow$ (i). The right $S^{-1}R$ action if it exists is clearly forced by

$$s_1^{-1}ad(r)t^{-1}b = s_1^{-1}as^{-1}\omega b \tag{18}$$

for s, ω chosen as above. On the other hand, if (18) holds, this right action does extend the right R-action. One has to prove that (18) can be taken as a definition of the right $S^{-1}R$-action (compatible with the left action), i.e. it does not depend on choices. If we choose s', ω' such that $s'd(r) = \omega't$ then $s^{-1}\omega t = (s')^{-1}\omega't$. As t does not annihilate from the right, $s^{-1}\omega = (s')^{-1}\omega'$. Other cases are left to the reader. Hence $S^{-1}\Omega^1(R)$ is a bimodule; its elements are of the form $s^{-1}adb$.

To prove that it is sufficient, define d_S from d by the generalization of formula (15) by $\bar{\partial}$ and ∂ replaced by d_S and d and proceed with the rest of the proof as in **13.1** – all the calculations there make sense.

14 A Gabriel filter $\mathcal{L}_S$ for any $S \subset R$

14.1 A lattice is a poset $(W, \succ)$ such that for any two elements z_1, z_2 the least upper bound $z_1 \vee z_2$ and the greatest lower bound $z_1 \wedge z_2$ exist. In other words, the binary operations *meet* $\wedge$ and *join* $\vee$ are everywhere defined. A poset is **bounded** if it contains a maximum and a minimum element, which we denote 1 and 0 respectively. A ('proper') **filter** in a bounded lattice $(W, \succ)$ is a subset $\mathcal{L} \subset W$ such that $1 \in \mathcal{L}$, $0 \notin \mathcal{L}$, $(z_1, z_2 \in \mathcal{L} \Rightarrow z_1 \wedge z_2 \in \mathcal{L})$ and $(z \in \mathcal{L}, z' \succ z \Rightarrow z' \in \mathcal{L})$.

E.g. in any bounded lattice $(W, \succ)$, given $m \in W$, the set ^mW of all $n \succ m$ is a filter.

14.2 *Notation.* Given a left ideal J in R and a subset $w \subset R$ define

$$(J : w) := \{z \in R \,|\, zw \subset J\}$$

Then $(J : w)$ is a left ideal in R. If $w =: K$ is also a left ideal, then $(J : K)$ is 2-sided ideal. In particular, if $w = K = R$, then $(J : R)$ is the maximal 2-sided ideal contained in J. For $r \in R$ we write $(J : r)$ for $(J, \{r\})$.

Given subsets $v, w \subset R$, set $((J : v) : w)$ contains precisely all t_1 such that $t_1 w \subset (J : v)$, i.e. $t_1 wv \subset J$. Hence $((J : v) : w) = (J : wv)$.

14.3 Preorders on left ideals. Let $I_l R$ be the set of all left ideals in a ring R. It is naturally a preorder category with respect to the inclusion preorder. This category is a lattice. For the localization and spectral questions another partial order $\succ$ on $I_l R$ is sometimes better: $K \succ J$ (category notation: $J \to K$) iff either $J \subset K$, or there exist a finite subset $w \subset R$ such that $(J : w) \subset K$. Any filter in $(I_l R, \succ)$ is called a **uniform filter**.

14.4 Let R be a unital ring and $S \subset R$ a multiplicative set. Consider

$$\mathcal{L}_S := \{J \text{ left ideal in } R \,|\, \forall r, \, (J : r) \cap S \neq \emptyset\} \subset I_l R. \qquad (19)$$

We make the following observations:

- As $(R : r) = R$, $R \in \mathcal{L}_S$.

- Suppose $J, K \in \mathcal{L}_S$. Given $r \in R$, $\exists s, t$, such that $s \in (J : r) \cap S$ and $t \in (K : sr) \cap S$. Hence $tsr \in J \cap K$. The set S is multiplicative, hence $ts \in S$ and $ts \in (J : r) \cap (K : r) \cap S = (J \cap K : r) \cap S$. Thus $J \cap K \in \mathcal{L}_S$.

- $(J : r) \cap S \neq \emptyset$ then, a fortiori, $(K : r) \cap S \neq \emptyset$ for $K \supset J$.

- If $J \in \mathcal{L}_S$ then $\forall r \; (J : r) \cap S \neq \emptyset$. In particular, this holds with r replaced by rr'. Using $((J : r) : r') = (J : r'r)$ we see that $(J : r) \in \mathcal{L}_S$ for all $r \in R$.

- If $\forall r' \in R \; (J : r') \cap S \neq \emptyset$ and $((J' : j) : r) \cap S \neq \emptyset$ for all $j \in J$, $r \in R$, then $\exists s \in S$ such that $srj \in J'$ and $\exists s' \in S$ such that $s'r' \in J$. In particular for $r = 1$ and $j = s'r'$ we have $ss'r' \in J$. Now $ss' \in S$ and r' is arbitrary so $J' \in \mathcal{L}_S$.

These properties can be restated as the axioms for a **Gabriel filter** $\mathcal{L} \subset I_l R$ (synonyms "radical set", "radical filter", "idempotent topologizing filter"):

- (F1) $R \in \mathcal{L}$ and $\emptyset \notin \mathcal{L}$.

- (F2) If $J, K \in \mathcal{L}$, then $J \cap K \in \mathcal{L}$.

- (F3) If $J \in \mathcal{L}$ and $J \subset K$ then $K \in \mathcal{L}$.

- (UF) $J \in \mathcal{L} \Leftrightarrow (\forall r \in R, (J : r) \in \mathcal{L})$.

- (GF) If $J \in \mathcal{L}$ and $\forall j \in J$ the left ideal $(J' : j) \in \mathcal{L}$, then $J' \in \mathcal{L}_S$.

Axioms (F1-3) simply say that a set $\mathcal{L}$ of ideals in R is a filter in $(I_l R, \subset)$. Together with (UF) they exhaust the axioms for a uniform filter (cf. **14.3**). Axioms (GF) and (UF) imply (F2): If $j \in J$, $(I \cap J : j) = (I : j) \cap (J : j) = (I : j) \in \mathcal{L}$ by (UF). Since $\forall j \in J$ $(I \cap J : j) \in \mathcal{L}$, (GF) implies $I \cap J \in \mathcal{L}$. (GF) & (F1) imply (F3): $(\forall j \in J \subset K)$ $(K : j) = R \in \mathcal{L}$, hence $K \in \mathcal{L}$.

There are examples of Gabriel filters $\mathcal{L}$, even for commutative R, which are *not* of the form $\mathcal{L}_S$ for a multiplicative $S \subset R$. Moreover, for rings without unity (F1-3,UF,GF) still make sense, whence a good notion of a multiplicative set and filters $\mathcal{L}_S$ fails to exist.

Notice that if a multiplicative set S satisfies the left Ore condition, then $\mathcal{L}_S = \mathcal{L}'_S := \{J$ is left ideal $\mid J \cap S \neq \emptyset\}$. Namely, $(J : 1) \cap S = J \cap S$ for *any* S, hence $\mathcal{L}_S \subset \mathcal{L}'_S$; and the *left* Ore condition implies that given an element $s \in J \cap S$ and $r \in R$ we can find $s' \in S$, and $r' \in R$ with $s'r = r's \in r'J \subset J$, hence $s' \in (J : r) \cap S$; hence $\mathcal{L}'_S \subset \mathcal{L}_S$.

14.5 Exercise. Check that the intersection of any family of Gabriel filters is a Gabriel filter.

Remark: this is *not* always true for the union: (GF) often fails.

14.6 For given R-module M and a filter $\mathcal{L}$ in $(I_l R, \subset)$, the inclusions $J \hookrightarrow J'$ induce maps $\mathrm{Hom}_R(J', M) \to \mathrm{Hom}_R(J, M)$ for any M, hence we obtain an inductive system of abelian groups. The inclusion also induce the projections $R/J \to R/J'$ and hence, by composition, the maps $\mathrm{Hom}_R(R/J', M) \to \mathrm{Hom}_R(R/J, M)$. This gives another inductive system of abelian groups. If a filter $\mathcal{L}$ is uniform, we consider the same systems and limits of groups (without new morphisms), and use (UF) as ingenious device to define the R-module structure on them.

14.7 Proposition. *Let $\mathcal{L}$ be a uniform filter and M a left R-module.*

(i) The inductive limit of abelian groups taken over downwards directed family of ideals

$$H_\mathcal{L}(M) := \lim_{J \in \mathcal{L}} \mathrm{Hom}_R(J, M)$$

has a canonical structure of an R-module. $H_\mathcal{L}$ extends to an endofunctor.

(ii) The abelian subgroup

$$\sigma_\mathcal{L}(M) := \{m \in M \mid \exists J \in \mathcal{L}, Jm = 0\} \subset M$$

is a R-submodule of M.

(iii) If $f : M \to N$ is a map of R-modules, $\mathrm{Im}\, f|_{\sigma_\mathcal{L}(M)} \subset \sigma_\mathcal{L}(N)$, hence the formula $f \mapsto \sigma_\mathcal{L}(f) := f|_{\sigma_\mathcal{L}(M)}$ extends $\sigma_\mathcal{L}$ to a subfunctor of identity.

(iv) The inductive limit of abelian group $\sigma'_{\mathcal{L}}(M) := \lim_{J \in \mathcal{L}} \operatorname{Hom}_R(R/J, M)$
has a structure of a left R-module.

(v) If $1 \in R$ *then the endofunctors* $\sigma_{\mathcal{L}}$ *and* $\sigma'_{\mathcal{L}}$ *(on the categories of modules* M *with* $1_R m = m$, *where* $m \in M$) *are equivalent.*

Proof. (i) Given $f \in H_{\mathcal{L}}(M)$, represent it as f_J in $\operatorname{Hom}_R(J, M)$ for some $J \in \mathcal{L}$. By (UF), $\forall r \in R$, $(J : r) \in \mathcal{L}$. The rule $x \mapsto f_J(xr)$ defines a map $(rf)_{(J:r)}$ in $\operatorname{Hom}_R((J : r), M)$ which we would like to represent the class rf. Suppose we have chosen another representative f_I, then there is $K \in \mathcal{L}$, $K \subset I \cap J$, such that $f_I|_K = f_J|_K =: h$. Then $(K : r) \subset (I \cap J : r) = (I : r) \cap (J : r)$ and the map $x \mapsto h(xr) : K \to M$ agrees with $(rf)_{(J:r)}|_K$ and $(rf)_{(I:r)}|_K$ hence the class rf is well defined.

This is a left action: $((rr')f)_{(J:rr')}(x) = f_J(xrr') = (r'f)_{(J:r')}(xr) = (r(r'f))_{((J:r'):r)}(x) = (r(r'f))_{(J:rr')}(x)$. We used $((J : r') : r) = (J : rr')$.

(ii) Suppose $m \in \sigma_{\mathcal{L}}(M)$, i.e. $Jm = 0$ for some $J \in \mathcal{L}$. For arbitrary $r \in R$ the ideal $(J : r) \in \mathcal{L}$ by (UF). Let $k \in (J : r)$. Then $kr \in J$, hence $krm = 0$. This is true for any such k, hence $(J : r)rm = 0$ and $rm \in \sigma_{\mathcal{L}}(M)$. As r was arbitrary, $R\sigma_{\mathcal{L}}(M) \subset \sigma_{\mathcal{L}}(M)$.

(iii) If $m \in \sigma_{\mathcal{L}}(M)$ then $0 = f(0) = f(Jm) = Jf(m)$ for some J in $\mathcal{L}$. Hence $f(m) \in \sigma_{\mathcal{L}}(N)$.

(iv) Let $r \in R$ and $f \in \lim \operatorname{Hom}_R(R/J, M)$. Take a representative $f_J \in \operatorname{Hom}_R(R/J, M)$. Let $(rf)_{(J:r)} \in \operatorname{Hom}_R(R/(J : r), M)$ be given by $(rf)_{(J:r)}(r' + (J : r)) = f_J(r'r + J)$. This formula does not depend on r' because changing r' by an element $\delta r' \in (J : r)$ changes $r'r$ by an element $(\delta r')r$ in $(J : r)r \subset J$. Suppose $f_I \sim f_J$. In this situation, with projections as connecting morphisms, this means that $f_I(x + I) = f_J(x + J)$ for all $x \in R$, and in particular for $x = r'r$, hence $(rf)_{(J:r)} \sim (rf)_{(I:r)}$ and rf is well defined.

Finally, $f \mapsto rf$ is a left R-action. Indeed, for all $r, r', t \in R$,

$$
\begin{aligned}
((rr')f)_{(J:rr')}(t + (J : rr')) &= (f_J)(trr' + J) \\
&= (r'f)_{(J:r')}(tr + (J : r')) \\
&= (r(r'f))_{((J:r'):r)}(t + ((J : r') : r)) \\
&= (r(r'f))_{(J:rr')}(t + (J : rr'))
\end{aligned}
$$

If R and M are unital, then $1_R f = f$ as well.

(v) To make the statement precise, we should first extend $\sigma_{\mathcal{L}}$ to a functor by defining it on morphisms as well ($\sigma'_{\mathcal{L}}$ is obviously a functor as the formula on object is explicitly written in terms of a composition of functors applied on M). As $\sigma_{\mathcal{L}}(M) \subset M$, it is sufficient to show that $f(\sigma_{\mathcal{L}}(M)) = \sigma_{\mathcal{L}}(f(M))$ and then define $\sigma_{\mathcal{L}}(f) := \sigma_{\mathcal{L}} \circ f$. Element $f(m) \in f(\sigma_{\mathcal{L}}(M))$ iff $Jm = 0$ for some $J \in \mathcal{L}$. This is satisfied iff $f(Jm) = Jf(m) = 0$, i.e. $f(m) \in \sigma_{\mathcal{L}}(f(M))$.

The equivalence $\nu : \sigma'_{\mathcal{L}} \Rightarrow \sigma_{\mathcal{L}}$ is given by $\nu_M([f_J]) := f_J(1_R+J) \in \sigma_{\mathcal{L}}(M)$ (because $Jf_J(1_R + J) = f_J(0 + J) = 0$), with inverse $m \mapsto [\theta_J^{(m)}]$ where $\theta_J^{(m)} : r + J \mapsto rm$ and any $Jm = 0$ (such J exists and the formulas for $\theta_J^{(m)}$ for different J agree, hence a fortiori define a limit class). Starting with $[f_J]$ with $m := f_J(1_R + J)$ and $\theta_J^{(m)} : r + J \mapsto rf_J(1_R + J) = f_J(r + J)$, hence $\theta_J^{(m)} = f_J$. Other way around, start with $m \in \sigma_{\mathcal{L}}(M)$, then $\nu_M(\theta_J^{(m)}) = \theta_J^{(m)}(1_R + J) = 1_R m = m$. Hence we see that each ν_M is an isomorphism of modules. The reader may check that ν, ν^{-1} are natural transformations.

14.8 If $\mathcal{A}$ is any abelian category, then a subfunctor σ of the identity (i.e. $\sigma(M) \subset M$ and $\sigma(f)|_{\sigma(M)} = f|_{\sigma(M)}$, cf. **1.4**) with the property $\sigma(M/\sigma(M)) = 0$ is called a **preradical** in $\mathcal{A}$. A preradical σ in $R - \mathrm{Mod}$ is left exact iff $J \subset K$ implies $\sigma(J) = \sigma(K) \cap J$. A **radical** is a left exact preradical.

14.9 Proposition. *If $\mathcal{L}$ is Gabriel filter, $\sigma_{\mathcal{L}}$ is an **idempotent radical** in the category of left R-modules, i.e. it is a radical and $\sigma_{\mathcal{L}}\sigma_{\mathcal{L}} = \sigma_{\mathcal{L}}$.*

14.10 To any Gabriel filter $\mathcal{L}$, one associates a localization endofunctor $Q_{\mathcal{L}}$ on the category of left modules by the formula

$$Q_{\mathcal{L}}(M) := H_{\mathcal{L}}(M/\sigma_{\mathcal{L}}(M)) = \lim_{J \in \mathcal{L}} \mathrm{Hom}_R(J, M/\sigma_{\mathcal{L}}(M)). \qquad (20)$$

Left multiplication by an element $r \in R$ defines a class $[r] \in Q_{\mathcal{L}}(R)$. There is a unique ring structure on $Q_{\mathcal{L}}(R)$, such that the correspondence $i_{\mathcal{L}} : r \mapsto [r]$ becomes a ring homomorphism $i_{\mathcal{L}} : R \to Q_{\mathcal{L}}(R)$.

Notice that (20) generalizes the RHS of Deligne's formula, **10.4a**.

14.11 Not only every Gabriel filter defines an idempotent radical, but also ([62]):

Proposition. *Every radical defines a Gabriel filter by the rule*

$$\mathcal{L}_\sigma := \{J \text{ left ideal in } R \,|\, \sigma(R/J) = R/J\}.$$

More generally, if M is a left R-module and σ is a radical, define

$$\mathcal{L}_{M,\sigma} := \{L \text{ left } R\text{-submodule in } M \,|\, \sigma(M/L) = M/L\}.$$

Then $\mathcal{L}_M := \mathcal{L}_{M,\sigma}$ satisfies the following properties

- *(GT1) $M \in \mathcal{L}_M$.*

- *(GT2) If $L, K \in \mathcal{L}_M$, then $L \cap K \in \mathcal{L}$.*

- *(GT3) If $L \in \mathcal{L}_M$, $K \subset M$ a left submodule and $L \subset K$, then $K \in \mathcal{L}_M$.*

- *(GT4) If $J \in \mathcal{L}_M$ and $K \in \mathcal{L}_J$ the left submodule $K \in \mathcal{L}_M$.*

14.11a When we restrict to the *idempotent radicals*, then the rule $\sigma \mapsto \mathcal{L}_\sigma$ gives a *bijection* between the idempotent radicals and Gabriel filters.

15   Localization in abelian categories

The language of Gabriel filters is not suited for some other categories where additive localization functors are useful. Subcategories closed with respect to useful operations (e.g. extensions of objects) are often used as the localization data, particularly in abelian and triangulated categories. We confine ourselves just to a summary of basic notions in abelian setting and comment on the connection to the language of Gabriel filters, as a number of references is available ([21, 42, 43, 62, 103, 107]).

15.1 Let $\mathcal{A}$ be an additive category. Let $\mathcal{P}$ be a full subcategory of $\mathcal{A}$. Define the left and right orthogonal to $\mathcal{P}$ to be the full subcategories $^\perp \mathcal{P}$ and $\mathcal{P}^\perp$ consisting of all objects $A \in \mathcal{A}$ such that $\mathcal{A}(P, A) = 0$ (resp. $\mathcal{A}(A, P) = 0$) for all $P \in \mathcal{P}$. The zero object is the only object in $\mathcal{P} \cap {}^\perp \mathcal{P}$. It is clear that taking (left or right) orthogonal reverses inclusions and that $\mathcal{P} \subset {}^\perp(\mathcal{P}^\perp)$ and $\mathcal{P} \subset ({}^\perp \mathcal{P})^\perp$. We leave as an exercise that $\mathcal{P}^\perp = ({}^\perp(\mathcal{P}^\perp))^\perp$ and $^\perp \mathcal{P} = {}^\perp(({}^\perp \mathcal{P})^\perp)$.

15.2 A **thick subcategory** of an abelian category $\mathcal{A}$ is a strictly full subcategory $\mathcal{T}$ of $\mathcal{A}$ which is closed under extensions, subobjects and quotients. In other words, an object M' in a short sequence $0 \to M \to M' \to M'' \to 0$ in $\mathcal{A}$ belongs to $\mathcal{T}$ iff M and M'' do.

Given a pair $(\mathcal{A}, \mathcal{T})$ where $\mathcal{A}$ is abelian and $\mathcal{T} \subset \mathcal{A}$ is thick, consider the class

$$\Sigma(\mathcal{T}) := \{ f \mid \mathrm{Ker}\, f \in \mathrm{Ob}\, \mathcal{T}, \text{ and } \mathrm{Coker}\, f \in \mathrm{Ob}\, \mathcal{T} \}.$$

The quotient category $\mathcal{A}/\mathcal{T}$ is defined as follows. $\mathrm{Ob}\, \mathcal{A}/\mathcal{T} = \mathrm{Ob}\, \mathcal{A}$ and $\mathrm{Mor}\, \mathcal{A}/\mathcal{T} := \mathrm{Mor}\, \mathcal{A} \coprod \Sigma^{-1}(\mathcal{T})$, where $\Sigma^{-1}(\mathcal{T})$ is the class of formal inverses of morphisms $f \in \Sigma$; impose the obvious relations. $\mathcal{A}/\mathcal{T}$ is additive in a unique way making the quotient functor additive. In fact ([42, 43]), it is abelian.

Proposition. (GROTHENDIECK [58]) *Let $\mathcal{T}$ be a thick subcategory in $\mathcal{A}$ and $\Sigma(\mathcal{T})$ as above. Then Σ is a left and right calculus of fractions in $\mathcal{A}$ and $\mathcal{A}[\Sigma(\mathcal{T})]^{-1}$ is naturally isomorphic to $\mathcal{A}/\mathcal{T}$.*

A thick subcategory $\mathcal{T}$ is a **localizing subcategory** if the morphisms which are *invertible* in the quotient category $\mathcal{A}/\mathcal{T}$ are precisely the images of the morphisms in $\Sigma(\mathcal{T})$.

Every exact localization functor $T^* : \mathcal{A} \to \mathcal{B}$ (i.e. an exact functor with fully faithful right adjoint T_*) of an abelian category $\mathcal{A}$ is the localization at the localizing subcategory Σ consisting of those morphisms f such that $T^* f$ is either a kernel or cokernel morphism of an invertible morphism in $\mathcal{B}$.

If $T^* : \mathcal{A} \to \mathcal{B}$ is any exact localization functor, then set $\mathcal{T} := \operatorname{Ker} T^*$ to be the full subcategory of $\mathcal{A}$ generated by all objects X such that $T^*(X) = 0$. Then T^* factors uniquely as $\mathcal{A} \xrightarrow{\mathcal{Q}^*} \mathcal{A}/\mathcal{T} \to \mathcal{B}$ where $\mathcal{Q}^*$ is the natural quotient map.

More than one thick subcategory may give the same quotient category, and that ambiguity is removed if we consider the corresponding localizing subcategories instead ([103]).

A composition of localization functors corresponds to **Gabriel multiplication** • on thick subcategories. For any two subcategories $\mathcal{B}$, $\mathcal{D}$ of an abelian category $\mathcal{A}$ one defines $D \bullet B$ to be the full subcategory of $\mathcal{A}$ consisting of precisely those A in $\mathcal{A}$ for which there is an exact sequence $0 \to B \to A \to D \to 0$ with B in $\operatorname{Ob}\mathcal{B}$ and D in $\operatorname{Ob}\mathcal{D}$. In categories of modules one can redefine Gabriel multiplication in terms of radical filters, cf. ([113]).

15.3 In this article, we often view exact localizations (and quotient categories, cf. **10.4b**) as categorical analogues of open spaces. Their complements should then be the complementary data to the quotient categories, and such data are localizing subcategories. A more precise and detailed discussion of those subcategories, which may be considered as subschemes and closed subschemes, may be found in [79], Part I and [113, 124]. Cf. the notion of a (co)reflective subcategory in **8.9**.

Thus, in our view, it is geometrically more appealing to split the data of a category to a localizing subcategory and a quotient category, than into two *sub*categories. However, the latter point of view is more traditional, under the name of "torsion theory' and has geometrically important analogues for triangulated categories. A **torsion theory** ([21, 62]) in an abelian category $\mathcal{A}$ is a pair $(\mathcal{T}, \mathcal{F})$ of subcategories of $\mathcal{A}$ closed under isomorphisms and such that $\mathcal{F}^\perp = \mathcal{T}$ and ${}^\perp \mathcal{T} = \mathcal{F}$.

For any idempotent radical σ in $\mathcal{A}$ (**14.9**), the class $\mathcal{T}_\sigma$ of σ-**torsion objects** and the class $\mathcal{F}_\sigma$ of σ-**torsion free objects** are defined by formulas

$$\mathcal{T}_\sigma = \{M \in \operatorname{Ob}\mathcal{A} \mid \sigma(M) = M\}, \quad \mathcal{F}_\sigma = \{M \in \operatorname{Ob}\mathcal{A} \mid \sigma(M) = 0\}.$$

This pair $(\mathcal{T}_\sigma, \mathcal{F}_\sigma)$ is an example of a torsion theory and $\mathcal{T}_\sigma$ is a thick subcategory of $\mathcal{A}$. Not every torsion theory corresponds to a radical, but hereditary theories do. That means that a subobject of a torsion object is torsion. The

Cohn localization of the next section is not necessarily hereditary, but it is always a torsion theory as shown there.

16   Quasideterminants and Cohn localization

Notation. Let $M_m^n(R)$ be the set of all $n \times m$ matrices over a (noncommutative) ring R, so that $M_n(R) := M_n^n(R)$ is a ring as well. Let I, J be the ordered tuples of row and column labels of $A = (a_j^i) \in M_m^n(R)$ respectively. For subtuples $I' \subset I, J' \subset J$ and $A = (a_j^i) \in M_m^n(R)$, denote by $A_{J'}^{I'}$ the submatrix of $A_J^I := A$ consisting only of the rows and columns with included labels; e.g. $A_{\{j\}}^{\{i\}} = a_j^i$ is the entry in i-th row and j-th column. When I is known and $K \subset I$, then $|K|$ is the cardinality of K and $\hat{}$ is the symbol for omitting, i.e. $\hat{K} = I \backslash K$ is the complementary $(|K| - |I|)$-tuple.

We may consider the r-tuple $\tilde{I} = (I_1, \ldots, I_r)$ of sub-tuples which partitions the n-tuple $I = (i_1, i_2, \ldots, i_n)$, i.e. I_k are disjoint and all labels from I are included; then $|\tilde{I}| := r$. Given $\tilde{I}, \tilde{J}$ form the corresponding **block matrix** in $M_{\tilde{J}}^{\tilde{I}}$ out of A, i.e. the $|\tilde{I}| \times |\tilde{J}|$ matrix $A_{\tilde{J}}^{\tilde{I}}$ whose entries are matrices $A_{\tilde{J}_l}^{\tilde{I}_k} := A_{J_l}^{I_k}$ cut-out from A by choosing the selected tuples. Forgetting the partition gives the canonical bijection of sets $M_{\tilde{J}}^{\tilde{I}} \to M_J^I$. The multiplication of block matrices is defined by the usual matrix multiplication formula $(AB)_{J_j}^{I_i} = \sum_{l=1}^r A_{K_l}^{I_i} A_{J_j}^{K_l}$ if AB *and* the sizes of subtuples for columns of A and rows of B match. One can further nest many levels of partitions (block-matrices of block-matrices …). Some considerations will not depend on whether we consider matrices in R or block matrices, and then we'll just write M_J^I etc. skipping the argument. More generally, the labels may be the objects in some abelian category $\mathcal{A}$, and entries $a_j^i \in \mathcal{A}(i, j)$; I will be the sum $\oplus_{i \in I} i$, hence $A : I \to J$. Ring multiplication is replaced by the composition, defined whenever the labels match.

Observation. *Multiplication of block matrices commutes with forgetting (one level) of block-matrix structure.* In other words we may multiply in stages (if working in $\mathcal{A}$ this is the associativity of $\oplus$). Corollaries:

(i) *if $\tilde{I} = \tilde{J}$ then $M_{\tilde{I}}(R) := M_{\tilde{I}}^{\tilde{I}}(R)$ is a ring.*

(ii) *We can invert matrices in stages as well ('heredity').*

(iii) *The same for linear equations over noncommutative rings.*

Any pair $(i, j) \in I \times J$ determines partitions $\tilde{I} = (i, \hat{i})$ and $\tilde{J} = (j, \hat{j})$. For each A in M_J^I it induces a 2×2 block-matrix $A_{\tilde{J}}^{\tilde{I}}$. Reader should do the exercise of inverting that block matrix (with noncommutative entries), in terms of the inverses of blocks. As we will see, the (i, j)-quasideterminant of A is the inverse of the (j, i)-entry of A^{-1} if the latter is defined; though it may be defined when the latter is not.

16.1 The (i, j)-th quasideterminant $|A|_{ij}$ of A is

$$\boxed{|A|_{ij} = a_j^i - \sum_{k \neq i, l \neq j} a_l^i (A_j^{\hat{i}})_{lk}^{-1} a_j^k} \tag{21}$$

provided the right-hand side is defined (at least in the sense of evaluating a rational expression, which will be discussed below). In alternative notation, the distinguished labels ij may be replaced by a drawing of a box around the entry a_j^i as in

$$\begin{vmatrix} a_1^1 & a_2^1 & a_3^1 \\ a_1^2 & a_2^2 & a_3^2 \\ a_1^3 & \boxed{a_2^3} & a_3^3 \end{vmatrix} \equiv \begin{vmatrix} a_1^1 & a_2^1 & a_3^1 \\ a_1^2 & a_2^2 & a_3^2 \\ a_1^3 & a_2^3 & a_3^3 \end{vmatrix}_{32}$$

At most n^2 quasideterminants of a given $A \in M_n(R)$ may be defined.

16.2 *If all the n^2 quasideterminants $|A|_{ij}$ exist and are invertible then the inverse A^{-1} of A exists in $A \in M_n(R)$ and*

$$(|A|_{ji})^{-1} = (A^{-1})_j^i. \tag{22}$$

Thus we also have

$$|A|_{ij} = a_j^i - \sum_{k \neq i, l \neq j} a_l^i |A_j^{\hat{i}}|_{kl}^{-1} a_j^k \tag{23}$$

16.3 Sometimes the RHS of (21) makes sense while (23) does not. So for subtle existence questions one may want to be careful with alternative formulas for quasideterminants. Some identities are often proved using alternative forms, so one has to justify their validity. Different expressions differ by **rational identities** ([1, 27]), and under strong assumptions on the ring R (e.g. a skewfield which is of ∞ dimension over the center which is also infinite), the rational identities induce a well-behaved equivalence on the algebra of rational expressions and the results of calculations extend in an expected way to alternative forms once they are proved for one form having a nonempty domain of definition ([1, 27, 123])

16.4 On the other hand, the existence of an inverse A^{-1} does *not* imply the existence of quasideterminants. For example, the unit 2×2 matrix $\mathbf{1}_{2 \times 2}$ over field $\mathbb{Q}$ has only 2 quasideterminants, not 4. Or, worse, the matrix

$\begin{pmatrix} 3 & 2 \\ 2 & 3 \end{pmatrix}$ over the commutative ring $\mathbb{Z}[\frac{1}{5}]$ is invertible, but no single entry is invertible, and in particular no quasideterminants exist.

16.5 Quasideterminants are *invariant under permutation* of rows or columns of A if we appropriately change the distinguished labels.

16.6 Suppose now we are given an equation of the form

$$Ax = \xi$$

where $A \in M_n(R)$ and x, ξ are n-tuples of indeterminates and free coefficients in R respectively (they are column "vectors"). Then one can attempt to solve the system by finding the inverse of the matrix A and multiply the equation by A^{-1} from the left, or one can generalize the Cramer's rule to the noncommutative setup.

Define thus $A(j, \xi)$ as the $n \times n$ matrix whose entries are the same as of A except that the j-th column is replaced by ξ. Then the noncommutative **left Cramer's rule** says

$$\boxed{|A|_{ij} x^j = |A(j, \xi)|_{ij}}$$

and the right-hand side does not depend on i.

To see that consider first $n = 2$ case:

$$a_1^1 x^1 + a_2^1 x^2 = \xi^1$$
$$a_1^2 x^1 + a_2^2 x^2 = \xi^2$$

Then

$$
\begin{aligned}
|A|_{11} x^1 &= a_1^1 x^1 - a_2^1 (a_2^2)^{-1} a_1^2 x^1 \\
&= (\xi^1 - a_2^1 x^2) - a_2^1 (a_2^2)^{-1} a_1^2 x^1 \\
&= \xi^1 - a_2^1 (a_2^2)^{-1} a_2^2 x^2 - a_2^1 (a_2^2)^{-1} a_1^2 x^1 \\
&= \xi^1 - a_2^1 (a_2^2)^{-1} \xi^2 = |A(1, \xi)|_{11}.
\end{aligned}
$$

The general proof is exactly the same, just one has to understand which indices are included or omitted in the sums involved:

$$
\begin{aligned}
|A|_{ij} x^j &= a_j^i x^j - \sum_{k \neq j, l \neq i} a_k^i (A_{\hat{j}}^{\hat{i}})_{kl}^{-1} a_j^l x^j \\
&= (\xi^i - \sum_{h \neq j} a_h^i x^h) - \sum_{k \neq j, l \neq i} a_k^i (A_{\hat{j}}^{\hat{i}})_{kl}^{-1} a_j^l x^j \\
&= \xi^i - \sum_{h \neq j, k \neq j, l \neq i} a_k^i (A_{\hat{j}}^{\hat{i}})_{kl}^{-1} a_h^l x^h - \sum_{k \neq j, l \neq i} a_k^i (A_{\hat{j}}^{\hat{i}})_{kl}^{-1} a_j^l x^j \\
&= \xi^i - \sum_{1 \leq h \leq n, k \neq j, l \neq i} a_k^i (A_{\hat{j}}^{\hat{i}})_{kl}^{-1} a_h^l x^h \\
&= \xi^i - \sum_{1 \leq h \leq n, k \neq j, l \neq i} a_k^i (A_{\hat{j}}^{\hat{i}})_{kl}^{-1} a_h^l x^h \\
&= |A(j, \xi)|_{ij}.
\end{aligned}
$$

Similarly consider equation $\sum_k y^k B_k^l = \zeta^l$. Apparently the individual coefficients multiply y^k from the right, but the combinatorics of matrix labels is organized as if we multiply By (alas, otherwise the rule of writing upper indices for rows would force us to write such equations upside-down!). The canonical antiisomorphism $R \to R^{\mathrm{op}}$ clearly sends any quasideterminant into the quasideterminant of the *transposed matrix*. Hence the left Cramer's rule implies the **right Cramer's rule**

$$\boxed{y^j |B^T|_{ji} = |(B(j,\zeta))^T|_{ji}.}$$

16.7 Row and column operations. Ordinary determinants do not change if we add a multiple of one row to another, and similarly for the columns.

We have to distinguish between left and right linear combinations.

If $|A|_{ij}$ is defined and $i \neq l$, then it is unchanged under left-row operation

$$A^l \to A^l + \sum_{s \neq l} \lambda_s A^s$$

Proof. We may assume $i = 1$. Define the row matrix

$$\vec{\lambda} = (\lambda_2, \ldots, \lambda_n).$$

Then $\vec{\lambda} T = \sum_{s \neq k} \lambda_s T^s$ for any matrix T with row-labels $s = 2, \ldots, n$. Then $\Lambda T = \sum_{s \neq k} \lambda_s T^s$. Assume the matrix A is in the block-form written as

$$\begin{pmatrix} a & \vec{b} \\ \vec{c}^T & D \end{pmatrix}$$

with a of size 1×1. Then

$$\begin{vmatrix} \boxed{a + \vec{\lambda}\vec{c}^T} & \vec{b} + \vec{\lambda}D \\ \vec{c}^T & D \end{vmatrix} = a + \vec{\lambda}\vec{c}^T - (\vec{b} + \vec{\lambda}D) D^{-1} \vec{c}^T$$

$$= a - \vec{b}D^{-1}\vec{c}^T.$$

If we multiply the l-th row from the left by an invertible element μ then the quasideterminant $|A|_{ij}$ won't change for $i \neq l$ and will be multiplied from the left by μ if $i = l$. Actually, more generally, left multiply the i-th row by μ and the block matrix consisting of other rows by invertible square matrix Λ (i.e. other rows can mix among themselves, and scale by different factors):

$$A \to \begin{pmatrix} \mu & 0 \\ 0 & \Lambda \end{pmatrix}$$

Then $|A|_{ij}$ gets left-multiplied by μ:

$$\left| \begin{array}{cc} \boxed{\mu a} & \mu \vec{b} \\ \Lambda \vec{c}^T & \Lambda D \end{array} \right|_{11} = \mu a - \mu \vec{b} \, (\Lambda D)^{-1} \Lambda \vec{c}^T$$

$$= \mu \, (a - \vec{b} D^{-1} \vec{c}^T) = \mu \, |A|_{ij}.$$

16.8 Jacobi's ratio theorem. ([46]) *Let A be a matrix with possibly noncommutative entries such that the inverse $B = A^{-1}$ is defined. Choose some row index i and some column index j. Make a partition of the set of row indices as $I \cup \{i\} \cup J$ and a partition of the set of column indices as $I' \cup \{j\} \cup J'$, with the requirements $\operatorname{card} I = \operatorname{card} I'$ and $\operatorname{card} J = \operatorname{card} J'$. Then*

$$(|A_{I \cup \{i\}, I' \cup \{j\}}|_{ij})^{-1} = |B_{J' \cup \{j\}, J \cup \{i\}}|_{ji}.$$

Proof. ([71]) The block decomposition of matrices does not change the multiplication, i.e. we can multiply the block matrices and then write out the block entries in detail, or we can write the block entries of the multiplicands in detail and then multiply and we get the same result. In particular, as $A = B^{-1}$, the block-entries of A can be obtained by block-inversion of B.

After possible permutation of labels, we may find the block-entry of the matrix $A = B^{-1}$ at the intersection of rows $I \cup \{i\}$ and columns $I' \cup \{j\}$ by means of block-inverting the block matrix

$$A = \left(\begin{array}{cc} A_{I \cup \{i\}, I' \cup \{j\}} & A_{I \cup \{i\}, J'} \\ A_{J, I' \cup \{j\}} & A_{J, J'} \end{array} \right)$$

Then $A_{I \cup \{i\}, I' \cup \{j\}} = (B_{I \cup \{i\}, I' \cup \{j\}} - B_{I \cup \{i\}, J'} (B_{J J'})^{-1} B_{J, I \cup \{i\}})^{-1}$ or, equivalently,

$$(A_{I \cup \{i\}, I' \cup \{j\}})^{-1} = B_{I \cup \{i\}, I' \cup \{j\}} - B_{I \cup \{i\}, J'} (B_{J J'})^{-1} B_{J, I' \cup \{j\}}$$

This is a matrix equality, and therefore it implies the equality of the (i, j)-th entry of both sides of the equation. We obtain

$$((A_{I \cup \{i\}, I' \cup \{j\}})^{-1})_{ij} = b_{ij} - \sum_{k \in J', l \in J} b_{i,k} (B_{J J'})^{-1}_{kl} b_{lj}.$$

Finish by applying the formula $(|C|_{ji})^{-1} = (C^{-1})_{ij}$ at LHS.

16.9 Muir's law of extensionality. ([46, 48, 49]) Let an identity $\mathcal{I}$ between quasiminors of a submatrix A_J^I of a generic matrix A be given. Let $K \cap I = \emptyset$, $L \cap J = \emptyset$ and $K = L$. If every quasiminor $|A_V^U|_{uv}$ of A_J^I in the identity $\mathcal{I}$ is replaced by the quasiminor $|A_{V\cup L}^{U\cup K}|_{uv}$ of $A_{J\cup L}^{I\cup K}$ then we obtain a new identity $\mathcal{I}'$ called the extensional to $\mathcal{I}$.

16.10 Quasitelescoping sum. Let $A = (a_j^i)$ be a generic $n \times n$ matrix. For any $k > 2$, and $i, j \in \{1, k-1\}$ consider the quasiminor

$$|A_{j,k,k+1,\dots,n}^{i,k,k+1,\dots,n}|_{ij}.$$

The quasitelescoping sum involves such minors:

$$QT(A_{1,\dots,n}^{1,\dots,n}) = \sum_{k=3}^{n} |A_{k-1,k,\dots,n}^{1,k,\dots,n}|_{1,k-1} |A_{k-1,k,\dots,n}^{k-1,k,\dots,n}|_{k-1,k-1}^{-1} |A_{1,k,\dots,n}^{k-1,k,\dots,n}|_{k-1,1}$$

Then, by Muir's law and induction on n, we obtain

$$QT(A_{1,\dots,n}^{1,\dots,n}) = a_1^1 - |A|_{11}. \tag{24}$$

For $n = 3$ this is simply the identity obtained by extending by the third row and column the identity expressing the expansion of the 2×2 upper left quasiminor. Suppose now we have proved (24) for n. Take an $(n+1) \times (n+1)$-matrix A. Then, by induction, this is true for the submatrix

$$A_{\hat{2}}^{\hat{2}} = A_{1,3,\dots,n}^{1,3,\dots,n}.$$

But

$$\begin{aligned}
QT(A_{1,\dots,n}^{1,\dots,n}) &= QT(A_{1,3,\dots,n}^{1,3,\dots,n}) + |A_{2,3,\dots,n}^{1,3,\dots,n}|_{1,2} |A_{2,3,\dots,n}^{2,3,\dots,n}|_{2,2}^{-1} |A_{1,3,\dots,n}^{2,3,\dots,n}|_{2,1} \\
&= a_1^1 - |A_{1,3,\dots,n}^{1,3,\dots,n}|_{11} + |A_{2,3,\dots,n}^{1,3,\dots,n}|_{1,k-1} |A_{2,3,\dots,n}^{2,3,\dots,n}|_{k-1,k-1}^{-1} |A_{1,3,\dots,n}^{2,3,\dots,n}|_{k-1,1} \\
&= a_1^1 - |A_{1,2,3,\dots,n}^{1,2,3,\dots,n}|_{11}
\end{aligned}$$

where the last two summands were added up, using the identity which represents the expansion of 2×2 upper left corner of A and extending the identity by rows and columns having labels $3, \dots, n$.

16.11 Homological relations. Start with the identity

$$(a_1^1 - a_2^1(a_2^2)^{-1}a_1^2)(a_1^2)^{-1} = -(a_2^1 - a_1^1(a_1^2)^{-1}a_2^2)(a_2^2)^{-1}.$$

which in the quasideterminant language reads

$$\begin{vmatrix} \boxed{a_1^1} & a_2^1 \\ a_1^2 & a_2^2 \end{vmatrix} (a_1^2)^{-1} = - \begin{vmatrix} \boxed{a_1^1} & a_2^1 \\ a_1^2 & a_2^2 \end{vmatrix} (a_1^2)^{-1}$$

and extend the latter applying Muir's law, adding the rows $3, \ldots, n$ of A to each minor in the expression. Renaming the indices arbitrarily we obtain the **row homological relations**:

$$\boxed{|A|_{ij}|A^{\hat{i}}_{j'}|^{-1}_{i'j} = -|A|_{ij'}|A^{\hat{i}}_{\hat{j}}|^{-1}_{i'j'}} \tag{25}$$

for $j \neq j'$. Similarly, starting with the identity

$$(a_2^1)^{-1}(a_1^1 - a_2^1(a_2^2)^{-1}a_1^2) - (a_2^2)^{-1}(a_1^2 - a_2^2(a_2^2)^{-1}a_1^2),$$

we obtain the **column homological relations**

$$\boxed{|A^{\hat{i'}}_{\hat{j}}|^{-1}_{ij'}|A|_{ij} = -|A^{\hat{i}}_{\hat{j}}|^{-1}_{i'j'}|A|_{i'j}.} \tag{26}$$

16.12 Laplace expansion for quasideterminants. Start with the identity

$$\sum_j a^i_j (A^{-1})^j_k = \delta^i_k.$$

If $i \neq k$ and A^{-1} exists, then substituting $(A^{-1})^j_k = |A|^{-1}_{kj}$ this becomes

$$\sum_j a^i_j |A|^{-1}_{ij} = 1.$$

Multiply this equation from the right by $|A|_{il}$ and split the sum into the part with $j \neq l$ and the remaining term:

$$a^i_l + \sum_{j \neq l} a^i_j |A|^{-1}_{ij} |A|_{il} = |A|_{il}$$

and apply the row homological relations (25) to obtain the following Laplace expansion for the (i, j)-th quasideterminant by the k-th row:

$$\boxed{a^i_l - \sum_{j \neq l} a^i_j |A^{\hat{i}}_{\hat{l}}|^{-1}_{kj} |A^{\hat{i}}_{\hat{j}}|_{kl} = |A|_{il}} \tag{27}$$

Similarly, multiplying from the left the equation $\sum_i |A|^{-1}_{ij} a^i_j = 1$ by $|A|_{lj}$ and splitting the sum into two terms we obtain

$$a^l_j + \sum_{i \neq l} |A|_{lj} |A|^{-1}_{ij} a^i_j = |A|_{lj},$$

which after the application of the column homological relations (26) gives the following Laplace expansion for the (i,j)-th quasideterminant by the k-th column:

$$a_j^l - \sum_{j \neq l} |A_{\hat{j}}^{\hat{i}}|_{lk} |A_{\hat{j}}^{\hat{i}}|_{ik}^{-1} a_l^i = |A|_{lj} \tag{28}$$

Notice that the summation sign involves $(n-1)$ summands whereas the similar summation in the recursive formula (23) for quasideterminants involves $(n-1)^2$ summands.

16.13 ([27, 28]) Let R be an associative unital ring, and Σ a given set of square *matrices* of possibly different (finite) sizes with entries in R. The map $f : R \to S$ of rings is Σ-inverting if each matrix in Σ is mapped to an invertible matrix over S. A Σ-inverting ring map $i_\Sigma : R \to R_\Sigma$ is called a **Cohn localization** (or equivalently a universal Σ-inverting localization) if for every Σ-inverting ring map $f : R \to S$ there exists a unique ring map $\tilde{f} : R_\Sigma \to S$ such that $f = \tilde{f} \circ i_\Sigma$.

A set Σ of matrices is called (upper) **multiplicative** if $1 \in \Sigma$ and, for any $A, B \in \Sigma$ and C of right size over R, $\begin{pmatrix} A & C \\ 0 & B \end{pmatrix}$ is in Σ. If Σ is the smallest multiplicative set of matrices containing Σ_0, then a map is Σ_0-inverting iff it is Σ-inverting. Inclusion $\Sigma_0 \subset \Sigma$ makes every Σ-inverting map $f : R \to S$ also Σ_0-inverting. Conversely, if each of the diagonal blocks can be inverted, a block-triangular matrix can be inverted, hence Σ_0-inverting maps are Σ-inverting.

The universal Σ-inverting localization can be constructed by the "inversive method", as follows. Represent R as a free algebra F on a generating set $\mathbf{f}$ modulo a set of relations I. For each quadratic matrix $A \in \Sigma$ of size $n \times n$, add n^2 generators (A, i, j) to $\mathbf{f}$. In this way we obtain a free algebra F' over some generating set $\mathbf{f'}$. All (A, i, j) for fixed A clearly form a $n \times n$-matrix A' over F'. Then $\Sigma^{-1} R = F'/I'$ where I' is the ideal generated by I and by all elements of matrices $AA' - I$ and $A'A - I$ for all $A \in \Sigma$. Then $i_\Sigma : R \to \Sigma^{-1} R$ is the unique map which lifts to the embedding $F \hookrightarrow F'$.

16.14 Warning. A naive approach to quotient rings, would be just adding new generators a' and relations $aa' = a'a = 1$ for each $a \in R \backslash \{0\}$ which needs to be inverted in the first place. In geometrical applications this could induce pretty unpredictable behaviour on modules etc. But suppose we just want to do this in an extreme special case: constructing a quotient skewfield. After inverting all the nonzero elements, we try inverting all their nonzero sums and so on. The problem arises that one may not know which elements from m-th step will be forced to be zero by new relations added a few steps later. So one should skip inverting some new elements, as they will become zeros after a few more steps of inverting other elements. There is no recipe for which elements to leave out at each step. For a given ring R, there may be none (no quotient field) or multiple possibilities for such a recipe. More precisely, given two embeddings $R \hookrightarrow K_i$ into skewfields $K_1 \neq K_2$, there may be *different* smallest subskewfields $L_i \hookrightarrow K_i$ containing R.

16.15 Proposition. *Let Σ be a multiplicative set of square matrices over R and let $f : R \to S$ be a Σ-inverting map. Let $S(i, \Sigma) \subset R$ consist of all the components of solutions over S of all the equations $f(A)x = f(b)$ where $A \in \Sigma$, b is a column-vector over R and x a column of unknowns.*

(i) $S(i, \Sigma)$ is a subring of S.

(ii) $S(i, \Sigma)$ coincides with the image of R_Σ under the unique map $\tilde{f} : R_\Sigma \to S$ for which $f = \tilde{f} \circ i_\Sigma$.

In particular, if f is 1-1 then $\tilde{f}$ is isomorphism and i_Σ is 1-1.

(i) If the components x_i and y_j of column vectors x and y over S are in $S(i, \Sigma)$, with $f(A)x = b$ and $f(B)y = c$, then, possibly after enlarging x, y, b, c by zeroes and A and B by diagonal unit blocks, we may always make $i = j$ and b and c of the same length. Then $f \begin{pmatrix} A & -A+B \\ 0 & B \end{pmatrix} \begin{pmatrix} x+y \\ y \end{pmatrix} = f \begin{pmatrix} b+c \\ c \end{pmatrix}$ and as the left-hand side matrix is in $f(\Sigma)$ by multiplicativity, then $x_i + y_i \in S(i, \Sigma)$, as claimed. For z a (row or column) vector consider the diagonal square matrix $\mathrm{diag}(z)$ with diagonal z. Then $\mathrm{diag}(z)(1, 1, \ldots, 1)^T = (z_1, \ldots, z_n)$. For a fixed i, there is a matrix P_i such that $P_i(y_1, \ldots, y_n)^T = (y_i, \ldots, y_i)^T$. Hence, $f \begin{pmatrix} B & -\mathrm{diag}(c)P_i \\ 0 & A \end{pmatrix} \begin{pmatrix} y \\ x \end{pmatrix} = f \begin{pmatrix} 0 \\ b \end{pmatrix}$ has as the j-th component $(f(B)^{-1}f(c))_j(f(A)^{-1}f(b))_i$. But our block-triangular matrix is in Σ, hence $x_i y_j$ is in $S(i, \Sigma)$. Similarly, had we worked with algebras over **k**, we could have considered all possible weights on the diagonal instead of just using the non-weighted diagonal $\mathrm{diag}(c)$ to obtain any possible **k**-linear combination of such.

(ii) The corestriction of i onto $S(i, \Sigma)$ is also Σ-inverting. Hence there is a unique map form R_Σ. But, by construction, there is no smaller ring than $S(i, \Sigma)$ containing $f(R_\Sigma)$. As $i(R_\Sigma)$ is a ring they must coincide. If the map is 1-1 it has no kernel hence $\tilde{f}$ is an isomorphism.

16.16 Proposition. (left-module variant of P. M. Cohn [28], 2.1) *If Σ is multiplicative, then $\exists!$ subfunctor $\sigma_\Sigma : R - \mathrm{Mod} \to R - \mathrm{Mod}$ of identity* *such that, as a subset, $\sigma_\Sigma(M)$ equals*

$$\{m \in M \mid \exists u = (u_1, \ldots, u_n)^T \in M^{\times n}, \exists i, \, m = u_i \text{ and } \exists A \in \Sigma, \, Au = 0\}$$

for every $M \in R - \mathrm{Mod}$. Moreover, σ_Σ is an idempotent preradical.

   Proof. 1. $\sigma_\Sigma(M)$ *is an R-submodule of M.* It is sufficient to show that for any $r \in R$, $m, m' \in \sigma_\Sigma(M)$ the left linear combination $m + rm' \in \sigma_\Sigma(M)$. Choose $A, B \in \Sigma$, $Au = 0, Bv = 0$, $u \in M^{\times k}, v \in M^{\times l}$, $m = u_i, m' = v_j$. We may assume $k = l$, $i = j$, hence $m + rm' = (u + rv)_i$, by adjusting A, B, u, v. For example, $\tilde{A} := \mathrm{diag}\,(I_s, A, I_t) \in \Sigma$ and $\tilde{u} := (0_s, u_1, \ldots, u_k, 0_t)^T \in M^{s+k+t}$ satisfy $\tilde{A}\tilde{u} = 0$ with $m = \tilde{u}_{i+s}$.
Then $\begin{pmatrix} A & -Ar \\ 0 & B \end{pmatrix} \in \Sigma$ and $\begin{pmatrix} A & -Ar \\ 0 & B \end{pmatrix}\begin{pmatrix} u + rv \\ v \end{pmatrix} = \begin{pmatrix} 0 \\ 0 \end{pmatrix}$.

   2. $M \mapsto \sigma_\Sigma(M)$ *extends to a unique subfunctor of identity.* If $m = u_i \in \Sigma$ for some i and $A(u_1, \ldots, u_k)^T = 0$ then $A(f(u_1), \ldots, f(u_k))^T = 0$ whenever $f : M \to N$ is R-module map. As $f(m) = f(u_i)$ this proves that $f(\sigma_\Sigma(M)) \subset \sigma_\Sigma(f(M))$ as required.

   3. $\sigma_\Sigma(M)$ *is a preradical:* $\sigma_\Sigma(M/\sigma_\Sigma(M)) = 0$. If $m \in \sigma_\Sigma(M)$, then $\exists u_1, \ldots, u_k \in M$, $\exists A \in \Sigma$, $\exists p \leq k$, such that $A\vec{u}^T = 0 \bmod \sigma_\Sigma(M)$ and $m = u_p$ where $\vec{u} := (u_1, \ldots, u_k)$. Hence $\exists v_1, \ldots, v_k \in \sigma_\Sigma(M)$ such that $A\vec{u}^T = (v_1, \ldots, v_k)^T$ and there are matrices $B_1, \ldots, B_k$, where B_s is of size $h_s \times h_s$, and vectors $(w_{1s}, \ldots, w_{h_s s})$ of size h_s, such that $B_s(w_{1s}, \ldots, w_{hs})^T = 0$ for all s; and we have that $v_i = w_{s_i i}$ for some correspondence $i \mapsto s_i$. Let $\vec{w} = (w_{11}, \ldots, w_{h_1 1}, w_{12}, \ldots, w_{k s_k})$. Let the matrix $J = (J_j^i)$ be defined by $J_{s_i}^i = 1$ for each i and all other entries are 0. This matrix satisfies $J\vec{w}^T = \vec{v}$ by construction. Define also the block matrix $B := \mathrm{diag}\,\{B_1, \ldots, B_k\}$. Clearly $B\vec{w}^T = 0$ by construction and $B \in \Sigma$ by the multiplicativity of Σ. In this notation a summary of the above is encoded in the block identity

$$\begin{pmatrix} A & -J \\ 0 & B \end{pmatrix}(\vec{u}, \vec{v})^T = 0, \quad \begin{pmatrix} A & -J \\ 0 & B \end{pmatrix} \in \Sigma, \quad m = u_p.$$

   4. $\sigma_\Sigma(\sigma_\Sigma(M)) = \sigma_\Sigma(M)$. If $m = u_i$ for some i and $A(u_1, \ldots, u_k)^T = 0$ for some $A \in \Sigma$ with all $u_j \in \sigma_\Sigma(M)$, then in particular, all $u_j \in M$.

   Exercise. *Let Σ, Σ' be multiplicative sets of matrices over R.*

   If for every $A \in \Sigma$ there are permutation matrices $w, w' \in GL(k, \mathbb{Z})$ such that $wAw' \in \Sigma'$ then $\sigma_{\Sigma'}(M) \subset \sigma_\Sigma(M)$ for all M.

   16.17 Warning. σ_Σ is not necessarily left exact. Equivalently, the associated torsion theory is not always hereditary (i.e. a submodule of a σ_Σ-torsion module is not necessarily σ_Σ-torsion). Hereditary torsion theories correspond to Gabriel localizations.

16.18 Quasideterminants vs. Cohn localization. Quasideterminants are given by explicit formulas. It is sometimes more algorithmically manageable to invert them, than the matrices (if the inverse can not be expressed in terms of them anyway). The two procedures often disagree, as some simple (e.g. diagonal) matrices do not possess all the n^2 possible quasideterminants. One may combine the process, by first inverting quasideterminants which exist, and then performing the Cohn localization for the simpler matrix so obtained. The combination is not necessarily a Cohn localization.

Thus, let Σ be as before. For each $A \in \Sigma$ and pair (ij) such that $|A|_{ij}$ exists and is nonzero add a variable B_{ij} and require $B_{ij}|A|_{ij} = |A|_{ij}B_{ij} = 1$. One obtains a localization $j : R \rightarrow R_\Sigma^{q0}$. Then one inverts $j(\Sigma)$ by the Cohn method, which amounts to adding formal variables just for those entries which are not added before as quasideterminants, and adding relations for them. The result is some localization $i_\Sigma^q : R \rightarrow R_\Sigma^q$ which is Σ-inverting and clearly a quotient ring of the Cohn localization. If i_Σ^q is injective, it is just the Cohn localization.

There are obvious variants of this method, (cf. the reasoning in **16.14**, and recall that quasideterminants may be defined inductively by size). Some rings may be quotiented by ideals to get commutative or Ore domains. A quasideterminant may be proven to be nonzero, as its image in the quotient is nonzero, which is a good procedure for some concrete Σ (cf.[122], Th.7).

We have seen in Ch.8 that for the usual descent of quasicoherent sheaves one needs flatness, which is often lacking for Cohn localization. In the special case of Cohn localization at a 2-sided ideal, the flatness of the localization map i_Σ as a map of left modules implies the right Ore condition ([132]). Though in geometrical situations one inverts sets of matrices for which this theorem does not apply, flatness is not expected for useful non-Ore universal localizations. Less essential, but practically difficult, is to find the kernel of the localization map i_Σ (there is a criterium using the normal form mentioned below).

We would still like, in the spirit of an example [122], Th.7 (more accurately described in [123]), to be able to consider some global noncommutative spaces where the local coordinates are compared using nonflat Cohn localizations. Localizations between full categories of modules ('perfect localizations') are described by their underlying rings (the forgetful functor from the localized category to the category of modules over localized ring is an equivalence). Similarly, the knowledge of the restriction of the Cohn localization functor to the category of finitely generated (f.g.) projective modules is equivalent to the knowledge of the localization morphism on the level of rings. Of course, the theory is here not any more complicated if one inverts any multiplicative set of morphisms between f.g. projectives than only matrices. The localization functor for f.g. projectives has an explicit description (GERASIMOV-MALCOLMSON normal form ([50, 86])) (analogous to the description of Ore localization as $S \times R/ \sim$ where $\sim$ is from **4.13**) and has an interesting homological interpretation and properties ([97]). Thus while the torsion theory corresponding to Cohn localization is bad (nonhereditary, cf. **10.3,15.3,16.17**), other aspects are close to perfect localizations (thus better than arbitrary hereditary torsion theory). This suggests that there is hope for a geometry of "covers by Cohn localizations" if we find a way beyond the case of flat descent for full categories of modules.

Acknowledgements. My biggest thanks go to Prof. A. RANICKI, who patiently encouraged the completion of this article and gave me numerous small suggestions.

Many thanks to Prof. J. ROBBIN for educating me to improve my writing style, to be straight, explicit and to avoid a too frequent use of 'it is well-known that'. Thanks to Indiana University, Bloomington where a piece of my Wisconsin student 'preprint' started growing into this paper, and Prof. V. LUNTS of Indiana for insights into mathematics and his manuscripts. I thank the Rudjer Bošković Institute, Zagreb, Max Planck Institute, Bonn, and IHÉS, Bures-sur-Yvette, for allowing me to use a big chunk of my research time there to complete this survey. Finally, my apology to the reader that I needed much more space and time for this article to be polished and balanced than I could deliver.

Bibliography. We try to give a few of the most useful references for a reader who has similar geometric needs to the author. The bibliography is intentionally incomplete. For the benefit of the geometrically oriented reader, we used the following preference criteria: geometrically motivated, historically important, readable (for the author at least), irreplacable. The literature which is obscure to me in a major way is naturally not in the list. However I mention here some undoubtedly important alternative works by listing only Math. Reviews code. For ring theorists there is a monograph on torsion theories by GOLAN MR88c:16934 and on localization by JATAGEONKAR MR88c:16005 and by GOLAN MR51:3207. Many equivalent approaches to Gabriel localization have been multiply discovered (GOLDMAN (1969) [53], SILVER (1967) MR36:205, MARANDA (1964) MR29:1236 etc.) in various formalisms, e.g. torsion theories (the term is basically from DICKSON (1965) MR32:2472).

Despite their historical importance, we ignore these, and recommend the systematic treatment in Gabriel's thesis [42] as well as the books [62, 103, 130] and Ch. 6 of [23]. For abelian categories see [21, 39, 42, 44, 45, 62, 103, 125, 58, 145]; for localization in abelian categories see books [21, 39, 62, 103, 125]. Other longer bibliographies on localization are in [39, 62, 72, 103]. Neither the present article nor the bibliography survey noncommutative geometry beyond the localization aspects; rather consult [26, 30, 55, 74, 75, 89, 113, 117, 125, 127, 136, 139, 141] and bibliographies therein; quantum group literature (e.g. [83, 87]; CHARI, PRESSLEY MR95j:17010; MAJID MR90k:16008, MR97g:17016, MR2003f:17014; KLIMYK-SCHMÜDGEN MR99f:17017; VÁRILLY MR2004g:58006); and, for the physics, also [33]. Abbrev.: LMS = London Mathematical Society, MPI = Max Planck Inst. preprint (Bonn). & for Springer series: GTM

(Graduate Texts in Math.), LNM (Lecture Notes in Math.), Grundl.MW (Grundlehren Math.Wiss.).

References

[1] S. A. AMITSUR, *Rational identities and application to algebra and geometry*, J. Alg. 3 (1966), pp. 304–359.

[2] H. APPELGATE, M. BARR, J. BECK ET AL., *Seminar on triples and categorical homology theory*, ETH 1966/67, ed. B. Eckmann, LNM 80, Springer 1969.

[3] P. ARA, M. MARTIN, *Local multipliers of C^*-algebras*, Springer, London 2003.

[4] M. ARTIN, *Noncommutative deformations of commutative rings*, preprint 1994.

[5] M. ARTIN, J. J. ZHANG, *Noncommutative projective schemes*, Adv. Math. 109 (1994), no. 2, pp. 228–287.

[6] K. ASANO, *Über die Quotientenbildung von Schiefringen*, (German) J. Math. Soc. Japan 1, (1949), pp. 73–78.

[7] P. BALMER, *Presheaves of triangulated categories and reconstruction of schemes*, Math. Ann. 324, no. 3 (2002), pp. 557-580.

[8] M. BARR, *Composite triples and derived functors*, in [2], pp. 336–356.

[9] M. BARR, C. WELLS, *Toposes, triples and theories*, Grundl. MW 178, Springer 1985.

[10] J. BECK, *Distributive laws*, in [2], pp. 119–140.

[11] J. M. BECK, *Triples, algebras and cohomology*, Ph.D. Thesis, Columbia Univ. 1963.

[12] KAI BEHREND, *Differential graded schemes I: perfect resolving algebras*, `math.AG/0212225`; *II: The 2-category of dg schemes*, `math.AG/0212226`.

[13] S. BEILINSON, V. DRINFELD, *Chiral algebras*, AMS Colloq. Publ. 51 (2004) 352 pp.

[14] D. BEN-ZVI, T. NEVINS, *From solitons to many-body systems*, `math.AG/0310490`

[15] G. M. BERGMAN, *The diamond lemma for ring theory*, Adv.M. 29 (1978), 178–219.

[16] J. BERNSTEIN, V. LUNTS, *Equivariant sheaves and functors*, LNM 1578 (1994).

[17] L. A. BOKUT', P. S. KOLESNIKOV, *Gröbner-Shirshov bases: from inception to the present time*, Zap. Nauchn. Sem. POMI 272 (2000), Vopr. Teor. Predst. Algebr i Grupp. 7, 26–67, 345; transl. in J. Math. Sci. (N. Y.) 116 (2003), no. 1, 2894–2916.

[18] A. I. BONDAL, M. M. KAPRANOV, *Representable functors, Serre functors, and reconstructions* (Russian), Izv. Akad. Nauk SSSR Ser. Mat. 53 (1989), no. 6, 1183–1205, 1337; English translation in Math. USSR-Izv. 35 (1990), no. 3, pp. 519–541.

[19] A. BONDAL, D. ORLOV, *Derived categories of coherent sheaves*, Proceedings of the ICM, Beijing 2002, vol. 2, pp. 47–56; `math.AG/0206295`.

[20] A. BONDAL, M. VAN DEN BERGH, *Generators and representability of functors in commutative and noncommutative geometry*, Mosc. Math. J. 3 (2003), no. 1, 1–36, 258; `math.AG/0204218`.

[21] F. BORCEUX, *Handbook of categorical algebra*, 3 vols. Enc. of Math. and its Appl. 50, Cambridge Univ. Press, 1994.

[22] N. BOURBAKI, Éléments de mathématique: *Algèbre commutative, II*, 1964-9 Hermann, Paris, (Engl. transl. *Commutative Algebra* II, Addison Wesley 1972).

[23] I. BUCUR, A. DELEANU, *Introduction to the theory of categories and functors*, Pure Appl. Math. XIX, Wiley 1968 (and Russian annotated transl. Mir, 1972).

[24] T. BRZEZIŃSKI, S. MAJID, *Coalgebra bundles*, Comm. M.Phys. 191 (1998), 2, 467–492.

[25] S. CAENEPEEL, G. MILITARU, S. ZHU, *Frobenius and separable functors for generalized module categories and nonlinear equations*, LNM 1787, 2002.

[26] P. CARTIER, *A mad day's work: from Grothendieck to Connes and Kontsevich – the evolution of concepts of space and symmetry*, Bull. AMS 38, No.4, pp. 389–408, 2001.

[27] P. M. COHN, *Free rings and their relations*, Academic Press 1971.

[28] P. M. COHN, *Inversive localization in Noetherian rings*, Comm. Pure. Appl. Math. 26 (1973), pp. 679–691.

[29] P. M. COHN, *Algebra*, Vol. 3, 2nd ed. John Wiley and Sons, Chichester, 1991.

[30] A. CONNES, *Noncommutative geometry*, Academic Press, New York 1994.

[31] M. DEMAZURE, P. GABRIEL, *Groupes algébriques I*, 1970 (French), *Introduction to algebraic geometry and algebraic groups* (Engl.transl. Ch.1-3), N. Holland 1980.

[32] J. DIXMIER, *Enveloping algebras*, North Holland (1977) (enhanced version AMS 1997).

[33] M. DOUGLAS, N. NEKRASOV, *Noncommutative field theory*, Rev. Modern Phys. 73 (2001), no. 4, 977–1029.

[34] V. DRINFELD, *DG quotients of DG categories*, J. Algebra 272 (2004), pp. 643–691; math.KT/0210114.

[35] C. EHRESMANN, *Catégories et structures* (French), Paris, Dunod 1965, xvii+358 pp.

[36] S. EILENBERG, *Abstract description of some basic functors*, J. Indian Math. Soc. (N.S.) 24 (1960), 231–234.

[37] S. EILENBERG, J. C. MOORE, *Adjoint functors and triples*, Illinois J. Math. 9 (1965) pp. 381–398.

[38] V. P. ELIZAROV, *Kol'ca častnyh* (Russian), Alg. i logika 8 (4) 1969, pp. 381–424; *K obščej teorii kolec častnyh*, Sib. Mat. Ž. 11 (3), 1970 pp. 526–546; *Sil'nye prekručenija i sil'nye fil'try, moduli i kol'ca častnyh*, Sib. Mat. Ž. 14 (3), 1973, pp. 549–559.

[39] C. FAITH, *Algebra I,II*, Grundleheren Math. Wiss. 190, 191, Springer 1973,1976,1981.

[40] C. FAITH, *Rings and things and a fine array of twentieth century associative algebra*, Math. Surveys & Monographs 65, AMS 1999.

[41] O. GABBER, L. RAMERO, *Almost ring theory*, LNM 1800, Springer 2003, vi+307 pp.

[42] P. GABRIEL, *Des catégories abéliennes*, Bull. Soc. Math. France 90 (1962), 323–448;

[43] P. GABRIEL, M. ZISMAN, *Calculus of fractions and homotopy theory*, Springer 1967.

[44] S. I. GELFAND, YU. I. MANIN, *Homological algebra* (Russian), Cur. problems in math.. Fund. dir. 38, 5–240, Itogi Nauki i Tehniki, AN SSSR, VINITI 1989. (Engl. transl. Algebra, V, Encyclopaedia Math. Sci., 38, Springer, Berlin, 1994).

[45] S. I. GELFAND, YU. I. MANIN, *Methods of homological algebra*, Springer 1996; translated from the Russian original, Nauka 1988.

[46] I. M. GEL'FAND, V. S. RETAKH, *Determinants of matrices over noncommutative rings*, Fun. Anal. Appl. vol 25 (1991), no. 2, pp. 91–102. transl. 21 (1991), pp. 51–58.

[47] I. M. GEL'FAND, V. S. RETAKH, *A theory of noncommutative determinants and characteristic functions of graphs*, Fun. Anal. Appl. vol. 26 (1992), no.4, pp. 231–246.

[48] I. M. GEL'FAND, V. S. RETAKH, *Quasideterminants I*, Selecta Mathematica, New Series 3 (1997) no.4, pp. 517–546; `q-alg/9705026`.

[49] I. GELFAND, S. GELFAND, V. RETAKH, R. WILSON, *Quasideterminants*, `arXiv:math.QA/0208146`.

[50] V. N. GERASIMOV, *Localization in assoc. rings*, Sib. M. Zh. 23-6 (1982) 36–54, 205; *Inverting homomorphisms of rings*, Algebra i Logika 18-6 (1979) 648–663, 754.

[51] J. GIRAUD, *Cohomologie non abélienne*, Grundl. Math. Wiss. 179, Springer 1971.

[52] J. S. GOLAN, J. RAYNAUD, F. VAN OYSTAEYEN, *Sheaves over the spectra of certain noncommutative rings*, Comm. Alg. 4(5), (1976), pp. 491–502.

[53] O. GOLDMAN, *Rings and modules of quotients*, J. Algebra 13 (1969), pp. 10–47.

[54] K. R. GOODEARL, R. B. WARFIELD JR., *An introduction to noncommutative Noetherian rings*, LMS Student Texts, Vol. 16, Cambridge Univ. Press 1989.

[55] J. M. GRACIA-BONDÍA, J. VÁRILLY, H. FIGUEROA, *Elements of noncommutative geometry*, Birkhäuser Advanced Texts: Basler Lehbúcher, 2001, xviii+685 pp.

[56] A. GROTHENDIECK, *Éléments de géométrie algébrique. IV. Étude locale de schémas et de morphismes de schémas*, Publ. Math. IHES 32 (1967).

[57] A. GROTHENDIECK ET AL., *Revêtements étales et groupe fondamental*, Séminaire de Géometrie Algébrique du Bois Marie 1960–1961 (SGA 1), LNM 224, Springer (1971), retyped as: `arXiv:math.AG/0206203`.

[58] A. GROTHENDIECK, *Sur quelques points d'algèbre homologique*, Tôhoku Math. J.(2) 9 (1957) pp. 119–221.

[59] R. HARTSHORNE, *Algebraic geometry*, Springer, GTM 52 (1977) xvi+496 pp.

[60] R. HARTSHORNE, *Residues and duality* (& appendix by P. Deligne), Springer 1966.

[61] T. J. HODGES, S. P. SMITH, *Sheaves of noncommutative algebras and the Beilinson-Bernstein equivalence of categories*, Proc. AMS, 92, N. 3, 1985.

[62] P. JARA, A. VERSCHOREN, C. VIDAL, *Localization and sheaves: a relative point of view*, Pitman Res. Notes in Math. 339, Longman 1995.

[63] P. T. JOHNSTONE, *Topos theory*, LMS Monographs 10, Acad. Press 1977; *Sketches of an elephant: a topos theory compendium*, 3 vols, Oxford Univ. Press 2002.

[64] M. KAPRANOV, *Noncommutative geometry based on commutator expansions*, J. Reine Angew. Math. 505 (1998), pp. 73–118; `arXiv:math.AG/9802041`.

[65] B. KELLER, *Introduction to A-infinity algebras and modules*, Homology, Homotopy and Applications 3 (2001), no. 1, 1–35 (electronic).

[66] A. KAPUSTIN, A. KUZNETSOV, D. ORLOV, *Noncommutative instantons and twistor transform*, Comm. Math. Phys. 221 (2001), pp. 385–432; `arXiv:hep-th/0002193`.

[67] M. KONTSEVICH, *Deformation quantization for algebraic varieties*, `math.QA/0106006`.

[68] M. KONTSEVICH, A. L. ROSENBERG, *Noncommutative smooth spaces*, The Gelfand Math. Seminars, 1996–1999, Birkhäuser 2000, pp. 85–108; `arXiv:math.AG/9812158`.

[69] M. KONTSEVICH, A. L. ROSENBERG, *Noncommutative spaces*, MPI 2004 - 35; *Noncommutative spaces and flat descent*, MPI 2004 - 36; *Noncommutative stacks*, MPI 2004 - 37; preprints, Bonn 2004.

[70] M. KONTSEVICH, Y. SOIBELMAN, *Homological mirror symmetry and torus fibrations*, in: Symplectic geometry and mirror symmetry (Seoul, 2000), pp. 203–263, World Sci. Publ. (2001) math.SG/0011041.

[71] D. KROB, B. LECLERC, *Minor identities for quasi-determinants and quantum determinants*, Comm.Math.Phys. 169 (1995), pp. 1–23.

[72] J. LAMBEK, *Torsion theories, additive semantics and rings of quotients*, LNM 177.

[73] N.P. LANDSMAN, *Bicategories of operator algebras and Poisson manifolds*, Math. physics in math. & physics (Siena, 2000), Fields Inst. Comm. 30, 271–286, AMS 2001.

[74] N. P. LANDSMAN, *Math. topics between class. & quantum mechanics*, Springer 1998.

[75] O. A. LAUDAL, *Noncommutative algebraic geometry*, Rev. Mat. Iberoamericana 19 (2003), 509–580.

[76] D. A. LEITES, *Introduction to the theory of supermanifolds* (Russian), Usp. Mat. Nauk. 35 (1980), pp. 3–57.

[77] W. T. LOWEN, M. VAN DEN BERGH, *Def. theory of ab. categories*, math.CT/0405226; *Hochschild cohomology of abelian categories and ringed spaces*, math.KT/0405227.

[78] V. A. LUNTS, *Deformations of quasicoherent sheaves of algebras*, J. Alg. 259 (2003), pp. 59-86; arXiv:math.AG/0105006.

[79] V. A. LUNTS, A. L. ROSENBERG, *Differential calculus in noncommutative algebraic geometry*, Max Planck Institute Bonn preprints:
I. D-calculus on noncommutative rings, MPI 96-53;
II. D-calculus in the braided case. The localization of quantized enveloping algebras, MPI 96-76, Bonn 1996.

[80] V. A. LUNTS, A. L. ROSENBERG, *Differential operators on noncommutative rings*, Sel. Math. N. S. 3 (1997), pp. 335–359.

[81] V. A. LUNTS, A. L. ROSENBERG, *Localization for quantum groups*, Selecta Math. (N.S.) 5 (1999), no. 1, pp. 123–159.

[82] V. A. LUNTS, A. L. ROSENBERG, *Localization for quantum groups II*, manuscript.

[83] G. LUSZTIG, *Introduction to quantum groups*, Progr. in Math. 110, Birkhäuser, 1993.

[84] S. MAC LANE, *Categories for the working mathematician*, GTM 5, Springer 1971.

[85] S. MAC LANE, I. MOERDIJK, *Sheaves in geometry and logic*, Springer 1992.

[86] P. MALCOLMSON, *Construction of universal matrix localizations*, pp. 117–131, Advances in Non-comm. Ring theory. Proc. 12th G. H. Hudson Symp. Plattsburgh, 1981. LNM 951, Springer 1982.

[87] YU. I. MANIN, *Quantum groups & non-commutative geometry*, CRM, Montreal 1988.

[88] YU. I. MANIN, *Moduli, motives, mirrors*, pp. 53–73, Proc. 3rd Eur. Cong. Math. Barcelona July 10–14, 2000, Prog. Math. 201; math.AG/0005144.

[89] YU. I. MANIN, *Topics in noncommutative geometry*, Princeton Univ. Press 1991.

[90] H. FAUSK, PO HU, J. P. MAY, *Isomorphisms between left and right adjoints*, Theory Appl. Categ. 11 (2003), pp. 107–131 (electronic).

[91] J. C. McConnell, J. C. Robson, *Noncommutative noetherian rings*, Graduate Studies in Mathematics 30, A.M.S. (2001)

[92] F. Morel, *An introduction to $\mathbb{A}^1$-homotopy Theory*, Cont. Dev. in Alg. K-theory, M. Karoubi, A. O. Kuku, C. Pedrini eds. ICTP Lec. Notes Series 15 (2004).

[93] D. C. Murdoch, F. van Oystaeyen, *Noncommutative localization and sheaves*, J. Alg. 38 (1975), pp. 500–515.

[94] S. Montgomery, H.-J. Schneider, *Hopf crossed products, rings of quotients and prime ideals*, Adv. Math. 112 (1995), pp. 1–55.

[95] J. Ndirahisha, F. van Oystaeyen, *Grothendieck representations of categories and canonical noncommutative topologies*, K-theory 25 (2002), pp. 355-371.

[96] A. Neeman, *The Grothendieck duality theorem via Bousfield's techniques and Brown representability*, J. AMS 9 (1996), pp. 205–236.

[97] A. Neeman, A. Ranicki, *Noncommutative localization and chain complexes I. Algebraic K- and L-theory*, math.RA/0109118, 75 pp., Geometry and Topology 8 (2004), pp. 1385–1425.

[98] O. Ore, *Linear equations in noncommutative fields*, Ann. Math. 32 (1931), pp. 463–477.

[99] D. O. Orlov, *Quasi-coherent sheaves in commutative and non-commutative geometry*, Izvestiya: Math. 67:3 (2003), pp. 119–138.

[100] D. O. Orlov, *Derived categories of coherent sheaves and equivalences between them*, (Russian) Uspehi Mat. Nauk 58 (2003), no. 3, pp. 89–172.

[101] C. J. Pappacena, *Frobenius bimodules between noncommutative spaces*, math.QA/0304386, J. Algebra 275 (2005), pp. 675–731

[102] A. Polishchuk, M. Rothstein, *Fourier transform for D-algebras*, math.AG/9901009.

[103] N. Popescu, *Abelian categories with applications to rings and modules*, London Math. Soc. Monographs 3, Academic Press (1973).

[104] L. Popescu, N. Popescu, *Theory of categories*, Bucuresti & Alphen: Editura Academiei & Sijthoff & Noordhoff International Publishers, 1979.

[105] T. Porter, *S-categories, S-groupoids, Segal categories and quasicategories*, math.AT/0401274.

[106] N. Reshetikhin, A. Voronov, A. Weinstein, *Semiquantum geometry*, Algebraic geometry 5, J. Math. Sci. 82 (1996), no. 1, pp. 3255–3267; q-alg/9606007.

[107] A. L. Rosenberg, *Noncomm. local algebra*, Geom. Fun. Anal. 4 (5), 1994, 545–585.

[108] A. L. Rosenberg, *The spectrum of abelian categories and reconstruction of schemes*, in Rings, Hopf algebras, and Brauer groups (Antwerp/Brussels, 1996), pp. 257–274, Lec. Notes Pure Appl. Math. 197, Dekker, NY 1998. ('96 version: MPI-1996-108).

[109] A. L. Rosenberg, *Noncommutative schemes*, Comp. Math. 112 (1998), pp. 93–125.

[110] A. L. Rosenberg, *Noncommutative spaces and schemes*, MPI-1999-84, Bonn 1999.

[111] A. L. Rosenberg, *Spectra related with localizations*, MPI-2003-112, Bonn 2003.

[112] A. L. Rosenberg, *Non-commutative affine semischemes and schemes*, Seminar on supermanifolds No. 26, edited by D. Leites, Dept. of Math., Univ. of Stockholm 1988.

[113] A. L. ROSENBERG, *Noncommutative algebraic geometry and representations of quantized algebras*, MAIA 330, Kluwer 1995.

[114] A. L. ROSENBERG, *Spectra of noncommutative spaces*, MPI-2003-110, Bonn 2003.

[115] A. L. ROSENBERG, *Underlying spectra of noncommutative schemes*, MPI-2003-111.

[116] L. ROWEN, *Ring theory*, 2 vols, Acad. Press (Pure & Applied Math. 127,128), 1988.

[117] A. C. DA SILVA, A. WEINSTEIN, *Geometric models for noncommutative algebra*, UC Berkeley, AMS, 1999, 184 pp.

[118] Z. ŠKODA, *Coherent states for Hopf algebras*, `math.QA/0303357`.

[119] Z. ŠKODA, *Distributive laws for actions of monoidal categories*, `math.CT/0406310`.

[120] Z. ŠKODA, *Globalizing Hopf-Galois extensions*, preprint;
Quantum bundles using coactions and localization, in preparation.

[121] Z. ŠKODA, *Localized coinvariants I,II*, preprints.

[122] Z. ŠKODA, *Localizations for construction of quantum coset spaces*, in "Noncommutative geometry and quantum groups", P. M. Hajac, W. Pusz eds. Banach Center Publications vol.61, pp. 265–298, Warszawa 2003; `arXiv:math.QA/0301090`.

[123] Z. ŠKODA, *Universal noncommutative flag variety*, preliminary version.

[124] P. S. SMITH, *Subspaces of non-commutative spaces*, `math.QA/0104033`;
Integral non-commutative spaces, `math.QA/0104046`;
Maps between non-commutative spaces, `math.QA/0208134`.

[125] P. S. SMITH, *Non-commutative algebraic geometry*, preprint notes, U. Wash. 2000.

[126] Y. SOIBELMAN, *Quantum tori, mirror symmetry and deformation theory*, Lett. Math. Phys. 56 (2001), no. 2, 99–125.`math.QA/0011162`.
with P. BRESSLER, *Mirror symmetry and deformation quantization*, `hep-th/0202128`.
with V. VOLOGODSKY, *Non-commutative compactifications and elliptic curves*, `math.AG/0205117`.

[127] J. T. STAFFORD, *Noncommutative projective geometry*, Proceedings of the ICM, Beijing 2002, vol. 2, pp. 93–104; `math.RA/0304210`.

[128] J. T. STAFFORD, T. A. NEVINS, *Sklyanin algebras and Hilbert schemes of points*, `math.AG/0310045`.

[129] J. T. STAFFORD, M. VAN DEN BERGH, *Noncommutative curves and noncommutative surfaces*, Bull. AMS 38 (2001), pp. 171–216.

[130] BO STENSTRÖM, *Rings of quotients*, Grundlehren Math. Wiss. 217, Springer 1975.

[131] R. STREET, *The formal theory of monads*, J. Pure & Appl. Alg. 2 (1972), pp. 149–168;
& part II (with S. LACK), J. Pure Appl. Algebra 175 (2002), no. 1-3, pp. 243–265.

[132] P. TEICHNER, *Flatness and the Ore condition for rings*, Proc. A.M.S. 131 (2003), pp. 1977–1980;

[133] B. TOEN, *Homotopical and higher categorical structures in algebraic geometry*, `math.AG/0312262`; *Affine stacks* (Champs affines), `math.AG/0012219`.

[134] B. TOEN, G. VEZZOSI, *Algebraic geometry over model categories* etc. `math.AG/0110109,0207028,0210407,0212330`, `math.AT/0309145`.

[135] R. W. THOMASON, T. TROBAUGH, *Higher algebraic K-theory of schemes and of derived categories*, The Grothendieck Festschrift, vol. III, Progr. Math. 88, Birkhäuser (1990), pp. 247–435.

[136] M. VAN DEN BERGH, *Blowing up of non-commutative smooth surfaces*, Mem. Amer. Math. Soc. 154, 2001.

[137] A. VAN DEN ESSEN, *Alg. microlocalization and modules with regular singularities over filtered rings*, Handbook of algebra, v.1, ed. M. Hazewinkel, 813–840, N. Holland, 1996.

[138] A. VERSCHOREN, *Sheaves and localization*, J. Algebra 182 (1996), no. 2, pp. 341–346.

[139] F. VAN OYSTAEYEN, *Algebraic geometry for associative algebras*, Marcel Dekker 2000.

[140] F. VAN OYSTAEYEN, A. VERSCHOREN, *Reflectors and localization, application to sheaf theory*, Lec. Notes in Pure. Appl. Math. 41, M. Dekker 1979.

[141] F. VAN OYSTAEYEN, A. VERSCHOREN, *Noncommutative algebraic geometry. An introduction.* LNM 887, Springer 1981.

[142] F. VAN OYSTAEYEN, L. WILLAERT, *Grothendieck topology, coherent sheaves and Serre's theorem for schematic algebras*, J. Pure Appl. Alg. 104 (1995), pp. 109–122. *Cohomology of schematic algebras*, J. Alg. 185 (1996), pp. 74–84.

[143] A. VISTOLI, *Notes on Grothendieck topologies, fibered categories and descent theory.*

[144] C. E. WATTS, *Intrinsic characterization of some additive functors*, Proc. AMS 11 (1960), pp. 1–8.

[145] C. WEIBEL, *An introduction to homological algebra*, Cambridge Univ. Press 1994.

[146] R. WISBAUER, *Module and comodule categories – a survey*, Proc. of the Math. Conf. Birzeit/Nablus, 1998, World Scientific (2000), pp. 277–304.

[147] H. WOLFF, $\mathcal{V}$-*localizations and* $\mathcal{V}$-*monads*, J. Alg. 24 (1973), pp. 405–438.

[148] A. YEKUTIELI, *On deformation quantization in algebraic geometry*, math.AG/0310399.

[149] A. YEKUTIELI, J. L. ZHANG, *Dualizing complexes and perverse sheaves on noncommutative ringed schemes*, J. Algebra 256 (2002), pp. 556–567; math.QA/0211309.

Index

Jacobi's ratio theorem, 296
join, 285

kernel, 224

Laplace expansion for quasideterminants, 298
lattice, 285
left calculus of fractions, 235
left Cramer's rule, 294
left denominator set, 234
left Ore set, 234
left orthogonal subcategory, 290
Leibnitz identity, 284
localization functor, 233
localizing subcategory, 290
localizing to open subscheme, 270

meet, 285
middle unitary law, 260
module over monad, 253
monad, 253
     associated to adjunction, 254
monoid, 234
monoidal category
     symmetric, 265
morphism of monads, 255
Muir's law of extensionality, 296
multiplicative set
     of matrices, 299
multiplicative subset, 234
multiplicative system, 235
multiplicatively generated, 234

normal form, 303

Ore localization, 237

partitioned matrix, 292
pentagon axiom, 252
perfect localization, 302
Poisson algebra, 282
Poisson bracket, 282
preradical, 289
preserving limits, 224
prime ideal, 267
principal open set, 267
pseudofunctor, 258

quantization, 230
quantum groups, 303

quasicompact relative noncommutative scheme, 272
quasitelescoping sum, 297

radical, 289
radical filter, 286
rational identities, 293
reflecting limits, 224
regular element, 279
restriction of scalars, 270
right Cramer's rule, 295
right orthogonal subcategory, 290
ring of fractions, 237
row and column operations, 295
row homological relations, 298

saturated multiplicative subset, 280
semigroup, 234
semiquantum geometry, 228
separated filtration, 275
sieve, 266
simplicial endofunctor, 255
site, 266
subobject, 223
symbol map, 276

terminal object, 223
thick subcategory, 290

unit of adjunction, 254

Zariski topology, 267
zero object, 224

Institute Rudjer Bošković, P.O.Box 180, HR-10002 Zagreb, Croatia

Institut des Hautes Études Scientifiques,
Le Bois-Marie, 35 route de Chartres, F-91440 Bures-sur-Yvette, France

e-mail: zskoda@irb.hr

Printed in the United States
by Baker & Taylor Publisher Services